MEDIEVAL OPTICS AND THEORIES OF LIGHT IN THE WORKS OF DANTE

Front Cover: *Dante Gazing at a Large Sphere of Light*,
Yates Thompson. 36 f179.
By Permission of the British Library.

MEDIEVAL OPTICS AND THEORIES OF LIGHT IN THE WORKS OF DANTE

Simon A. Gilson

The Edwin Mellen Press
Lewiston•Queenston•Lampeter

Library of Congress Cataloging-in-Publication Data

Gilson, Simon A.
    Medieval optics and theories of light in the works of Dante / Simon A. Gilson.
        p. cm. --
    Includes bibliographical references and index.
    ISBN 0-7734-7808-6
    1. Dante Alighieri, 1265-1321--Knowledge--Optics. 2. Optics in literature. 3. Vision in
literature. 4. Light and darkness in literature. I. Title.

   PQ4432.O68 G55   2000
   851'.1--dc21

                                                   99-089195

hors série.

A CIP catalog record for this book is available from the British Library.

Copyright © 2000 Simon A. Gilson

All rights reserved. For information contact

<table>
<tr><td>The Edwin Mellen Press</td><td>The Edwin Mellen Press</td></tr>
<tr><td>Box 450</td><td>Box 67</td></tr>
<tr><td>Lewiston, New York</td><td>Queenston, Ontario</td></tr>
<tr><td>USA 14092-0450</td><td>CANADA L0S 1L0</td></tr>
</table>

The Edwin Mellen Press, Ltd.
Lampeter, Ceredigion, Wales
UNITED KINGDOM SA48 8LT

Printed in the United States of America

To Julie

Contents

Acknowledgements..ix

Abbreviations..xi

Preface..xiii

Introduction..1

Part One: Optical Science in Dante's Thought and Poetry

Chapter 1: The Science of "Perspective": Light, Vision, and Optics
in the Thirteenth Century..7

Chapter 2: Optics and Vision in Dante before the *Comedy*..........................39

Chapter 3: Aspects of Vision in the *Comedy*: Blinding, Optical Illusions,
and Visual Error..75

Chapter 4: Light Reflection, Mirrors, and Meteorological Optics
in the *Comedy*..109

Part Two: Theories of Light in Dante

Chapter 5: Dante and the "Metaphysics of Light": A Reassessment.............151

Chapter 6: Light in the Cosmos and in God's Creative Act.........................171

Chapter 7: Adaptations Drawn from Light in the Imagery
and Doctrine of the *Paradiso*..217

Conclusion...257

Appendix..261

Bibliography...263

Index of Longer Quotations from the Works of Dante................................291

Name and Subject Index..293

ACKNOWLEDGEMENTS

For the completion of this volume I am grateful to many friends and colleagues who have provided me with encouragement and support. I owe a particular debt to Pat Boyde who read several drafts and whose insight and enthusiasm during earlier stages of research was particularly helpful. I would also like to thank a number of individuals who either helped with, or commented on, drafts of one or more chapters, particularly Peter Armour, David Gibbons, Julie Gilson, Catherine Keen, Robin Kirkpatrick, Michio Fujitani, Daniel Ractliffe, and David Ruzicka. I would also like to record my gratitude to colleagues in the Departments of Italian at the Universities of Birmingham and Leeds for their generosity and friendship during the writing of this book. Chapters 1-4 of this study incorporate some parts of two earlier articles, "Dante's Meteorological Optics: Reflection, Refraction, and the Rainbow," *Italian Studies* 52 (1997), 51-62 and "Dante and the Science of 'Perspective': A Reappraisal," *Dante Studies* 115 (1997), forthcoming, and I am grateful to the respective editors of these journals, Brian Richardson and Christopher Kleinhenz, for permission to reprint the relevant sections. A short section of my "Light Reflection, Mirror Metaphors, and Optical Framing in Dante's *Comedy*: Precedents and Transformations," *Neophilologus* 83 (1999), 241-252 is reproduced in Chapter 4 by kind permission of Kluwer Academic Publishers. The cover illustration, Dante and Beatrice gazing at a large sphere of light, is taken from Giovanni di Paolo's illuminations to the *Paradiso* and reproduced from the Yates Thompson Ms. 36, f. 179 by permission of the British Library. I dedicate the work, with love, to my wife, Julie.

ABBREVIATIONS

The following abbreviations have been used in the notes and bibliography:

AHDLMA	Archives d'histoire doctrinale et littéraire du moyen âge
Beiträge	Beiträge zur Geschichte der Philosophie des Mittelalters
CCCM	Corpus Christianorum Continuatio Medievalis
CCSL	Corpus Christianorum Series Latina
CSEL	Corpus Scriptorum Ecclesiasticorum Latinorum
DS	Dictionnaire de spiritualité ascétique et mystique doctrine et histoire
ED	Enciclopedia dantesca
GSLI	Giornale storico della letteratura italiana
PG	Patrologia Graeca
PL	Patrologia Latina
RThAM	Recherches de théologie ancienne et médiévale
ST	Thomas Aquinas, *Summa theologiae*

All quotations from Dante's minor works are taken from the Ricciardi-Rizzoli editions listed in the bibliography, and quotations from the *Comedy* are from *La "Commedia" secondo l'antica vulgata*, ed. Giorgio Petrocchi, 4 vols. (Milan: Mondadori, 1966-1967). Standard abbreviations have been used to refer to the works of Dante. The Vulgate Bible (including Vulgate numbering of Psalms) is quoted from *Biblia sacra latina iuxta vulgatam versionem*, ed. Robert Weber, 2 vols. (Stuttgart: Württembergische Bibelanstalt, 1969) and standard abbreviations have been used. Aristotle is cited using references to the Bekker edition of his works and unless otherwise stated Latin quotations are taken from the medieval Latin translations by William of Moerbeke found in the Marietti editions of Thomas Aquinas' commentaries. The works of the Pseudo-Dionysius the xii

Aeropagite are quoted from the medieval Latin translations in Philippe Chevalier, *Dionysiaca: Receuil donnant l'ensemble des traditions latines des ouvrages attribués au Denys l'Aréopage*, 2 vols. (Paris: Desclée de Brouwer, 1937-1950).

Unless otherwise noted, all translations from the Latin are mine.

PREFACE

This is a comprehensive study of Dante's theories concerning the nature of light and vision and of the imagery and narrative situations derived from those theories by the poet in whose work the words "vedere" and "luce" (and their cognates) are among the most frequent and distinctive items.

Part One reconstructs with admirable clarity and just the right amount of detail the context of Dante's concepts and technical terms, which is to say the medieval science of perspective – one of the most important disciplines in the Middle Ages. This is the first time this very important area has been treated in its full complexity with constant reference to Dante, and Dr. Gilson's treatment of both the background and foreground will be welcomed by all his readers.

Drawing extensively on the wide range of sources available to the thirteenth century, the author offers an important revision of established scholarly opinion by demonstrating that Dante was not dependent on certain technical treatises on optics, as was previously assumed, since the concepts and terminology he used had already found their way into secondary sources such as the commentaries to the works of Aristotle. As well as placing the many strands of Dante's optical knowledge in their proper context, Part One also provides a close literary analysis of Dante's poetic transformations of this material in his writings, especially in the *Comedy*, and demonstrates how Dante's scientific interest in the processes of human vision (with all its preconditions, intermediate phases and limitations) animates the poem from beginning to end. Part One is rounded off with a thorough discussion of imagery deriving from the reflection of light, the nature and properties of mirrors, and related phenomena such as the rainbow and moon halo.

The same interest in Dante as both thinker and poet informs the whole of Part Two. Light remains the principal subject, but attention is now directed to the relevance of the tradition often called "light metaphysics." Dr. Gilson asks, for

xiv

example, how far do the writings of Grosseteste and Bonaventure, in their attempt to explicate biblical and patristic light-metaphors, contribute to our understanding of Dante's treatment of light as a philosophical and theological concept, rather than a description of the behaviour of light-rays. Here too Dr. Gilson offers a major revision of current critical opinion, by arguing that this widely-invoked critical category is not helpful in categorizing and discussing Dante's treatment of light imagery. In its place he offers a new approach to Dante's views about light, in which closer attention is paid to specific contexts and relevant authorities. By concentrating on the role of light in Dante's cosmos, on his views concerning the act of creation, and on his use of analogies to express theological concepts, Dr. Gilson achieves two important and innovative results. He drastically revises many previous claims concerning Dante's dependence on certain authors by showing more precisely the points at which Dante is borrowing from a traditional and widely-assimilated body of doctrine. And once he has shown what Dante's probable sources actually were, he is able to identify with much greater accuracy the many innovations that Dante as poet brings to his source material.

Taken as a whole, the volume brings together detailed historical scholarship with literary analysis to provide a thorough account of the place of optical science and medieval concepts of light in Dante's writings.

June 1999   Professor Patrick Boyde

INTRODUCTION

Light was a phenomenon which exercised such an enduring fascination over Dante that it occupies a role in his writings which seems to cut across the various images we have constructed of him as philosopher, theologian, venerator of antiquity, lover of nature, and poet. Light was, for Dante, a primary element in visual experience, a feature common to both the heavens and the earth, an object of scientific and philosophical inquiry, and the most fitting symbol for a Christian poet writing about God. He was enraptured by its brilliance and splendour, its power to radiate and diffuse itself, its effects on different surfaces, its expressive character, and its ability to suggest psychological and emotional responses.

As both poet and thinker, Dante is often regarded as the meeting-point and culmination of the diverse strands of a rich and extensive thirteenth-century literature about light which can be divided into several different categories. Optical specialists paid detailed attention to the various kinds of light rays (direct, reflected, and refracted) involved in vision. Natural philosophers, who also examined light's role in vision, considered its place in the constitution and operations of the cosmos and in natural phenomena such as the rainbow, heat, and colour. Theologians used analogies drawn from light as the basis for doctrines relating to God, the act of creation, divine action, and man's relation to the deity. Compilers of medieval encyclopaedias incorporated various aspects of all the above material; and, in a variety of distinctive ways, vernacular poets repeatedly

2

used light, optics, and vision as sources of imagery and structuring themes in their own compositions.

Even a cursory examination of Dante's writings attests to the centrality of this contemporary interest in luminous and optical phenomena, ranging from his early lyric production which celebrates the lady's luminosity to the *Convivio* in which he outlines a theory of vision, and onto the *Paradiso* with its references to optical lore and its vast proliferation of light imagery. Dante's recurrent concern with light and optics is often developed in such a technical vein that he has been associated with the latest developments in the science of optics or *perspectiva* and with several prominent thirteenth-century theologians and philosophers who used light in order to develop a set of doctrines which is today known as the "metaphysics of light."

One of the principal objectives of this book will be to show that Dante was not at the forefront of thirteenth-century thought on optics and light but that he relied instead on information which was available in more general medieval sources. To achieve this aim, it will be necessary to pay the closest possible attention to a wide variety of primary sources in order to reconstruct in as much detail as possible the background to specific areas of Dante's thought on light, optics, and vision. Of course, there is nothing new about this approach: it is the style of scholarship pioneered above all by Bruno Nardi, who has done more than any other scholar to ensure that notions fostered earlier this century of a "Thomist" Dante have now been supplanted, in contemporary Dante studies at least, by a sense of his intellectual eclecticism and syncretism. It is, however, my hope that a detailed historical study of the kind proposed here will allow us to get beyond useful but generic labels such as Dante the syncretist in order to appreciate more precisely how he relies upon, reuses, and on occasion departs from contemporary ideas related to light and optics in his prose works and especially in his poetry. With this aim in mind, the many strands of Dante's thought on light and vision will be situated in their historical contexts, but my ultimate purpose is neither to engage solely in the sometimes futile search for precise sources nor to concentrate exclusively on Dante the thinker. Apart from establishing an accurate set of historical contexts for Dante's own ideas, then, this book will also be concerned with investigating his innovative treatment of traditional materials through analysis of the formal and structural features of his writings, especially in a poetic work which is as rich, many-layered, and intricate as the *Comedy*. The *Comedy* has its own principles of organisation, its own syntheses, its own stylistic and narrative practices; and to ignore these is to fail to do justice to Dante's craft as a poet and to his originality as a thinker. It is with these considerations in mind – the need for detailed historical contextualisation, for caution and discrimination over the question of precise sources, and for a sense of the uniquely transforming quality of Dante's poetry – that this study aims to outline Dante's general debts to, and specific reworkings of, scientific, philosophical, and theological writings which deal with light, vision, and optics.

The study is divided into two parts. Part One (Chapters 1-4) deals with Dante's knowledge and use of the medieval science of optics or *perspectiva*. Despite a recent revival of critical interest in Dante and the medieval scientific tradition, medieval optics is a discipline which has received surprisingly little critical coverage.[1] Chapters 1-4 address this omission by offering a detailed

---

[1]For recent work on natural philosophy in Dante, see Patrick Boyde, *Dante Philomythes and Philosopher: Man in the Cosmos* (Cambridge: Cambridge University Press, 1981); idem, *Perception and Passion in Dante's "Comedy"* (Cambridge: Cambridge University Press, 1993); Maria Corti, *La felicità mentale: Nuove prospettive su Cavalcanti e Dante* (Turin: Einaudi, 1986). On Dante's astronomy, see Ideale Capasso, *L'astronomia nella Divina Commedia* (Pisa: Domus Galilaeana, 1967); Corrado Gizzi, *L'astronomia nel poema sacro*, 2 vols. (Naples: Loffredo, 1974). For astrology, see Richard Kay, *Dante's Christian Astrology* (Philadelphia: University of Pennsylvania Press, 1992). Dante's medical references are examined by Patrizia Bertini Malgarini, "Il linguaggio medico e anatomico nelle opere di Dante," *Studi danteschi* 61 (1989), 29-108; Mario Mattioli, *Dante e la medicina* (Naples: Edizioni scientifiche italiane, 1965); Nancy G. Siraisi, "Dante and the Art and Science of Medicine Reconsidered," in *The Divine Comedy and the Encyclopaedia of the Arts and Sciences*, ed. Giuseppe di Scipio and Aldo Scaglione (Amsterdam and Philadelphia: John Benjamins, 1988), pp. 223-245. For two collections of essays covering a wide, but not comprehensive, range of subjects, see *The Divine Comedy and the Encyclopaedia of the Arts and Sciences*, ed. di Scipio and Scaglione; *Dante e la*

4

analysis of Dante's relationship to the medieval optical tradition and investigating his creative deployment of its terminology and doctrines, especially in the *Comedy*. The first chapter provides a general introduction to the medieval optical tradition, establishing the background for a detailed consideration, in Chapter 2, of his use of the visual pyramid, light terminology, and visual theory in the *Rime*, the *Vita Nuova*, and the *Convivio*. Chapter 3 addresses the narrative implications of the optical heritage on the ways Dante presents the visual act in the *Comedy*, with particular attention to the motifs of dazzling and blinding as well as to passages involving optical illusions. Chapter 4 reconsiders Dante's treatment of reflection and refraction and it also examines the poet's use of scientific ideas in imagery which is based on the law of light reflection, mirrors, and meteorological phenomena with optical causes such as the rainbow and the moon-halo.

Part Two retains light as its unifying theme but focuses upon Dante's use of concepts of light in his minor works and especially in the *Paradiso*. Chapter 5 presents a critique of the "metaphysics of light," the dominant critical category which is used to assess Dante's doctrinal treatment of light. It outlines instead a more context-related approach to the intellectual matrices which underpin Dante's pronouncements and poetic disquisitions about the role of light in creation, cosmology, the angelic intelligences, the transmission of causal power, and the Empyrean. Chapter 6 examines Dante's views on the role of light in his cosmogony and cosmology by comparing his own elaborations with a wide range of medieval views. The final chapter considers Dante's intellectual relationship to theological writings about light, offers a broad inquiry into the poet's use of analogies drawn from light in the *Paradiso*, and gives specific consideration to his doctrine of the Empyrean.

---

scienza, ed. Patrick Boyde and Vittorio Russo, 2 vols. (Ravenna: Longo, 1996). A general introductory work is Beniamino Andriani, *Aspetti della scienza in Dante* (Florence: Le Monnier, 1981).

In the above ways, then, this book as a whole aims to contribute not only to the understanding of Dante's intellectual orientations and poetic practices, but also to the history and reception of ideas related to optics and light in the Middle Ages.

PART ONE

Optical Science in Dante's Thought and Poetry

CHAPTER 1

The Science of "Perspective": Light, Vision, and Optics in the Thirteenth Century

In the thirteenth and fourteenth centuries, the medieval science of optics or *perspectiva* became the subject of sustained intellectual inquiry, attained a high level of technical sophistication, and found its way onto university *curricula* throughout Europe. By 1600, the Latin term *perspectiva* had acquired a more limited meaning and was widely understood as *perspectiva artificialis*, the geometrical technique used by artists to create illusory spatial effects on a plane surface. But before 1450, *perspectiva* only ever referred to the medieval science of optics, a science which dealt with a variety of issues associated with the visual process, examining not only the preconditions and geometry of vision, but also the anatomy of the eye and the processes by which rational faculties in the brain reached a final judgement of the visual object. The medieval treatises on *perspectiva*, which date from the 1260s and are known as the *perspectivae*, analysed vision according to whether the visual act occurred by direct, reflected, or refracted light rays, and these writings therefore provided detailed discussions of light and colour effects, light reflection and mirrors, and phenomena involving refracted light in the atmosphere.

The principal contribution to Dante's relationship to the science of *perspectiva* so far has come from an historian of art, Alessandro Parronchi, in an

8

essay published in the 1959 edition of *Studi danteschi*.[1] Having distinguished between medieval and Renaissance "perspective," Parronchi reviewed previous scholarship and outlined his own approach which endeavoured to establish the parallels between Dante's optics and several of the later thirteenth-century *perspectivae*. In the course of his essay, he provided a detailed account of the many possible connections between Dante and these works, beginning with the *Convivio* and the *canzone* "Amor, che movi tua vertù da cielo" and paying detailed attention to the *Comedy*. After analysing Dante's descriptions of the visual act and optical illusions (direct vision), his use of light reflection and mirror imagery (reflected vision), and his comparisons based on meteorological phenomena (refracted vision), he summarised his findings and drew the following conclusions:

Se Dante, come ci sembra provato, raggiunse dunque una più esatta coscienza del fenomeno della visione a traverso lo studio dei trattati di *perspectiva*, le singole rispondenze potranno esser distribuite in diversi gradi [...] potremo rilevare come spesso i testi dei trattati offrano soltanto la spiegazione scientifica del passo [...] altre volte sembrano aver suggerito le immagini; qualche rara volta sembra abbiano offerto al poeta i termini stessi dell'espressione. Infine, in tre casi, e cioè nel trattato III cap. ix del *Convivio*, nella canzone, "Amor, che movi tua vertù da cielo," e nel canto XXIII del *Paradiso*, la *scientia perspectiva* pare aver fornito una traccia su cui s'è volta l'intera composizione.[2]

---

[1] Alessandro Parronchi, "La perspettiva dantesca," *Studi danteschi* 36 (1959), 5-103; reprinted in *Studi su la dolce prospettiva* (Milan: Martello, 1964), pp. 3-90.

[2] Parronchi, "La perspettiva dantesca," p. 102. For Parronchi's detailed range of correspondences between the *Comedy* and the three main subject areas of medieval optical treatises, see: (i) direct vision (pp. 47-67); (ii) reflected vision (pp. 67-81); and (iii) refracted vision (pp. 82-102). In his *Enciclopedia dantesca* entry, he argued that "[d]ai versi della *Commedia* si può estrarre un compendio dell'intera materia della perspettiva, restituendole l'ordine che essa ha nei trattati" (IV, p. 438).

In recent years, Parronchi's work has met with widespread approval, and it is now generally accepted that Dante was directly acquainted with the *perspectivae*.[3] And yet, despite the general support and enthusiasm for Parronchi's essay, it is important to appreciate that he did not provide a detailed historical background to Dante's knowledge of medieval optics. Parronchi noted that few attempts had been made to establish the "necessaria storicizzazione" of Dante's optical references, but he did not undertake such a study himself because he was primarily concerned with establishing correspondences. The chapters in Part One attempt to remedy this situation by relating Dante's optical and visual thought as closely as possible to their ancient and medieval traditions. With the assistance of the critical editions and commentaries provided by historians of medieval optics, especially David Lindberg, the remainder of this chapter will outline the intellectual patrimonies that came to form part of the medieval optical tradition.[4] For

[3]Egidio Guidubaldi was one of the first Dantists to endorse Parronchi's general thesis and to extend its field of inquiry to other thirteenth-century optical specialists; see his *Dante Europeo*, 3 vols. (Florence: Olschki, 1965-1968), II, pp. 237-254, 337-361, 383-407; III, pp. 267-318, 297-318. For recent supporters of Parronchi's essay, see Peter Dronke, *Dante and Medieval Latin Traditions* (Cambridge: Cambridge University Press, 1986), p. 36; idem, "Tradition and Innovation in Medieval Western Colour-Imagery," *Eranos Jahrbuch* 41 (1972), nn. 106 and 144, pp. 88, 105; Paul Hills, *The Light of Early Italian Painting* (New Haven and London: Yale University Press, 1987), p. 14 and n. 29, p. 150; Malgarini, "Il linguaggio medico e anatomico," pp. 32-33; Bortolo Martinelli, *"Esse* ed *essentia* nell'Epistola a Cangrande (capp. 19-23)," *Critica letteraria* 12 (1984), nn. 80-81, pp. 661-662; Gianni Oliva, *Per altre dimore: Forme di rappresentazione e sensibilità medievale in Dante* (Rome: Bulzoni, 1991), pp. 115-116; Monica Rutledge, "Dante, the Body and Light," *Dante Studies* 113 (1995), pp. 151-152; Claudio Varese, "Parola e immagine figurativa nei canti del Paradiso Terrestre," *Rassegna della letteratura italiana* 94 (1990), pp. 40-41; Cesare Vasoli, *Il Convivio*, in *Opere minori*, tomo I, parte II (Milan and Naples: Ricciardi, 1988), pp. 130-131; see also introd. p. lxxiii.

[4]See esp. David C. Lindberg's *Theories of Vision from al-Kindi to Kepler* (Chicago and London: Chicago University Press, 1976) and his other articles listed below in the notes and bibliography. Several of Lindberg's more important articles are collected in his *Studies in the History of Medieval Optics* (London: Variorum, 1983). On medieval optics, see also A. Mark Smith, "Getting the Big Picture in Perspectivist Optics," *Isis* 72 (1981), 568-589; idem, introd. to *Witelonis Perspectivae liber quintus* (Wroclaw: Ossolineum, 1983), pp. 18-44, 58-72; Katherine H. Tachau, *Vision and Certitude in the Age of Ockham: Optics, Epistemology, and the Foundations of Semantics, 1250-1345* (Leiden and New York: E.J. Brill, 1988); Graziella Federici Vescovini, *Studi sulla prospettiva medievale* (Turin: Giappichelli, 1965). For good short surveys of the development and subject-matter of medieval optics, see A.C. Crombie, "Expectation, Modelling and Assent in the History of Optics: Part 1. Alhazen and the Medieval

10

convenience of exposition, I will follow Lindberg in dividing medieval optics into geometrical, physical, and medical traditions.[5] A general introduction to each of these three areas will help to provide a clearer indication of the ideas that shaped medieval thinking about light, colour, the eye, and the process of vision before 1300. This introduction will pay particular attention to the reception and assimilation of Greek and Arabic ideas on optics in the thirteenth century. Historians of science have been primarily interested in the *perspectivae*, the most sophisticated and influential category of thirteenth-century optical writing. However, these treatises were not the only contemporary works which discussed optical matters; and, if we are to do justice to the extensive medieval literature on optics before Dante, we need to draw attention to a wide range of other sources. Once this historical conspectus is complete, it will then be possible to make fuller and more accurate estimates of Dante's relationship not only to the *perspectivae* but also to the various other strands of the medieval optical tradition. We will then be in a better position from which to explore the ways in which he invokes and redeploys this heritage in dealing with optical and visual phenomena in his own works of prose and poetry.

Geometrical Optics: Euclid and Ptolemy

The Greek geometer and optician, Euclid (fl. 300 B.C.), was the first writer to use geometrical principles to explain the perception of space. In the *Optica*, he accounted for the relative sizes, shapes, and *loci* of objects in the visual field by

---

Tradition," *Studies in the History and Philosophy of Science* 21 (1990), 605-632; David C. Lindberg, "The Science of Optics," in *Science in the Middle Ages*, ed. David C. Lindberg (Chicago and London: Chicago University Press, 1978), pp. 338-369.

[5]For Lindberg's trichotomy which divides ancient, Islamic, and medieval optics into geometrical, physical, and medical traditions, see *Theories of Vision*, passim, but esp. pp. 1, 57.

means of rectilinear light rays.[6] As Euclid stated in the opening propositions of his *Optica*, these straight-line rays issued from the eyes to form a visual cone which had its apex at the centre of the eye and its base on the surface of the visual object:

Ponatur ab oculo eductas rectas lineas ferri spacio magnitudinum immensarum. Et sub visibus contentam figuram conum esse verticem quidem in oculo habentem basim vero ad terminos conspectorum.[7]

By measuring the distance travelled by straight-line rays before they intersected a given object, Euclid was able to ascertain its location in the visual field. Moreover, the angle which the visual cone subtended in the eye could be calculated and used to determine the size and shape of visual objects: a distant object, by narrowing the base of the cone, would subtend a smaller angle, and appear proportionately smaller to the viewer. Euclidean optics was so successful in dealing with spatial reality that even Greek thinkers who had quite different views on the nature and direction of the visual process applied visual rays to optical inquiries.[8]

It is now recognised that the Alexandrian mathematician and astronomer, Ptolemy (fl. 127-147 A.D.), did most to refine Euclid's analysis of vision and to establish the tradition of geometrical optics. Ptolemy modified Euclid's account of

---

[6]For ancient sources on the principle of the rectilinear transmission of light, see e.g. Plato, *Parmenides*, 137E; Pseudo-Aristotle, *Problemata*, XI, 49, 904b 12, Latin trans. by Bartholomew of Messina (c. 1258-1266) in *Aristotele: Problemi di fonazione e di acustica*, ed. Gerardo Marenghi (Naples: Libreria scientifica editrice, 1962), p. 115.

[7]Propositions I and II of Euclid's *Optica* quoted from the critical edition from medieval manuscripts in Wilfred R. Thiesen, *"Liber de visu*: The Greco-Latin Translation of Euclid's *Optics,*" *Mediaeval Studies* 41 (1979), p. 62: "Let it be assumed that straight lines drawn from the eye pass through a great extent of space, and that within what is seen is a cone the vertex of which is at the eye and the base of which is at the limits of the objects seen."

[8]Aristotle, for example, accepted Euclidean extramitted visual rays in certain contexts, despite his advocacy of an intramission theory of vision, see *De caelo*, II, 8, 290a 17-25; *Meteorologica*, III, 4, 373b 3-10.

12

vision by stressing the varying sensitivities in the visual pyramid, especially the acuity of its central ray(s):

Et quia visibilis radius est super capud unius puncti sibi proprii, id quod videtur per medium radiorum visus, illum videlicet qui est super axem, magis debet videri quam illud quod in lateribus per laterales aspicitur [...] oportet ut aspectus eius quod elongatur a capite piramidis, fiat cum radio minori in virtute quam illius cuius distantia est moderata.[9]

Ptolemy also introduced other elements into his conception of vision such as the perception of colour, and he attached some importance to the role of mental powers in processing visual information.[10] More importantly still, he established refraction as a new subject area in geometrical optics by paying close attention to the way in which light rays changed direction as they passed through media of varying densities. After Ptolemy, it became standard to divide optical science into three branches according to whether the rays determining vision were direct, reflected, or refracted.[11] Like Euclid, however, Ptolemy's work on optics was

[9]*Optica*, lib. II, c. 20, ed. Albert Lejeune, in *L'optique de Claude Ptolémée dans la version latine d'après l'arabe de l'émir Eugène de Sicile*, new edn (New York and Leiden: E.J. Brill, 1989), pp. 20-21: "and since every visual ray reaches a single point which is proper to it, what is seen by the central visual ray, namely, that which is on the axis, must be more clearly visible than that which, on the sides, is seen by lateral rays [...] it is necessary that the vision of that which is distant from the vertex of the cone takes place with a ray of lesser strength than the vision of that which is less distant." For a more detailed discussion of Euclidean and Ptolemaic optics, see Albert Lejeune, *Euclide et Ptolémée: Deux stades de l'optique géométrique grecque* (Louvain: Bibliothèque de l'Université de Louvain, 1948); Lindberg, *Theories of Vision*, pp. 11-17; A. Mark Smith, *Ptolemy's Theory of Visual Perception: An English Translation of the "Optics" with Introduction and Commentary* (Philadelphia: American Philosophical Society, 1996), introd. pp. 14-49.

[10]See A. Mark Smith, "The Psychology of Visual Perception in Ptolemy's *Optics*," *Isis* 79 (1988), 189-207.

[11]This tripartite distinction was adopted in Islamic optical treatises, see e.g. Alhazen's *De aspectibus* in which one finds the following subdivisions: (direct vision), Bks I-IV; (reflected vision), Bks V-VI; (refracted vision), Bk VII. The distinction is repeatedly found (but with differing terminology) in medieval works, see Bartholomew the Englishman, *De rerum*

primarily concerned with explicating the causes of perceptual illusions; and a large part of his *Optica* is devoted to geometrical explanations of visual errors (Book II, props. 84-142). The order of Ptolemy's treatise helps to give us a clearer idea of the subject matter of ancient geometrical optics. Book One is unfortunately no longer extant, but the remainder of the treatise is divided as follows. Part one, which is concerned with optics proper, deals with direct vision and provides scientific explanations for visual errors (Book II). The second branch of optics, known as catoptrics, treats vision by reflected rays and examines problems of image formation in various types of mirrors, plane, convex, concave, conical, and pyramidal (Books III and IV). The third part studies the refraction of light both in bodies of different density and in the atmosphere (Book V).

Both Euclid's and Ptolemy's optical works were available to the Latin West from the early twelfth century onwards in Latin translations from Greek and Arabic texts.[12] And Euclidean-Ptolemaic optics was in fact to become so popular in the late Middle Ages and the Renaissance that the geometrical analysis of vision not only formed the foundation of medieval optics but also insinuated itself into medieval astronomical treatises, encyclopedias, Aristotelian commentaries, theological writings, and, from the mid-fifteenth century, works on painterly "perspective."[13] The work of Euclid and Ptolemy was, however, far less

---

proprietatibus, lib. III, c. 17, ed. Wolfgang Richter (Frankfurt, 1601; facsimile, Frankfurt-am-Main: Minerva, 1964), pp. 62-63; Pseudo-Aquinas, *In De meteorologicorum*, lib. III, lect. 3, § 272, ed. Raymund M. Spiazzi (Turin and Rome: Marietti, 1952), pp. 619-620; William of Conches, *De philosophia mundi*, lib. IV, cc. 26-27 (*PL* 172, cols 96-97).

[12]Euclid's work on optics was translated in the late eleventh century, twice from the Arabic and once from the Greek. On the translations and extant manuscripts, see David C. Lindberg, *A Catalogue of Medieval and Renaissance Optical Manuscripts* (Toronto: Pontifical Institute of Mediaeval Studies, 1975), pp. 209-210; idem, *Theories of Vision*, pp. 209-213.

[13]"Radi[i] nostr[i] visuales" are mentioned in chapter 1 of Sacrobosco's extremely popular astronomical treatise, *De sphaera* (c. 1215), see *The Sphere of Sacrobosco and Its Commentators*, ed. and trans. Lynn Thorndike (Chicago: Chicago University Press, 1949), p. 81. For references to the geometry of vision in thirteenth-century encyclopaedias, Aristotelian commentaries, and theological writings, see the references given below in Ch. 1, nn. 54-55; Ch. 2, nn. 24-26; Ch. 3, nn. 56-57. In the 1430s, the architect and humanist, Leon Battista Alberti, used the optics of the visual pyramid as the basis for constructing pictorial space; see his seminal

14

satisfactory in explaining other aspects of the visual process. Euclid did not, for instance, discuss how images were received into the eye, and he gave no analysis of the physical nature of the rays themselves. While Ptolemy paid some attention to the nature of colour and rays, his conception of the science remained, like Euclid's, an account of optical phenomena based upon straight lines and angles. Neither writer attempted to explain how images were sent from the object to the eye of a percipient.

Physical Theories of Vision: Plato and Aristotle

A response to these more "physical" questions can be found in many early Greek philosophers, but the most formative theories of vision for medieval thinkers were those elaborated by Plato and Aristotle. In the *Timaeus*, Plato (d. 347 B.C.) proposed a doctrine which combined extramission (i.e. visual rays emitted from the eye) with the idea that seeing involved some sort of inward reception or intramission. He argued that a stream of visual fire went out from the eye, coalesced with external light, and then, having reached a sensory object, returned its form to the eye.[14] This version of extramission was available to medieval Europe in Chalcidius' fourth-century translation and commentary, and it came to exercise a dominant influence on all major philosophers and theologians before

---

exposition (c. 1435) of *perspectiva artificialis* in *De pictura*, lib. I, cc. 5-8. ed. and trans. Cecil Grayson, in *"On Painting" and "On Sculpture": The Latin Texts of "De Pictura" and "De statua"* (London: Phaidon, 1972), pp. 38-44. On painterly "perspective," see Samuel Y. Edgerton, Jr., *The Renaissance Rediscovery of Linear Perspective* (New York: Harper Row, 1975); Hills, *The Light of Early Italian Painting*, pp. 64-72; Martin Kemp, *The Science of Art: Optical Themes in Western Art from Brunelleschi to Seurat* (New Haven and London: Yale University Press, 1990), part 1.

[14]For Plato's theory, see *Timaeus*, 45B-D in the Latin translation by Chalcidius, in *Timaeus a Calcidio translatus commentarioque instructus*, ed. J.H. Waszink and P.J. Jensen (London and Leiden: Warburg Institute, 1962), pp. 41-43. For "physical" theories of vision in other early Greek thinkers, see J.I. Beare, *Greek Theories of Elementary Cognition from Alcmaeon to Aristotle* (Oxford: Clarendon Press, 1906), pp. 22-37, 99-102; Lindberg, *Theories of Vision*, pp. 3-6, 91-94.

1200.[15] The section of the *Timaeus* available to the Middle Ages (until 53B) also contained a short encomium to the visual sense and some discussion of the properties of mirrors. Chalcidius' commentary supplemented this information with observations about optical illusions, rival ancient theories of vision, the tripartite division of optics, and the anatomy of the eye.[16]

By the second half of the thirteenth century, Aristotle's explanation of vision had begun to supplant Plato's as the dominant visual model in the universities of the Latin West.[17] In his *De anima*, and its accompanying treatise, *De sensu*, Aristotle put forward an intramission theory of vision in which he applied the philosophical concepts of act and potency to light, colour, and the all-important intervening medium. He conceived light as a state of the transparent medium by virtue of which this medium became transparent in act and could be seen through; and he regarded colour as a quality inherent in the surface of bodies which was made active by light. The transparent medium, or diaphanous as it was called, was the nature or state found in all bodies, including air, water, and the fifth element, the aether, of which the heavens were composed.[18] The diaphanous

---

[15]Chalcidius's commentary establishes three preconditions for vision; see his *Comm. in Timaeus*, c. 245 (Waszink, p. 256): "lumen caloris intimi per oculos means [...] lumen extra positum [...] lumen quoque, quod ex corporibus visibilium specierum fluit" ("the light of inner heat emanating through the eyes [...] external light [...] and also light of visible species which is given off by bodies"). On the later development of Platonic visual theories amongst members of the so-called "Chartrian school," see Tullio Gregory, *Anima Mundi: La filosofia di Guglielmo di Conches e la scuola di Chartres* (Florence: Sansoni, 1955), p. 173. The extramission theory never completely lost its appeal (see Ch. 2, n. 70), and it was especially popular with medieval and Renaissance lyric poets who repeatedly describe the lady's power over the lover with references to her effulgent eye rays.

[16]On Plato's praise of vision and the properties of convex and concave mirrors, see *Timaeus*, 47A-C; 51-52B (Waszink, pp. 44-45, 49-51). Further elaborations of visual and optical doctrine by Chaldidius include: (i) visual errors, *Comm. in Timaeus*, c. 271 (pp. 248-251); (ii) ancient visual theory and the tripartite distinction of optics, c. 272 (p. 250); (iii) eye anatomy, c. 280 (p. 257).

[17]For Aristotle's theory of vision, see *De anima*, II, 7, 418a 24-419b 2; *De sensu*, 3, 439a 7-440b 25. For a more complete discussion, see Beare, *Greek Theories*, pp. 56-92; Lindberg, *Theories of Vision*, pp. 6-9; Smith, "Getting the Big Picture," pp. 571-572.

[18]*De anima*, II, 7, 418b 5-10; *De sensu*, 3, 439a 21-25.

16

occupied a fundamental place in Aristotle's explanation of vision, for only coloured forms of visual objects were seen by the eye, but to reach the organ of sight, these forms had to traverse the medium. Light was not seen as such, but it was required to bring the medium to a state of actual transparency.[19] Although rather elliptical, Aristotle's discussion of the external phases of vision did have one important advantage over many earlier and later theories insofar as it provided a coherent analysis of the visual act from the coloured object up to the eye.

In the *De anima*, he offered no detailed discussion of the anatomy of the eye and its relation to the seat of perception, but he did supply a fairly full account of the process of visual perception itself. The eye was said to perceive coloured forms directly, because colour was its immediate object of vision, or "proper sensible." Other non-visible properties such as shape, magnitude, and number (termed collectively the "common sensibles") were evaluated by a special faculty called the "general sensibility." As is well known, Aristotle believed that thinking took place through a process of abstraction based on sensory images.[20] The "general sensibility" (or *sensus communis* as it was later called by the schoolmen) assisted in the formation of concepts by passing a unified impression of a sensory object to the "imagination" (*phantasia*), the post-sensory faculty which received sensible forms and made them available to the intellect. In this way, Aristotle established a close connection between vision and the process of intellection, one which was to be developed by medieval writers and to prove influential well into the Renaissance.[21]

---

[19]References in order: (i) colour as the "proper sensible" of sight, *De anima*, II, 6, 418a 13; (ii) light actualises the potentially transparent; colour alters the actualised transparent, II, 7, 419a 9-12; (iii) eye receives coloured effect of the medium, II, 7, 419a 13-15.

[20]For Aristotle's discussion of the "common sensibles," see *De anima*, II, 6, 418a 7-25, esp. 17-20; III, 1, 425a 15-425b 10; *De sensu*, 4, 442b 4-10; 7, 449a 2-19. For the idea that universals are abstracted from sensible forms, see e.g. *De anima*, III, 7, 431a 14-17; 8, 432a 3-10; *De memoria*, I, 449b 30; *Analytica posteriora*, II, 19, 100a 16.

[21]For the cognitive aspects of Aristotelian visual theory and their influence on later optical writers, see Smith, "Getting the Big Picture"; Tachau, *Vision and Certitude*, passim. On the

In his writings on vision, Aristotle did not use geometrical principles to treat the crucial problem of how objects are seen less clearly at a distance. But he did possess a sound knowledge of Greek optics, and he employed visual rays and the cone to provide explanations for star scintillation in the *De caelo* and for other atmospheric phenomena in Book III of his *Meteorologica*.[22] Aristotle was also concerned with determining the value of optics as a scientific discipline. In so doing, he relied upon principles derived from what is now known as his theory of three grades of abstraction in which theoretical knowledge is divided into physics, mathematics, and metaphysics.[23] Physics is concerned with natural bodies in movement; mathematics with the numerical attributes that have been abstracted from these bodies; and first philosophy (Aristotle did not call it "metaphysics") with being *qua* being independent of matter and movement. In Book II of his *Physica*, Aristotle reworked the distinction between physics and mathematics to determine the theoretical value of optical science. While he allowed optics to make use of mathematical principles because vision is effected along straight lines, he argued that optical science is primarily concerned with the physical, rather than the mathematical, line:

Demonstrant autem et quae magis physica quam mathematica, ut perspectiva et harmonica et astrologia: e contrario enim quodammodo se habent ad geometriam. Geometria quidem enim physicam intendit lineam, sed non inquantum est physica:

---

historical development of Aristotelian ideas which related the process of vision to that of knowing, see David Summers, *The Judgment of Sense: Renaissance Naturalism and the Rise of Aesthetics* (Cambridge: Cambridge University Press, 1987), passim.

[22]See n. 8 above.

[23]The classical discussion is *Metaphysica*, VI, 1, 1026a 13-16. For Aristotle's other elaborations of this theory, see *Physica*, II, 2, 193b 31-194a 12; *De anima*, I, 409b 7-17; III, 430b 15-20. Boethius was influential in transmitting Aristotle's division of *speculatio* into *naturalis*, *mathematica*, and *theologica* in his *De Trinitate*, lib. II (*PL* 64, cols 1250-1251).

18

sed perspectiva quidem mathematicam lineam, sed non inquantum mathematica, sed inquantum est physica.[24]

In the *Posteriora analytica*, he also made it quite clear that, unlike geometry, optics is not a demonstrative discipline in its own right. In Book I, Chapter 13, Aristotle considered the way in which knowledge of the fact (*quia*) differs from knowledge of the reasoned fact (*propter quid*), and argued that *perspectiva* is not a *propter quid* science, since it relies on the *a-priori* principles of a superior discipline (geometry).[25]

The Medical Tradition of Optics

Besides geometrical and physical theories of sight, medieval writers inherited a set of medical writings that dealt with the anatomy and physiology of the visual sense. Galen (c. 130-201 A.D.) was the leading figure in establishing this tradition, and greatly enriched the knowledge of eye anatomy bequeathed to him by Greek writers. By dividing the eye up into a number of humours and tunics, he provided the first rational, even if erroneous, understanding of the anatomy of the eye. He attributed an important visual function to one of these humours, the glass-like *crystallinus*, which he located in the anterior part of the eye. According to Galen –

---

[24]*Physica*, II, 2, 194a 7-12, quoted from *In De Physicorum*, ed. M-P. Maggiòlo (Turin: Marietti, 1965), p. 239: "Sciences that are more physical than mathematical, such as perspective, harmonics, and astronomy, also provide demonstrations, and they are in a way the converse of geometry. Geometry deals with the physical line but not insofar as it is physical as such, whereas perspective deals with the mathematical line, but as a physical entity, rather than a mathematical one." See further R.D. McKirahan, "Aristotle's Subordinate Sciences," *British Journal for the History of Science* 11 (1978), 197-220.

[25]*Posteriora analytica*, I, 13, 78b 32-79a 13.

and all medieval writers accepted his view – the crystalline humour was the sentient organ of vision.[26]

The role played by the nerves was also a significant feature of Galenic physiology, especially as far as vision was concerned. Galen believed that the nerves that led from the brain were filled with a luminous and highly active spirit (*pneuma psychikon*) which flowed into the crystalline humour of each eye. The excess light deposited by these nerves was central to the Galenic (and Stoic) doctrine of vision, since a flow of light issuing from the eyes excited the ambient air and illuminated the object of vision. By grafting Ptolemaic visual geometry onto his own theory, Galen maintained that this flux of luminous pneuma formed a cone of sentient air, and acted as a sort of extended nerve to relay sensations to the crystalline humour. The visual process was completed when the nerves transported sensory impressions from the crystalline back along the nerve-network to the brain. The brain, regarded by Galen as the principal seat of perception,[27] was divided into four pneuma-filled chambers, all of which had an active role in evaluating visual data. By directing attention to the pneumatic composition of the nerves and cerebral ventricles, Galen thus helped to provide a more detailed explanation of the mechanisms by which a visual impression could be transmitted from an object of vision to the eye and from there to the brain.[28] Even though the recovery of Galen's works did not begin to take place until the late thirteenth

---

[26]Galen, *De usu partium*, lib. X, c. 1, ed. and trans. Margaret T. May, in *Galen on the Usefulness of the Parts of the Body*, 2 vols. (Ithaca: Cornell University Press, 1968-1972), II, p. 463. On Galen's physiology of vision, see Rudolph E. Siegel, *Galen on Sense Perception* (Basel and New York: Karger, 1970), pp. 40-45. On Galen's medical doctrines in general, see David C. Lindberg, *The Beginnings of Western Science: The European Scientific Tradition in Philosophical, Religious, and Institutional Context, 600 B.C. to A.D. 1450* (Chicago and London: University of Chicago Press, 1992), pp. 125-131.

[27]According to Aristotle, the heart was the seat of all sensation and consciousness, see e.g. *De gen. animalium*, II, 6, 743b 25.

[28]Galen, *De usu partium*, lib. VIII, c. 6 (May, I, pp. 398-407). On Galen's visual doctrine, see Lindberg, *Theories of Vision*, pp. 10-11, 38, 40-41; Siegel, *Galen on Sense Perception*, pp. 45-123.

20

century, his ideas were widely available from the early twelfth century onwards in other sources, especially Latin translations of Arabic works.[29]

The three principal traditions discussed so far did on occasion interfere with one another, but it is fair to say that they remained largely independent enterprises with different aims and different authorities.[30] As such, all three traditions were refined and re-elaborated by Arabic philosophers to whom detailed attention must now be paid, because they made important innovations which thirteenth-century writers frequently used in developing their own ideas on vision and optics.[31]

From Arabic Optics to Thirteenth-Century *Perspectiva*

Arabic physicians and medical writers, such as Hunayn ibn Ishaq (809-877), re-elaborated the anatomical details present in the Galenic tradition, and gave more analytical descriptions of the parts of the eye. At the end of the eleventh century, Constantine the African (1015-1087) translated Hunayn's *Ten Treatises* as the *De*

---

[29]On the textual tradition of Galen's works before 1350, see Enzo Volpini, "Galeno, Claudio," in *ED* III, pp. 85-86.

[30]Galen, for example, placed his detailed conception of eye anatomy within the mathematical laws of perspective, see *De usu partium*, lib. X, cc. 12-15 (May, II, pp. 490-499); and the mathematician Ptolemy adopted Aristotle's "common sensibles" and his theory of colour perception, see *Optica*, lib. II, cc. 2 and 5 (Lejeune, pp. 12-13). From its inception, however, optics involved disciplines with quite different methodological approaches. For the problems associated with the hybrid nature of optics, see A.C. Crombie, *Science, Optics and Music in Medieval and Early Modern Thought* (London: Hambledon, 1990), pp. 176-184; David C. Lindberg, "The Intromission-Extramission Controversy in Islamic Visual Theory: Alkindi versus Avicenna," in *Studies in Perception: Interrelations in the History and Philosophy of Science*, ed. Peter K. Machamer and Robert G. Turnbull (Columbus, OH: Ohio State University Press, 1978), pp. 137-159.

[31]On the introduction of Arabic optics to medieval Europe, see David C. Lindberg, "The Western Reception of Arabic Optics," in *Encyclopaedia of the History of Arabic Science*, 3 vols. ed. Roshdi Rashed (London and New York: Routledge, 1996), II, pp. 716-729.

oculis, and this translation became one of the earliest and most popular sources of eye anatomy available to the Latin West.[32]

Alkindi (d. c. 873) was, however, the first Islamic thinker to contribute to the overall assimilation of Greek and Hellenist optical traditions; and his direct influence on thirteenth-century opticians is especially noteworthy. In a work known in Latin translation as *De aspectibus*, Alkindi reformed the Euclidean and Ptolemaic idea of the visual cone, by asserting that visual rays did not form discrete lines within this cone, but emanated from every point on the surface of the eye. This notion of dividing the visual field into a series of points was, as we shall see, one of the fundamental governing principles in the work of later Arabic and medieval optical specialists.[33]

In another influential treatise, the *De radiis* (or *Theorica artium magicorum*), Alkindi elaborated a theory of universal radiation based on the action of light rays; and it is this theory which helps to explain how some thirteenth-century authors were later to raise optics to the status of a universal science. In the *De radiis*, Alkindi argued that rays emanated in all directions not only from the eyes, but from all things in the cosmos, including words:

[...] manifestum est quod res huius mundi sive sit substantia sive accidens radios facit suo modo ad instar siderum [...] Hoc ergo pro vero assumentes dicimus quod

---

[32]Constantine also translated the *Pantegni* by 'Ali ibn al- 'Abbas. Several other Islamic works containing anatomical details were translated by Gerard of Cremona (d. 1167) in the mid-twelfth century; for further details, see Lindberg, "The Western Reception," pp. 718-719.

[33]Alkindi established the principle in his *De aspectibus*, prop. 13, ed. Axel Anthon Björnbo and Sebastian Vogl, "Alkindi, Tideus, und Pseudo-Euclid. Drei optische Werke," *Abhandlungen zur Geschichte der mathematischen Wissenschaften* 26/3 (1912), p. 22: "Non ergo restat, nisi ut lumen proveniat per corpus luminosum in toto aere ab eo contento, et ut omnis locus, a quo possibile est produci lineam rectam [...] illuminetur a lumine corporis luminosi" ("Hence, it can only be that light comes from a luminous body into all its surrounding air, and that every place from which it is possible to draw a straight line [...] is illuminated by the light of the luminous body"). On Alkindi's visual theory, see David C. Lindberg, "Alkindi's Critique of Euclid's Theory of Vision," *Isis* 62 (1971), 469-489; idem, *Theories of Vision*, pp. 18-32; Vescovini, *Studi sulla prospettiva*, pp. 38-52.

22

omnem quod actualem habet existentiam in mundo elementorum radios emittit in omnem partem, qui totum mundum elementarem replent suo modo. Unde est quod omnis locus huius mundi radios continet omnium rerum in eo actu existentium.[34]

The idea that light radiation is the universal cause of terrestrial phenomena clarifies the principal reason why the authors of the *perspectivae* came to accord optics a pre-eminent role in all scientific investigations.[35] For, if natural action could be reduced to straight-line rays of light, then presumably the science which dealt with these lines (i.e. geometrical optics) could be used to explain the causal agents involved in all natural phenomena. The laws of nature were directly analogous to the laws of optics. The first thirteenth-century writer to make these ideas his own was the Bishop of Lincoln, Robert Grosseteste, who, in the 1220s and 1230s, composed several scientific treatises that express a heightened sense of the value of geometrical configurations in the study of nature. Grosseteste's conviction that the causal system operated according to geometrical patterns of light rays is present most notably in his short treatises, *De lineis, angulis et figuris* and *De natura loci*:

Utilitas considerationis linearum, angulorum et figurarum est maxima, quoniam impossibile est sciri naturalem philosophiam sine illis. Valent autem in toto universo et partibus eius absolute [...] Omnes enim causae effectum naturalium

---

[34]Alkindi, *De radiis*, c. 3, ed. M.-Th. d'Alverny and F. Hudry, "Alkindi, *De Radiis*," *AHDLMA* 41 (1974), p. 224: "it is clear that everything in this world, whether substance or accident, produces rays in the manner of stars [...] and so we maintain that truly everything which has actual existence in the world of elements emits rays in every part and these rays fill in their way all the world of elements. Hence, every place in the world contains rays from everything that has actual existence."

[35]The idea of causation through emanation, frequently expressed by analogy to light radiation, is present in Plotinus and other Neoplatonic writers (see below Ch. 6, pp. 175-183).

habent dari per lineas, angulos et figuras. Aliter enim impossibile est sciri propter quid in illis.[36]

The most respected historians of Grosseteste's thought agree that, in this way, he established geometrical optics as the primary natural science, rather than adopting the more traditional Aristotelian view of optics as subordinate to geometry.[37] Roger Bacon (c. 1214-1292), one of the authors of the *perspectivae*, developed these ideas on the mathematical configuration of natural forces into a more systematic doctrine known as the "multiplication of species,"[38] and in this context his works contain several passages that praise the universal value of optics.

---

[36]Grosseteste, *De lineis*, ed. Ludwig Baur, *Die philosophischen Werke des Robert Grosseteste, Bischofs von Lincoln*, in *Beiträge* 9 (1912), pp. 59-60: "The usefulness of considering lines, angles, and figures is very great, since it is impossible to understand natural philosophy without them. They are useful in relation to the universe as a whole and its individual parts [...] Now, all causes of natural effects must be expressed by means of lines, angles, and figures, for otherwise it is impossible to grasp their explanation," trans. in *A Source Book in Medieval Science*, ed. Edward Grant (Cambridge, MA: Harvard University Press), p. 385. See also *De natura locorum* (Baur, pp. 65-66): "varietur omnis actio naturalis [...] per varietatem linearum, angulorum et figurarum" ("Every natural action varies [...] according to the variation of lines, angles, and figures").

[37]See A.C. Crombie, *Robert Grosseteste and the Origins of Experimental Science, 1100-1700* (Oxford: Clarendon Press, 1953), esp. p. 131; Bruce S. Eastwood, "Mediaeval Empiricism: The Case of Grosseteste's Optics," *Speculum* 43 (1968), pp. 307-308; James McEvoy, *The Philosophy of Robert Grosseteste* (Oxford: Clarendon Press, 1982), p. 210; Vescovini, *Studi sulla prospettiva*, pp. 9-10; William A. Wallace, *Causality and Scientific Explanation*, 2 vols. (Ann Arbor: University of Michigan Press, 1972), I, p. 63.

[38]According to Bacon, "species" emanate from all bodies and are a form of energy responsible for all change in the sublunar world, see *Opus tertium*, c. 26, ed. J.S. Brewer, in *Opera quaedam hactenus inedita* (London: Longman, 1859), p. 99: "Hae quidem species faciunt omnem mundi alterationem et corporum nostrorum et animarum" ("These species produce all change in the world, and they also produce all change in both our bodies and our souls"). For Bacon's doctrine of species, see David C. Lindberg, *Roger Bacon's Philosophy of Nature: A Critical Edition with English Translation, Introduction, and Notes, of "De multiplicatione specierum" and "De speculis comburentibus"* (Oxford: Clarendon Press, 1983), introd. pp. liii-lxxi. On the geometricisation of natural forces in Bacon, see *Opus maius*, lib. IV, dist. 2, c. 1, ed. J.H. Bridges, in *The "Opus Majus" of Roger Bacon*, 3 vols. (Oxford: Oxford University Press, 1897-1900), I, p. 112: "Omnis autem multiplicatio vel est secundum lineas, vel angulos, vel figuras" ("And all multiplication is either by lines, or angles, or figures"). See also *Opus tertium*, cc. 31-36, in *Opera*, ed. Brewer, pp. 107-117.

24

For example, in the *Opus tertium* which was completed before 1267, he wrote that:

Et necesse est omnia sciri per hanc scientiam [sc. perspectiva], quia omnes actiones rerum fiunt secundum specierum et virtutum multiplicationem ab agentibus huius mundi in materias patientes; et leges huiusmodi multiplicationum non sciuntur nisi ab perspectiva.[39]

Another highly influential Islamic writer on vision and optics was Avicenna (980-1037), perhaps better known as a *medicus* and metaphysician, who was responsible for reaffirming, in a modified form, the validity of Aristotelian visual theory. In Book VI of his *Kitab al Shifa* (or *Liber de anima* as it was known in the medieval Latin translation), he critically sifted all the arguments in favour of the extramission theory, and, one by one, he discredited them in support of Aristotle's intramission thesis.[40] Avicenna did not receive Aristotle's visual doctrine passively and he introduced many elements from other philosophies and technical disciplines into his own theory of vision. He accepted, for example, Stoic-Galenic ideas on psychic pneuma and the interconnecting system of nerves that linked sense organs with the brain.[41] And he also revised and extended the fairly limited classifications

[39]Bacon, *Opus tertium*, c. 11 (Brewer, p. 37): "Everything must be known through this science since all natural change takes place according to the multiplication of species and virtues from agents into the potentiality of matter, and the laws of this kind of multiplication are only known by perspective." See also idem, *Opus maius*, lib. V, p. i, dist. 1, c. 1 (Bridges, II, p. 3): "[Perspectiva] est flos philosophiae totius et per quam, nec sine qua, aliae scientiae sciri possunt" ("Perspective is the crowning glory of all philosophy, and through it, not without it, can all other sciences be known"); idem, *Opus minus* (Brewer, p. 32).

[40]For Avicenna's detailed criticism of emission theory, see *Liber de anima*, pars III, c. 5, ed. S. Van Riet, in *Avicenna latinus: liber de anima seu sextus de naturalibus I-III* (Louvain and Leiden: E.J. Brill, 1972), II, pp. 212-234. On the importance of Avicenna's visual theory to the Latin West, see Lindberg, *Theories of Vision*, pp. 43-52.

[41]In addition to Constantine's *De oculis* and Abbas' *Pantegni*, Avicenna's *Canon totius medicinae*, trans. Gerard of Cremona (d. 1187), provided an important source of eye anatomy. The *Liber canonis* and Averroës' *Colliget* (trans. 1250) later came to be used as university textbooks.

of the internal senses devised by Aristotle and Galen, providing a detailed elaboration which was widely discussed by the schoolmen. Avicenna posited five internal senses or *vires interiores*, each of which had a more precisely defined role in evaluating the sensible qualities of a visible form. Within this scheme, he now placed the "common sense" and several other internal senses that enabled man to receive, refine, and adjudicate sensible forms.[42] What is more, he endowed the "imagination" with the power to compare and combine forms and thereby accorded this faculty a more active role than Aristotle had done.[43]

Avicenna also enlarged the fairly limited role which Aristotle had assigned to light in the process of vision. Although he accepted the Aristotelian notion that light had a preparatory function, Avicenna no longer treated this phenomenon merely as the state of the medium which enabled colours to reach the eye. Instead, he gave light greater consideration in its own right, arguing that it was inherently visible and propagated by rays. In so doing, he provided a set of technical distinctions for light which was to become standard terminology in the literature on optics and vision from the mid-thirteenth century onwards. Avicenna categorised light according to whether it was found in its source, between the source and objects, or in contact with a given object:

Sunt autem his tres intentiones [...] quarum una est qualitas quam apprehendit visus in sole et igne [...] secunda est id quod resplendet ex his, scilicet splendor qui

---

[42] Avicenna discusses the internal senses (*sensus communis, imaginativa, aestimativa, cogitativa,* and *memorativa*) in *Liber de anima*, pars IV, c. 1, ed. S. Van Riet, in *Avicenna latinus: liber de anima seu sextus de naturalibus IV-V* (Louvain and Leiden: E.J. Brill, 1968) pp. 1-11. The seminal essay on the internal senses, including the Avicennan classification, is Harry Austryn Wolfson, "The Internal Senses in Latin, Arabic, and Hebrew Philosophic Texts," *Harvard Theological Review* 28 (1935), 69-133, esp. pp. 95-122. On medieval elaborations, see also Mary J. Carruthers, *The Book of Memory: A Study of Memory in Medieval Culture* (Cambridge: Cambridge University Press, 1990), pp. 47-60.

[43] On Avicenna's extension of the Aristotelian doctrine of the "imagination," see J. Portelli, "The 'Myth' that Avicenna Reproduced Aristotle's Concept of the 'Imagination' in *De anima*," *Scripta mediterranea* 3 (1982), 122-134.

26

videtur cadere super corpora et detegitur in eis albedo aut nigredo aut viriditas; tertia est quae apparet super corpora [...] si autem hoc fuerit in corpore acquirenti hoc ex alio corpore, vocabitur radiositas, si vero fuerit in corpore quod habet hoc ex se ipso, vocabitur radius. Si autem una earum, scilicet ea quae habet illud ex seipsa, lux, et utilitas eius sit lumen. Hoc autem quod vocamus lucem, sicut id quod habet sol et luna, est id quod videtur per seipsum.[44]

Despite the intense atmosphere of optical endeavour in the Muslim world, no single tradition nor any one individual had yet brought together all the optical traditions (geometrical, physical, and medical) into a single, coherent, and intellectually satisfying theory of vision. This was to be the achievement of an Arabic optician and physicist, Ibn-al Haytham, known to the West as Alhazen (965-1039). Alhazen created a brilliantly incisive synthesis of all previous traditions by making three highly important and quite original innovations. First, accepting the humours and tunics of Stoic eye anatomy, he used his understanding of optics, especially refraction, to provide a more sophisticated explanation of how visible forms are transmitted through the transparent parts of the eye. To do this, he showed that the central perpendicular ray was unrefracted by the tunics and humours as it passed into the eye: this ray was the principal determinant of vision:

[...] nihil pertransibit per diaphanitatem tunicarum visus secundum rectitudinem, nisi illud, quod erit super lineam rectam elevatam super superficiem visus

---

[44]*Liber de anima*, pars III, c. 1 (Van Riet, II, p. 171): "There are these three intentions [...] of which one is the quality which sight perceives in the sun and in fire [...] the second is that which shines back due to these [sun and fire], that is, the splendour which is seen to fall on bodies and white or black or green is revealed in them; the third is what appears above the surface of bodies [...] if this takes place in a body which acquires light from another then it will be called radiance, but if in a self-luminous body, it will be called a ray. If this takes place in one of the phenomena that have light inherently, it will be called *lux* and its offshoot will be called *lumen*. This which we call light, such as the sun and the moon have, is that which is visible by itself." The importance of the new and non-Aristotelian emphasis on the inherent visibility of light is noted by Smith in "Getting the Big Picture," p. 578.

secundum angulos rectos, et illud, quod fuerit super aliam, refringetur, et non pertransibit recte.[45]

Second, Alhazen modified Aristotle's intramitted forms. Unlike Aristotle, he did not regard these forms as coherent wholes, but conceived of them instead as an array of points:

Dicimus ergo quod, quando visus fuerit oppositus alicui rei visibili, veniet ex quolibet puncto superficies rei visae forma et coloris et lucis, quae sunt in ea, ad totam superficiem visus.[46]

As Lindberg has shown, this was one of the most original contributions in his theory; the Kindean principle of point forms enabled Alhazen to endow the intramission model of vision with the mathematical accuracy of ray geometry. Third, Alhazen did not restrict himself to the reception of forms in the eye. He also examined, as Aristotle, Galen, Avicenna, and others had done previously, the psychological processes required for a more reliable degree of visual perception.

---

[45]Alhazen, *De aspectibus*, lib. I, c. 5, sec. 18, ed. Friedrich Risner, in *Opticae thesaurus Alhazeni Arabis libri septem. Eundem liber de Crepusculis et Nubium Ascensionibus. Item Vitellonis Thuringopoloni libri X* (Basel, 1572), p. 10: "only that ray which follows a rectilinear path at right angles to the surface of the eye will pass perpendicularly through the transparent tunics; and that ray that were to follow another trajectory is refracted and will not pass undeviated." See also *De aspectibus*, lib. II, c. 1, sec. 8 (Risner, p. 29).

[46]Alhazen, *De aspectibus*, lib. I, c. 5, sec. 15 (Risner, p. 8): "Thus, we say that when sight is opposite some visible object all the forms of colour and light come from every point on the surface of a visible object to the entire surface of the eye." On Alhazen's innovation in this respect, see Lindberg, *Theories of Vision*, pp. 59-60: "Alhazen was the first writer to utilize the analysis of the visible object into point sources, each of which sends forth its ray, as the basis of an intromission theory of vision." On Alhazen's synthesised theory, see Lindberg, "Alhazen's Theory of Vision and Its Reception in the West," *Isis* 58 (1967), 321-341; idem, *Theories of Vision*, pp. 58-86. For a comprehensive treatment of Alhazen's visual theory, see A.I. Sabra, *The Optics of Ibn al-Haytham: Books I-III on Direct Vision*, 2 vols. (London: Warburg Institute, 1989); the introduction, glosses, concordances, and indexes in vol. II are most complete. On Alhazen's optics, see also Sabra's essays (II-XI) reprinted in his *Optics, Astronomy and Logic: Studies in Arabic Science and Philosophy* (Aldershot: Variorum, 1994).

28

Once again, he made important modifications, revising ancient and contemporary theories. The visible form was transmitted by a network of subtle spirits to a common nerve, where it was judged by a mental power called the *virtus distinctiva*. Another act of mental adjudication (called *intuitio*) then certified the impression, and yielded a faithful representation of the external object.[47]

The principal optical work in which Alhazen expounded these ideas, the *Kitab al-manazir*, was translated into Latin around the end of the twelfth century. Known to the Latin West as *De aspectibus*, *Optica*, or *Perspectiva*, this work became the main source of physical and physiological optics for the *perspectivae*. These Latin treatises on optics and vision were composed in the 1260s and 1270s by three authors: a Silesian, called Witelo (fl. 1250-1275); and two English Franciscans: John Pecham (c. 1230-1292); and Roger Bacon whose theory of "multiplication of species" and praise of optics have already been mentioned. For the sake of convenience, I will henceforth refer to these writers as the "perspectivists" and categorise their work as "perspectivist" optics.

Although the "perspectivists" relied on a wide variety of newly-translated optical sources (Euclid, Tideus, Ptolemy, Hero of Alexander, the Pseudo-Euclid, and Alkindi), it has been convincingly shown that their principal inspiration was Alhazen's elegant synthesis of eye anatomy, geometrical optics, and visual psychology.[48] Following Alhazen's example, they fashioned wide-ranging treatises on vision and, like him, blended optics and visual science into a

---

[47]On the quasi-rational operations in Alhazen's visual theory, see A.I. Sabra, "Sensation and Inference in Alhazen's Theory of Visual Perception," in *Studies in Perception*, ed. Machamer and Turnbull, pp. 160-185; Vescovini, *Studi sulla prospettiva*, pp. 113-135.

[48]Crombie has argued that Grosseteste's work profoundly influenced Bacon, Witelo, and Pecham, see *Robert Grosseteste*, passim, but esp. pp. 165-167, 213-217. Lindberg rightly centres attention on Alhazen, although he accepts that Grosseteste was a source of inspiration, see David C. Lindberg, "Lines of Influence in Thirteenth-Century Optics: Bacon, Witelo and Pecham," *Speculum* 46 (1971), p. 62; idem, *Theories of Vision*, pp. 94-102.

comprehensive discipline.[49]   Witelo compiled a voluminous ten-book treatise known as the *Perspectiva* (1268);[50] Pecham wrote an extremely popular work, called the *Perspectiva communis* (1270-1275);[51] and Bacon produced a number of works (completed before 1267) which contain extended sections on optics and include: the *Opus maius*, the *Opus tertium*, and the *De multiplicatione specierum*.[52] The technical details and overall conception of the visual process in these works correspond to Alhazen's treatment very closely, and discussion of the relevant doctrines will be provided in examining the Dantean material. But before we can begin to consider properly Dante's relationship to medieval optics, it is necessary to take into account several other thirteenth-century sources to which Dantists have paid either little or no attention.

Thirteenth-Century Optics: Secondary Lines of Influence

Despite the importance of the *perspectivae*, it should not be assumed that the "perspectivists" were the only thirteenth-century writers to assimilate the Latin translations of Greek and Arabic optical works.[53] It has already been seen how

[49]For the interrelations between the *perspectivae* and the centrality of Alhazen, see Lindberg, "Lines of Influence," pp. 66-83. For the dominant influence of Alhazen on each of the "perspectivists," see Lindberg, "Alhazen's Theory of Vision," p. 331: "the theories of vision expressed in Bacon, Pecham, and Witelo are essentially the same as Alhazen's"; idem, *Theories of Vision*, pp. 109, 117-118.

[50]Quotations from Witelo's *Perspectiva* are also taken from the Risner edition (see n. 45).

[51]Although Parronchi did not mention Pecham, Guidubaldi has used the *Perspectiva communis* in his *Dante Europeo* (see n. 3). For a critical edition of the *Perspectiva communis*, see *John Pecham and the Science of Optics: "Perspectiva communis," edited with an Introduction, English Translation, and Critical Notes*, ed. David C. Lindberg (Madison, Milwaukee, and London: University of Wisconsin Press, 1970). Pecham also wrote a *Tractatus de perspectiva*, ed. David C. Lindberg (New York: Franciscan Institute Publications, 1972).

[52]See the editions cited in n. 38 above. The relevant sections on optics in Book V of Bacon's *Opus maius* will be quoted from the critical edition by David C. Lindberg, *Roger Bacon and the Origins of "Perspectiva" in the Middle Ages: A Critical Edition and English Translation of Bacon's "Perspectiva" with Introduction and Notes* (Oxford: Clarendon, 1996).

[53]The Appendix provides a list of these works with approximate dates of translation.

30

earlier in the century Robert Grosseteste privileged optical science in his writings on geometrical forces in nature. Optical references are ubiquitous in Grosseteste's works and he also wrote original treatises on the rainbow and colour, two important natural phenomena involving light. Grosseteste was a key figure in developing what is often referred to as a "metaphysics of light," a complex matrix of philosophical and theological ideas related to light, which will be examined in Part Two of this study. As far as Grosseteste's pioneering optical studies are concerned, he certainly set an important example to the "perspectivists" and made extensive use of Euclid's *Optica*, Alkindi's *De aspectibus* and *De radiis*, and Aristotle's *Meteorologica*, but it seems that he had no knowledge of Alhazen.

The first writer to use Alhazen in the thirteenth century was not a specialist in optics, but the compiler of one of the best-known medieval encyclopaedias, Bartholomew the Englishman (c. 1190-c. 1250). In his extremely popular *De rerum proprietatibus* (c. 1230-1240), Bartholomew provided explanations about the sense of sight and the properties of the eye, and he also included wide-ranging discussions of, amongst other things, the technical terms for light and its propagation, the visual pyramid, the nature of light, refraction, the rainbow, and colour theory.[54] In the 1240s and 1250s, a considerable body of optical lore was also brought together and incorporated into Vincent of Beauvais' *Speculum maius*.[55] Although it is extremely difficult to establish whether Dante knew these

---

[54]Selected references from Bartholomew's *De rerum proprietatibus*: (i) visual matters, including eye anatomy, refraction, and the visual pyramid, lib. III, c. 17 (Richter, pp. 62-66); (ii) nature and behaviour of light, technical terms for light, and the rainbow, lib. VIII, c. 40-45 (Richter, pp. 425-435). In Book 3, Chapter 17, Bartholomew refers to Alhazen as the "phylosophum" and "auctor Perspective," using him as an *auctoritas* in discussing the tripartite division of optics, refraction, the nine preconditions of the visual act, and the more psychological aspects of vision.

[55]For important discussions of optics, light, and vision in Vincent of Beauvais, see *Speculum quadruplex sive speculum maius*, 4 vols. (Douai, 1624; reprint, Frankfurt-am-Main: Minerva, 1964): (i) Avicennan light distinctions, lib. II, c. 52, vol. I, cols 112-113; (ii) role of light in vision, lib. II, c. 33, col. 99; (iii) behaviour of light, lib. II, c. 35, col. 100; (iv) visual pyramid, lib. II, c. 78, col. 128; lib. XXV, cc. 36 and 45, cols 1799, 1804; (v) light reflection, lib. II, c. 83, col. 132; (vi) mirrors and optical illusions, lib. IV, c. 77, col. 280; (vii) rainbow and its colours, lib. IV, c. 74, col. 278. The optical lore which was incorporated into medieval encyclopaedias in

works, it is well worth making reference to them, since they indicate what was known and available in more general sources after 1250.[56]

Writers in the medieval tradition of Aristotelianism also drew heavily on translated optical works. From the early 1250s, commentators of the *libri naturales* began to amplify and enrich their commentaries and paraphrases with more detailed discussions of optical matters, often drawing on Arabic sources. Albert the Great (c. 1200-1280), writing in the 1240s and 1250s, before the "perspectivist" treatises of the 1260s and 1270s, occupies a central place in this process. Albert had a highly important role in re-elaborating and transmitting optical science and visual theory to the Latin West. His paraphrases of the Aristotelian corpus as well as his independent treatises contain a considerable number of learned digressions on optics in which he cites the optical works of Euclid, Alkindi, Avicenna, and even Alhazen.[57] Thomas Aquinas also provides many substantive explanations of optical matters, even though he is less inclined to stray from the Aristotelian text into the type of expository digression so favoured by Albert. Both Albert and Aquinas support a refined version of Aristotle's intramission model of vision and, in so doing, make continual reference to the

---

the 1240s and 1250s helped to supplement the relatively scarse knowledge of this subject found in the Latin encyclopaedic tradition; on this earlier tradition, see *A Source Book*, ed. Grant, pp. 376-384; Lindberg, *Theories of Vision*, pp. 87-89. More generally on thirteenth-century encyclopedias, see Pierre Michaud-Quantin, "Les petites encyclopédies du XIIIe siècle," *Cahiers d'histoire mondiale* 9 (1966), 580-596; *L'enciclopedismo medievale*, ed. Michelangelo Picone (Ravenna: Longo, 1994).

[56]Dante's own intellectual formation has often been related to medieval encylopaedias and *specula*, see Maria Simonelli, "Convivio," in *ED* II, p. 202; Cesare Vasoli, *Il Convivio*, introd., pp. xxvii-xxviii; idem, "Il *Convivio* di Dante e l'enciclopedismo medievale," *L'enciclopedismo medievale*, ed. Picone, pp. 363-381; idem, "Dante e l'immagine enciclopedica del mondo nel *Convivio*," *Studi sulle imago mundi: Centro di studi sulla spiritualità medievale* 22 (1983), 37-73. For an important critical re-evaluation, see also Zygmunt G. Baranski, "Dante fra 'sperimentalismo' e 'enciclopedismo'," *L'enciclopedismo medievale*, ed. Picone, pp. 383-404.

[57]Amongst his writings on Aristotle, Albert's major discussions of visual theory are: *De anima*, lib. II. tr. 3, cc. 7-10, ed. Clemens Stroick (Münster: Aschendorff, 1968), pp. 108-114; *De sensu et sensato*, tr. 1, cc. 5-16, ed. Peter Jammy, 21 vols. (Lyon, 1651), V, pp. 3b-20b. Chapters 5-11 of his *De sensu et sensato* are all digressions on visual and optical matters. On Albert's visual theory and his place in thirteenth-century optics, see Lindberg, *Theories of Vision*, pp. 104-107.

32

technical terms, ideas, and examples advanced by Avicenna (in his *Liber de anima*) and Averroës (in his Epitome on *Parva naturalia* and Long Commentary on *De anima*).

Since Parronchi's work was published, it would seem that no one has studied the Aristotelian commentaries as a possible influence on the whole apparatus of Dante's optical and visual thought. This type of approach is especially pertinent to Dante's "scientific" thought, because we know indubitably that he had frequent recourse to both the Albertine and Thomist commentaries, often using them in conjunction to satisfy his desire for critical exactitude.[58] As we shall see, Dante's own ideas on light, vision, and the eye frequently correspond more closely to this tradition of commentary and paraphrase than to the "perspectivist" tradition.

There is, however, a separate tradition which is more specifically theological, and this comprises commentaries on Scripture, commentaries on the Pseudo-Dionysius and on Peter Lombard's *Sentences*, and also works of theology, some of them quite popular. On the assumption that physical and spiritual light possessed similar qualities and behaved in similar ways, analogies between natural and divine light had provided an early and important means of representing God and the Trinity and explaining divine action. Around the middle of the thirteenth century, medieval theologians started to use technical ideas, derived from contemporary theories of light and optics, to deal with subjects such as creation, the Empyrean heaven, the action of grace, and the senses of the glorified body.

---

[58]Enrico Berti has demonstrated Dante's use of both Thomist and Albertine commentaries on *De anima* (*ED* II, p. 325), *De caelo* (II, p. 331), *De generatione animalium* (II, pp. 335-336), *De generatione et corruptione* (II, pp. 336-337), *Physica* (II, p. 934), *Metaphysica* (III, p. 924), *Meteorologica* (II, pp. 364-365), *De sensu* (II, pp. 387-388). Dante also seems to have made use of Albert's commentary on *De partibus animalium*, see *ED* II, 378. It is also likely that Dante had recourse to some of Averroës' paraphrases, especially the Long Commentary on *De anima* (cf. *Mon.* I, iii, 9) and possibly on *Meteorologica*. See also L. Minio-Paluello, "Dante's Reading of Aristotle," in *The World of Dante*, ed. Cecil Grayson (Oxford: Clarendon Press, 1980), pp. 61-79.

With the assimilation of Arabic sources, many theologians enriched their comparisons with detailed optical information about mirrors, geometrical optics, and the behaviour of light. The popularity of this kind of comparison even led certain writers to structure theological treatises almost entirely around motifs and doctrines related to the study of light and optics. The role of light in thirteenth-century theological writings will receive more detailed treatment in Chapter 7, but what is important to emphasise here is the extent to which such secondary lines of influence brought optics to a non-specialist audience. One pertinent example of this is offered by the widespread use of optical examples in medieval sermons.[59] Once again, it is difficult to know which works Dante might have used, but even a very limited exposure would have brought him into contact with discussions of light and mirrors in theological writings.

Poets were also becoming receptive to many of the optical doctrines and ideas that were entering into a general synthesis by the second half of the thirteenth century. The most extensive development of optical themes in a literary work before Dante is found in Jean de Meun's section of the *Roman de la Rose*, where Nature delivers a lengthy digression on optical matters in which she mentions Alhazen.[60] What is more, it would appear that optics is more than a peripheral concern in the *Roman*, since it has recently been claimed that optical doctrines directly influenced the structural design of both Guillaume's and Jean's sections of the poem.[61]

---

[59]On the use of optics in homiletic literature, see Peter Brown, *Chaucer's Visual World: A Study of His Poetry and the Medieval Optical Tradition*, 2 vols. unpublished D.Phil. dissertation, York University, 1981, I, pp. 147-166; David L. Clark, "Optics for Preachers: The 'De oculo morali' by Peter of Limoges," *Michigan Academician* 9 (1977), 329-343.

[60]For Jean de Meun's discussion of the properties and effects of mirrors (magnifying, reducing, distorting, deceiving, and burning), see *Roman de la Rose*, ll. 18014-18030, 18123-18186, ed. Félix Lecoy, 3 vols. (Paris: Champion, 1965-1970), III, pp. 41-45. Meun refers to "perspective" as "ceste merveilleuse science" (l. 18250, p. 48) and mentions "Alhacem" and his "livre des Regarz" (ll. 18004-18006, pp. 40-41).

[61]For the optics in this digression and its relation to the poem as a whole, see Patricia J. Eberle, "The Lovers' Glass: Nature's Discourse on Optics and the Optical Design of the *Romance of the*

34

In addition to the traditional *topoi* in early French and Italian poetry that celebrated the eye rays of the lady and the radiant effects of her beauty, more specialised optical themes were also used by Italian lyric poets in the generation before Dante.[62] There is a nascent optics of the lady's image in much of the Italian lyric tradition before Dante; and Giacomo da Lentini's sonnet "Or come pote sì gran donna entrare" offers a very good early example of the preoccupation with the lady's image, her radiance, and the optical implications of her mysterious passage into the lover's eyes and heart:

> Or come pote sì gran donna entrare
> per gli ochi mei che sì piccioli sone?
> e nel mio core come pote stare,
> che 'nentr'esso la porto là onque i' vone?
> Lo loco là onde entra già non pare,
> ond'io gran meraviglia me ne dòne;
> ma voglio lei a lumera asomigliare,
> e gli ochi mei al vetro ove si pone.
>
> Lo foco inchiuso, poi passa difore
> lo suo lostrore, sanza far rotura:
> così per gli ochi mi pass'a lo core,
>
> no la persona, ma la sua figura.
> Rinovellare mi voglio d'amore,

Rose," *University of Toronto Quarterly* 46 (1977), 241-262. See also Suzanne Conklin Akbari, "Medieval Optics in Guillaume de Lorris' *Roman de la Rose,*" *Medievalia et Humanistica* n.s. 21 (1994), 1-15; Bonnie Pavlis Baig, *Vision and Visualization: Optics and Light Metaphysics in the Imagery and Poetic Form of Twelfth- and Thirteenth-Century Secular Allegory, with Special Attention to the "Roman de la Rose,"* unpublished Ph.D. dissertation, University of California, Berkeley, 1982.

[62]On luminous, visual, and optical *topoi* in the Romance and Provençal literary traditions, see Flavio Catenazzi, *L'influsso dei provenzali sui temi e immagini della poesia siculo-toscana* (Brescia: Morcelliana, 1977), pp. 68-70, 74-77; Ruth H. Cline, "Heart and Eyes," *Romance Philology* 25 (1972), pp. 263-267, 289-297; Edgar de Bruyne, *Études d'esthétique médiévale,* 3 vols. (Bruges: De Tempel, 1946; reprint, Geneva, 1975), III, pp. 14-16; Jean Frappier, *Histoire, mythes et symboles: Étude de littérature française* (Geneva: Droz, 1976), pp. 149-167, 181-198.

poi porto insegna di tal crïatura.[63]

As central elements in the love passion, the lady's eye beams, her image, and their various phases and effects were repeatedly described by recourse to optical and luminous comparisons.[64] From the Sicilians onwards such motifs were often the starting-point for finely-wrought conceits about "occhi micidiali," learned comparisons to light sources in nature, and witty analogies.[65] As Lentini's sonnet also reveals, the lover's eyes also received special attention and another important early example of this concern is found in his *tenzone* with Jacopo Mostacci and Pier della Vigna.[66]

---

[63]Text in *Giacomo da Lentini: Poesie*, ed. Roberto Antonelli (Rome: Bulzoni, 1979), XXII, p. 291.

[64]Lentini's sonnet "Sì come il sol manda la sua spera" is structured upon the comparison between the effect of the lady's eye rays in penetrating the lover and that of a light ray passing through glass, see XXI, ed. Antonelli, p. 286. For an earlier example of this phenomenon in the vernacular religious lyric, see the composition "Ave, donna santissima," 2, ll. 19-26, in ed. Gianfranco Contini, *Poeti del Duecento*, 2 vols. (Milan and Naples: Ricciardi, 1960), II, pp. 15-16: "Quasi come ['n] la vitrera, / quando i rai del sol la fera, / dentro passa quella spera / ch'è tanto splendidissima, / altresì per tua mundizia / venne 'l sol de la iustizia / in te, donna de letizia, / sì fosti preclarissima." For further examples of this image in other medieval sources, see also Ch. 7, nn. 16-17, 67.

[65]For "occhi micidiali," see Pier della Vigna's *canzonetta*, "Uno piasente isguardo," II, ll. 14-15, (Contini, I, p. 123): "cogli oc[c]hi suo' micidari, / e quelli oc[c]hi m'hanno conquiso e morto." This theme was often developed through the motif of the basilisck (a mythical beast believed to kill with its sight; cf. Hugh of St Victor, *De bestiis et aliis rebus*, lib. III, c. 41, in *PL* 176, col. 177), see Stefano Protonotaro da Messina's *canzone*, "Assai mi placeria," ll. 40-45 (Contini, I, p. 138). Cf. Iacopone da Todi's "O femmene, guardate – a le mortal' ferute," 9, esp. ll. 3-5 (Contini, II, p. 91) where it is maintained that "El basalisco serpente occide om col vedere" and this motif is linked to the power of a woman's "aspetto" to make Christ lose souls. For variations on related *topoi* in the Sicilians, see Lentini, XIII, esp. ll. 5-6 (Contini, I, p. 81): "Più luce sua beltate e dà sprendore / che non fa 'l sole né null'autra cosa"; Guido delle Colonne's *canzone*, "Gioiosamente canto," esp. ll. 14-15 (Contini, I, p. 99): "la vostra fresca cera, / lucente più che spera"; cf. IV, ll. 36-39 (Contini, I, p. 105); Chiaro Davanzati, XII, ll. 1-6 (Contini, I, p. 428): "La splendïente luce, quando apare, / in ogne scura parte dà chiarore; / cotant'ha di vertute il suo guardare, / che sovra tutti gli altri è 'l suo splendore: / così madonna mi face alegrare, / mirando lei, chi avesse alcun dolore."

[66]*Poeti del Duecento*, ed. Contini, I, p. 90: "Amor è un[o] desio che ven da core / per abondanza di gran piacimento; / e li occhi in prima genera[n] l'amore / e lo core li dà nutricamento." See also ll. 7-8: "ma quell'amor che stringe con furore / de la vista de li occhi ha nas[ci]mento."

36

Amongst Italian poets who were even more formative influences on the young Dante, the two Guidos, Guinizzelli and Cavalcanti, also stand out for their elaboration of light and optical motifs. Guinizzelli reached a new level of refinement, both technical and spiritual, in his presentation of the lady's luminosity.[67] His *canzone* "Al cor gentil rempaira sempre Amore" is perhaps the most famous thirteenth-century lyric composition in which light imagery is used to express technical ideas.[68] Dante's "primo amico," Cavalcanti, also draws together and develops a wide variety of luminous and optical themes, often in a highly technical vein, in his poetic production. Cavalcanti celebrates the lady's beauty in terms of light, describes her "figura" entering the lover's eyes, and pays close attention to the wounds her eye beams inflict ("ferire," "fedire") on the lover.[69] The theme of sight is stressed in many of his compositions and in some cases seems to provide a structuring motif for individual sonnets. Elsewhere in his *Canzoniere*, the philosophical underpinnings of his concern with vision are evident; he uses, for example, scholastic light imagery and perceptual theory in the philosophical *canzone* "Donna me prega." But Cavalcanti's most important innovation of all in this respect was to re-elaborate the medical doctrine of spirits for his own poetic

---

[67]Selected references to optical and visual motifs in Guinizzelli include: (i) "colpo" through eyes to heart: I, esp. ll. 11-14 (Contini, II, pp. 450-451); VI, l. 9 (Contini, II, p. 468): "Per li occhi passa come fa lo trono"; (ii) lady's luminosity: I, ll. 35-40 (Contini, II, p. 452): "la notte, s'aparisce, / come lo sol di giorno dà splendore, / così l'aere sclarisce: / onde 'l giorno ne porta grande 'nveggia, / ch'ei solo avea clarore, / ora la notte igualmente 'l pareggia"; VII, ll. 1-6, p. 469; X, l. 3, p. 472; XII, ll. 9-11, p. 474; XVII, l. 11, p. 479. Both aspects are adroitly brought together in VIII, ll. 9-11, (Contini, II, p. 470): "Apparve luce, che rendé splendore, / che passao per li occhi e 'l cor ferìo, / ond'io ne sono a tal condizïone."

[68]Text in *Poeti del Duecento*, ed. Contini, II, pp. 460-464, esp. ll. 11-14, but see also ll. 16-20, 31-42.

[69]References in order: (i) light of lady: II, l. 3 (Contini, II, p. 493): "risplende più che sol vostra figura"; IV, l. 2 (Contini, II, p. 495): "che fa tremar di chiaritate l'âre"; (ii) passage of the lady's "figura": XIII, ll. 6-7 (Contini, II, p. 506): "che' deboletti spiriti van via: / riman figura sol en segnoria"; (iii) wounding eyes: IX, ll. 12-14, 23 (Contini, II, pp. 500-501): "la qual degli occhi suoi venne a ferire / in tal guisa, ch'Amore / ruppe tutti miei spiriti a fuggire [...] Per gli occhi fere la sua claritate."

purposes in order to emphasise the sensory spirits of sight which carry the lady into the lover's eyes and heart.[70]

In all these ways, the second half of the thirteenth century was a period of intense intellectual involvement with the newly-translated optical and visual heritage, one which was shared by optical specialists, encyclopaedists, natural philosophers, theologians, and poets. It is this atmosphere of assimilation, synthesis, and reelaboration that forms the essential background to the study of Dante's own use of optics. The chapters that now follow in Part One of this study will attempt to draw on all these thirteenth-century sources, and their antecedents, in order to provide a more comprehensive assessment of the medieval optical tradition in Dante's writings.

---

[70]On sight and the visual spirits as structuring themes, see VI (Contini, II, p. 497): "Deh, spiriti miei, quando mi vedete"; XIII (Contini, II, p. 506): "Voi che per li occhi mi passaste 'l core." Suggestions of Aristotelian light theory (used negatively with reference to Love) and psychological concepts are found in "Donna me prega," ll. 15-17, 67-68 (Contini, II, pp. 522-529). For Cavalcanti's use of perceptual theory, see esp. XIX, ll. 22-24 (Contini, II, p. 513): "ch'entra [sc. l'imagine] per li [occhi] miei sì debilmente / ch'oltra non puote color discovrire / che 'l 'maginar vi si possa finire." On the visual spirits, see esp. XXVIII, l. 1 (Contini, II, p. 530): "Pegli occhi fere un spirito sottile."

CHAPTER 2

Optics and Vision in Dante before the *Comedy*

Before investigating Dante's treatment of optical imagery and visual doctrines in the *Comedy*, it is necessary to consider his knowledge and use of several key aspects of the medieval optical and visual heritage in his earlier writings. The most important single text in this respect is the *Convivio* in which Dante twice mentions the science of "perspective" by name, uses the visual pyramid and the rectilinear ray to explain variations in the clarity of vision, employs technical terms for the propagation of light, and offers a lengthy digression on the processes of visual sensation and perception.[1] However, it is also well worth considering his handling of vision and recourse to optical knowledge in the *Vita Nuova* and the most optically-inspired of all his *canzoni*, "Amor, che movi tua vertù da cielo." In this way, it should be possible to determine more exactly what Dante understood by the term *perspectiva*, on what basis he formulated his thought about light and vision, and how he redeployed related ideas, doctrines, and imagery in earlier prose works and poetic compositions.

---

[1]References in order: (i) mentions of "perspettiva," *Con.* II, iii, 6; II, xiii, 27; (ii) ray geometry applied to vision, II, ix, 4-6; III, iii, 13; (iii) light terminology, III, xiv, 5-6; (iv) digression on visual theory, III, ix, 6-16.

40

The *Vita Nuova*

Chapter ii of the *Vita Nuova*, one of the most intricately wrought of the entire *libello*, describes Beatrice's first appearance to Dante and the effects she has upon him. In a sequence of three sentences with nearly identical rhetorical patterning Dante tells of the locations and the reactions of each of his "spirits" to her physical presence.[2] The second of these sentences which involves the animal spirit is of most interest for our purposes, since it draws particular attention to the sensory spirits ("li spiriti sensitivi") which act as intermediaries between the eye and the brain ("l'alta camera") in the visual act:

In quello punto lo spirito animale, lo quale dimora ne l'alta camera ne la quale tutti li spiriti sensitivi portano le loro percezioni, si cominició a maravigliare molto, e parlando spezialmente a li spiriti del viso, sì disse queste parole: «Apparuit iam beatitudo vestra». (ii, 5)

As was seen in Chapter 1, the medical doctrine of spirits, and especially that of the "spiriti del viso" was available to Dante in a variety of sources, including the *Canzoniere* of Guido Cavalcanti. There seems to be little doubt that Dante does indeed draw directly on Cavalcanti in a later prose passage in chapter xi of the *Vita Nuova* which deals with the overpowering of Dante's visual spirits by Beatrice and contains distinctively Cavalcantian echoes with its notion of the rout of the spirits and its "deboletti spiriti del viso":[3]

---

[2]For the sources and development of the *spiritus* doctrine, see E. Bertola, "La dottrina dello «spirito» in Alberto Magno," *Sophia* 19 (1951), 306-312; idem, "Le fonti medico-filosofiche della dottrina dello «spirito»," *Sophia* 26 (1958), 48-61; James S. Bono, "Medical Spirits and the Medieval Language of Life," *Traditio* 40 (1984), 91-130. On its use in the early Italian lyric, see Massimiliano Chiamenti, "The Representation of the Psyche in Cavalcanti, Dante and Petrarch: The 'Spirit'," *Neophilologus* 82 (1998), 71-81.

[3]On visual spirits in Cavalcanti, see the references given in Ch. 1, nn. 69-70.

41

E quando ella fosse alquanto propinqua al salutare, uno spirito d'amore, distruggendo tutti li altri spiriti sensitivi, pingea fuori li deboletti spiriti del viso, e dicea loro: «Andate a onorare la donna vostra»; ed elli si rimanea nel luogo loro. E chi avesse voluto conoscere Amore, fare lo potea mirando lo tremare de li occhi miei. (xi, 2)

One further prose passage, in chapter xiv, also describes the routing of Dante's visual spirits, and given its subject-matter it is significant that it precedes the last Cavalcantian-inspired sonnet in the *libello* ("Con l'altre donne mia vista gabbate"):[4]

Allora fuoro sì distrutti li miei spiriti per la forza che Amore prese veggendosi in tanta propinquitade a la gentilissima donna, che non ne rimasero in vita più che li spiriti del viso; e ancora questi rimasero fuori de li loro istrumenti, però che Amore volea stare nel loro nobilissimo luogo per vedere la mirabile donna. (xiv, 5)

After Dante's first perception of Beatrice, the God of Love takes control of Dante's soul and appears to him in a "maravigliosa visione" which weakens his natural spirit and leads to his silence before all interlocutors. One important narrative strand in the first half of the *Vita Nuova* is Dante's concern with Beatrice's power over his own sight and the role of his "imagination" in arousing visions and dreams related to her.[5] In chapter v a technical

---

[4]On the heightened Cavalcantian echoes in *Vita Nuova*, XIII-XVI before the shift to praise style, see Teodolinda Barolini, *Dante's Poets: Textuality and Truth in the "Comedy"* (Princeton, NJ: Princeton University Press, 1984), pp. 136-138; J.F. Took, *Dante: Poet and Philosopher: An Introduction to the Minor Works* (Oxford: Clarendon Press, 1990), p. 49.

[5]On this theme and related issues, see Margherita de Bonfils Tempier, "La prima visione della «Vita Nuova» e la dottrina dei tre spiriti," *Rassegna della letteratura italiana* 76 (1972), 303-316; Robert Hollander, "*Vita Nuova*: Dante's Perceptions of Beatrice," *Dante Studies* 92 (1974), 1-18; Nicolò Mineo, *Profetismo e apocalittica in Dante: Strutture e temi profetico-*

42

understanding of the visual process is implicit in the narrative sequence that relates how Dante was able to view Beatrice sitting in a church, but, because another lady was in his direct line of sight, he was mistakenly thought to have been staring not at Beatrice but at this lady. In language that is reminiscent of the extramission theory of vision and can be found elsewhere in his lyric poetry, Dante refers to his sight as appearing to come to a halt on the lady:

Uno giorno avvenne che questa gentilissima sedea in parte ove s'udiano parole de la regina de la gloria, ed io era in luogo dal quale vedea la mia beatitudine; e nel mezzo di lei e di me per la retta linea sedea una gentile donna di molto piacevole aspetto, la quale mi mirava spesse volte, maravigliandosi del mio sguardare, che parea che sopra lei terminasse. (v, 1)

This is the first reference in Dante's writings to the idea that vision takes place along a straight line, a principle which, as we have already seen, was a fundamental doctrine of medieval optics and whose medieval sources will be examined later in this chapter.

"Amor, che movi tua vertù da cielo"

The previous chapter outlined some of the principal ways in which early Italian lyric poets dealt with the phenomena of vision and optics when describing the lady's image or "figura," her eyes, eye rays, and the various processes of visual sensation and perception involved in the genesis of the love passion. Given the numerous antecedents and importance of visual motifs in this lyric production, it is hardly surprising that suggestions of visual theory and traces of optical

---

apocalittici in Dante: dalla Vita Nuova alla Divina Commedia (Catania: Università di Catania, 1968), pp. 103-141.

knowledge are to be found in the young Dante's own lyric poetry. In several of his early sonnets and *canzoni*, Dante develops conventional motifs such as the God of Love appearing in the lady's eyes, the entry of her "figura" into the lover's eyes and heart, and the force of her radiant gaze with all its ennobling, wounding, and murderous potentialities.[6] In a rather more technical vein, there are extended comparisons between the sun and the lady in the doctrinal *canzoni* addressed to the "Donna Gentile." And it is one of these doctrinal *canzoni*, "Amor, che movi tua vertù da cielo," where Dante offers his most sustained, sophisticated, and original development of optical and visual themes in a poetic work before the *Comedy*. It has been suggested that the *canzone* might have received an optical commentary in the *Convivio*, but whether this is the case or not, it is clear that the way in which Dante here adapts optical and visual motifs deserves detailed consideration in its own right.[7]

The opening lines of the first stanza tell of Love's capacity to move "vertù" from heaven in a way analogous to that in which the sun sends its light ("splendore": l. 2) into terrestrial objects:

> Amor, che movi tua vertù da cielo
> come 'l sol lo splendore,
> che là s'apprende più lo suo valore
> dove più nobiltà suo raggio trova;
> e come el fuga oscuritate e gelo,
> così, alto segnore,
> tu cacci la viltate altrui del core. (1-7)

---

[6]For the God of Love in the lady's eyes, see *Rime* LVII, 1-2; *VN* xxi, 2. On the lady's "figura" in Dante's eyes, see *Rime* LXVII, 43-44, 81; LXXX, 9, 15. Eyes and eye rays are notably developed in *Rime* LXV, 5, 10-11; LXVII, 7-13, 49; LXIX, 5-8; XC, 28; XCI, 17; CII, 5, 43; CIII, 74-75; cf. *Con.* III, x, 4.

[7]Parronchi was undoubtedly justified in arguing that this *canzone* provides evidence of Dante's concern with the subject-matter of medieval optics, see "La perspettiva dantesca," p. 44.

44

While "splendore" is not here being used in its technical sense, Dante does introduce the appropriate technical term for the presence of light in the medium in line 4 ("raggio") in order to describe the way in which the reception of Love's virtue varies according to the nobility of the recipient. One of the first effects of Love is to chase away faintheartedness, and this process is compared to a ray of sunlight banishing darkness and cold (l. 5). The notion that Love elicits nobility and, like the sun, vanquishes darkness is further refined in lines 11-15 which now add a greater emphasis on Love's capacity to activate worthy conduct. Here, Dante draws deftly on the Aristotelian doctrine that light is a precondition of vision insofar as it actualises the transparent medium and allows colour to be seen. The poet then illustrates this idea with a simile: as a picture in darkness can neither be seen nor delight the spectator with its colour and its ordered arrangement, so too without Love all human potential to do good is unrealised:

> sanza te è distrutto
> quanto avemo in potenzia di ben fare,
> come pintura in tenebrosa parte,
> che non si può mostrare
> né dar diletto di color né d'arte. (11-15)

After describing the effects of Love by analogies to the varying reception of the sun's rays and to light's central role in activating vision, in the second stanza Dante moves on to consider Love's effect on his own eyes and his heart. The luminous power of Love is still the central motif of the *canzone*, and the violent intensity with which it wounds Dante's heart is emphasised by the prominent metrical positions of the verb "ferire" (l. 16) and the noun "luce" (l. 17). The action of light on Dante's heart is again compared to a "raggio" of sunlight, but the poet now alludes to this by means of another highly compressed simile which is based upon the commonplace cosmological doctrine that the stars receive their light from the sun (1. 17).[8] The wounding light of Love brings about the capture of Dante's soul and gives birth to a desire which leads him to gaze upon all beautiful things, feeling greater joy where there is greater beauty:[9]

> Feremi ne lo cor sempre tua luce,
> come raggio in la stella,
> poi che l'anima mia fu fatta ancella
> de la tua podestà primeramente;
> onde ha vita un disio che mi conduce
> con sua dolce favella
> in rimirar ciascuna cosa bella
> con più diletto quanto è più piacente. (16-23)

At this point Dante makes use of a far more sophisticated series of optical analogies. His act of looking on beauty has led to his capture by a young lady who has entered his mind and ignited a fire there. To offer the reader an analogy for these phases of his enamourment and especially for the kindling effect of love, Dante borrows the idea that a body of water produces fire when light rays are concentrated as they pass through it. The radiance of the lady and her effect on Dante are made all the more intense because Love's rays ("li raggi

---

[8]For ancient and medieval elaborations of this doctrine, see Ch. 4, nn. 28-29; cf. *Il mare amoroso*, 1. 141, in *Poeti del Duecento*, ed. Contini, I, p. 492. For additional Dantean references, see Ch. 6, n. 76.

[9]For the notion that vision is the cause of love, see esp. Andreas Capellanus, *De amore*, lib. I, c. 1, ed. E. Trojet (München: Eidos Verlag, 1964), p. 1: "Amore est passio quaedam innata procedens ex visione et immoderata cogitatione formae alterius sexus" ("Love is a certain inherent passion which arises from the sight of, and immoderate cogitation on, the beauty of the opposite sex"). On the connection between sight and delight, see also Aristotle, *Metaphysica*, I, 1, 980a 21-23, Graeco-Latin trans. in *In De metaphysicorum*, ed. M.-R. Cathala, 2nd edn (Rome and Turin: Marietti, 1971), p. 5: "Omnes homines natura scire desiderant. Signum autem est sensuum dilectio. Praeter enim utilitatem, propter seipsos diliguntur, et maxime aliorum, qui est per oculos" ("By nature all mankind desires to know and a sign of this is the delight men take in the senses. Men take delight in the senses in themselves beyond what is practical and of all the senses this is especially so of the eyes"); cf. *Ethica*, IX, 5, 1167a 3-10. For the delight of sight in colours, see further Augustine, *Confessionum, libri XIII*, lib. X, c. 34, § 51, ed. Lucas Verheijen, in *CCSL* 26 (1981), p. 182; Hugh of St. Victor, *Eruditionis didascalicae*, lib. VII, c. 12 (*PL* 176, col. 821); cf. *Purg.* I, 13-16.

46

tuoi") were reflecting in her eyes at the very moment she entered into his mind (ll. 26-30):[10]

> Per questo mio guardar m'è ne la mente
> una giovane entrata, che m'ha preso,
> e hagli un foco acceso,
> com'acqua per chiarezza fiamma accende;
> perché nel suo venir li raggi tuoi,
> con li quali mi risplende,
> saliron tutti su ne gli occhi suoi. (24-30)

In the opening lines of the following stanza the poet turns to consider the next phase in his perception of the "giovane": the role played by a post-sensory faculty, "lo imaginar" (l. 33), in adorning her in his memory with the assistance of Love's powers (ll. 31-38). Since it is axiomatic that effects can be judged by their causes, her beauty is a demonstration of Love's value. Once again, Dante offers an analogy based on the effects of light to illustrate this concept: like the fire which is produced by the sun, the "giovane" is also the product of a higher cause (Love), an effect which varies depending on the person in whom Love's value is received:

> È sua beltà del tuo valor conforto,
> in quanto giudicar si puote effetto
> sovra degno suggetto,
> in guisa ched è 'l sol segno di foco;
> lo qual a lui non dà né to' virtute,
> ma fallo in altro loco
> ne l'effetto parer di più salute. (39-45)

---

[10]For the judicious interpretation of this verse as "come l'acqua risplendendo attraversata dai raggi suscita la fiamma," see Francesco Maggini, Review of Contini's *Poeti del Duecento*, in *GSLI* 116 (1940), pp. 43-44. On the concentration of light by water, see the references given in Ch. 4, nn. 54-55.

Thus, in the first forty-five lines of "Amor, che movi," Dante has moved from a grandiose macrocosmic vision of the sun sending out its rays into the Earth to a microcosmic understanding of the optical effects of Love's rays on his own eyes, heart, and mind. He alludes to a wide variety of optical and visual doctrines in order to provide analogies so as to clarify how these rays are received, how they actualise the lover's sight, and how they eventually lead to his mind being captured and set alight. As we can see, then, in the years preceding the *Convivio*, Dante was well versed in using concepts related to the study of light and optics, and this material provided him not only with a convenient reservoir of imagery and analogies, but also with a set of structuring themes upon which to develop his own poetic fictions.

Optical and Visual Theory in the *Convivio*

Dante and "perspettiva"

It was shown in the opening chapter that Aristotle's works offered no precedent for using the geometrical principles of *perspectiva* to analyse all physical phenomena, and that it was decidedly non-Aristotelian to conceive of optics as the primary natural science.[11] It would seem that this conception of optics, formulated by Grosseteste and elaborated by Roger Bacon, was shared by the other "perspectivists." Witelo writes of the pre-eminence of light in the rhetorical prologue to his *Perspectiva*; and in his prologue to *Perspectiva communis*, John Pecham resolutely declares that "Perspective is properly

---

[11]On the "non-Aristotelian" approach to natural science elaborated by Grosseteste and Bacon, see McEvoy, *The Philosophy of Robert Grosseteste*, p. 450; James A. Weisheipl, in *The Cambridge History of Later Medieval Philosophy: From the Rediscovery of Aristotle to the Disintegration of Scholasticism, 1100-1600*, ed. Norman Kretzmann (Cambridge: Cambridge University Press, 1983), p. 523.

48

preferred to all the traditional teachings of mankind."[12] By contrast, the commentaries of Averroës, Albert, and Aquinas present more orthodox views by referring to optics as a *scientia media*, a composite science which is more concerned with the physical than the mathematical.[13]

To what extent do Dante's two references to *perspectiva* reflect the differing conceptions of the optical enterprise expressed by Grosseteste and the "perspectivists," on the one hand, and by Aristotle and representatives of the Aristotelian tradition, on the other?

His first mention of "un'arte che si chiama perspettiva" in Book II of the *Convivio* shows that he regards optics as an auxiliary science, and not as a universal one. The site of the nine moveable spheres, Dante says, is made manifest and determined by "perspettiva, e [per] arismetrica e geometria, sensibilmente e ragionevolmente è veduto" (II, iii, 6). Optics is associated with astronomy, a discipline which itself rests on visual data, but there is absolutely no suggestion in this passage that what Dante understands by "perspettiva"

---

[12]See Witelo, *Perspectiva*, prologue (Risner, p. 1); Pecham, *Perspectiva communis*, prologue (trans. Lindberg, pp. 61-62); cf. idem, *Tractatus de perspectiva*, c. 1 (Lindberg, pp. 25-26). See also the texts quoted in Ch. 7, n. 25.

[13]For the elaboration of Aristotelian ideas on optics in medieval commentators, see Averroës, *Physica*, lib. II, summa secunda, c. 2, comm. 20, in *Aristotelis opera cum Averrois commentariis*, 12 vols. (Venice, 1562-1574; facsimile, Frankfurt-am-Main: Minerva, 1962), IV, p. 55vb; Albert, *Metaphysica*, lib. XII, tr. 1, c. 2, ed. Bernhard Geyer, 2 vols. (Münster: Aschendorff, 1964), II, p. 547; idem, *Physica*, lib. II, tr. 1, c. 8, ed. Paul Hossfeld, 2 vols. (Münster: Aschendorff, 1987-1993), I, pp. 90-91; Aquinas, *In De physicorum*, lib. II, lect. 3, § 164, ed. M.-P. Maggiòlo, reprint (Turin: Marietti, 1965), p. 84; idem, *In De metaphysicorum*, lib. III, lect. 7, § 412 (Cathala, p. 412); idem, *In Posteriorum analyticorum*, lib. I, lect. 25, §§ 210-211, ed. Raymund M. Spiazzi, 2nd edn (Turin: Marietti, 1964), pp. 231-232; and lect. 41, § 358 (p. 299): "quaedam autem scientiae sunt mediae, quae scilicet principia mathematica applicant ad materiam sensibilem" ("there are some middle sciences, which apply mathematical principles to sensible matter"); cf. idem, *ST*, Ia-IIae, q. 35, a. 8, resp., ed. P. Caramello, 3 vols. (Turin: Marietti, 1963), II, p. 171. See also Adriana Caparello, *La «perspettiva» in Sigieri di Brabante* (Vatican City: Libreria editrice Vaticana, 1987). For the different conceptions of optics in Albert and Bacon, see Jeremiah M.G. Hackett, "The Attitude of Roger Bacon to the *Scientia* of Albertus Magnus," in *Albertus Magnus and the Sciences: Commemorative Essays, 1980*, ed. James A. Weisheipl (Toronto: Pontifical Institute of Mediaeval Studies, 1980), pp. 53-72, esp. p. 65.

might hold the key to all knowledge of God's universe. The way in which he groups optics with mathematical disciplines, and then uses the adverb "sensibilmente" (i.e. by the evidence of the senses) to refer to "perspettiva," shows that he is adhering to the Aristotelian tradition represented by Averroës, Albert, and Aquinas. Like these commentators, Dante is aware that optics is a discipline which falls between the abstract and the concrete sciences and uses the demonstrative principles of geometry to correlate experiential facts.

The second reference to the science of "perspettiva," also in Book II, offers confirmation of this interpretation. In his extended discussion of the analogies between the planets and the seven liberal arts, Dante notes that geometry, the third discipline in the *quadrivium*, "è bianchissima, in quanto è sanza macula d'errore e certissima per sé e per la sua ancella, che si chiama Perspettiva" (II, xiii, 27). Optics is here related to geometry in a way that quite clearly corresponds to Aristotle's discussions in the *Physica* and the *Posteriora analytica*. In medieval elaborations of these works, optics was said to be a *scientia subalternata*, because it is subordinated to geometry, the *scientia subalternans*.[14] This principle of *subalternatio* (namely that certain disciplines are dependent on the principles of a higher science) can also be found in the

---

[14]See Albert, *Analytica posteriora*, lib. I, tr. 2, c. 18 and tr. 3, c. 7, ed. Peter Jammy, in *Opera omnia*, 21 vols. (Lyon, 1651), I, pp. 553a, 564a-566b; idem, *Metaphysica*, lib. I, tr. 2, c. 2 (Geyer, I, p. 19). See also Aquinas, *In Poster. analy.*, lib. I, lect. 15, § 131 (Spiazzi, pp. 197-198); lect. 25, § 211 (Spiazzi, p. 232): "Perspectiva enim subalternatur geometriae" ("Perspective is subordinated to geometry"); idem, *ST*, Ia, q. 1, a. 2, resp. (Caramello, I, p. 3): "Quaedam [sc. scientiae] vero sunt, quae procedunt ex principiis notis lumine superioris scientiae: sicut perspectiva procedit ex principiis notificatis per geometriam, et musica ex principiis per arithmeticam notis" ("There are certain sciences which proceed by means of principles known in the light of superior sciences, such as perspective which proceeds through principles given by geometry, and music which relies on principles of arithmetic"); idem, *ST*, Ia, q. 79, a. 9, resp. (Caramello, I, p. 390). For Aquinas' introduction of the term *scientia media* in this context, see Joan Gagné, "Du 'Quadrivium' aux 'scientiae mediae'," in *Arts libéraux et philosophie au moyen âge* (Paris and Montreal: Vrin, 1969), p. 982.

50

commentaries of Albert and Aquinas, although there is no sharp contrast with Grosseteste's views in this instance.[15]

It is extremely worthwhile to compare the position that Dante assigns to "perspettiva" on the periphery of the *quadrivium* with its place in other classifications, both Islamic and scholastic, of the liberal arts. Alkindi, for example, reveals his deep respect for optics by noting that a work on the subject is required to complete the *artes doctrinales*. A generation later, Alfarabi (c. 850-950), wrote two separate treatises, the shorter *De ortu scientiarum*, and the more detailed *De scientiis*, in which he placed optical science within the mathematical and demonstrative disciplines of an expanded *quadrivium*.[16] Around 1150, Dominic Gundissalinus (d. 1190) translated both these treatises, and inserted several sciences, including "perspective," between geometry and astronomy in his own work, *De divisione philosophiae*.[17] Unlike the writers in this special tradition, Dante does not place "perspettiva" directly within the quadrivial disciplines; he retains instead the original classification of seven disciplines handed down by Boethius in his *Institutio arithmetica* and popularised by Martianus Capella.

It seems clear, then, that Dante held a more limited view of the nature and utility of *perspectiva* than the aesthetically enticing and unified theory of the science formulated by Grosseteste, Bacon, and the other "perspectivists." In

[15]For Grosseteste's traditional use of Aristotelian ideas and his introduction of the term *subalternatio*, see W.R. Laird, "Robert Grosseteste on the Subalternate Sciences," *Traditio* 43 (1987), 147-169.

[16]For Alkindi's revealing comments, see his *De aspectibus* (Björnbo and Vogl, p. 4). On Alfarabi's inclusion of *perspectiva* within the *quadrivium*, see *De ortu scientiarum*, ed. Clemens Baeumker, in *Beiträge* 19/3 (1916), p. 20; idem, *De scientiis*, ed. Angel González Palencia, in *Catálogo de las ciencias* (Madrid, 1932), pp. 148-152. This special optical tradition is discussed by James A. Weisheipl, "Classification of the Sciences in Medieval Thought," *Mediaeval Studies* 27 (1965), pp. 68-72. On the place of *perspectiva* in medieval classifications of knowledge, see also Graziella Federici Vescovini, "La *perspectiva* nell'enciclopedia del sapere medievale," *Vivarium* 6 (1968), 35-45.

[17]Dominic Gundissalinus, *De divisione philosophiae*, ed. Ludwig Baur, in *Beiträge* 4/2-3 (1903), pp. 112-114.

51

fact, given the contexts of Dante's discussions, it seems reasonable to assume that his initial interest in the science arose from his astronomical studies.[18]

The rectilinear ray and the visual pyramid

In the *Convivio*, Dante's discussions of the visual pyramid and the rectilinear ray are perhaps the most notable aspects of his optical knowledge. The first passage in which Dante explicitly refers to the idea that vision takes place along a straight line gives evidence of his understanding of the entire visual process and is well worth quoting in full:

E qui si vuol sapere che avvenga che più cose ne l'occhio a un'ora possano venire, veramente quella che viene per retta linea ne la punta de la pupilla, quella veramente si vede, e ne la imaginativa si suggella solamente. E questo è però che 'l nervo per lo quale corre lo spirito visivo, è diritto a quella parte, e però veramente l'occhio l'altro occhio non può guardare, sì che esso non sia veduto da lui; ché, sì come quello che mira riceve la forma ne la pupilla per retta linea, così per quella medesima linea la sua forma se ne va in quello ch'ello mira: e molte volte, nel dirizzare di questa linea, discocca l'arco di colui al quale ogni arme è leggiere.[19]

In this scientific analysis of the love passion, Dante reveals his indirect debts not only to Euclidean-Ptolemaic straight-line optics, but also to the Galenic doctrine of nerves and visual spirits, and to the relevant parts of Aristotelian and

---

[18]By contrast, Parronchi suggests that Dante's interest in optics was stimulated by his membership of the "Arte dei medici e speziali," see "La perspettiva dantesca," p. 23; idem, "Perspettiva," in *ED* IV, p. 438.

[19]*Con.* II, ix, 4-5; cf. *Par.* XXVIII, 10-12. For the rectilinear pathway in vision, see also *Con.* III, iii, 13: "[...] come chi guarda col viso co[me] una retta linea, prima vede le cose prossime chiaramente; poi, procedendo, meno le vede chiare." For a further reference to the "radius rectus rei visibilis inter rem et oculum," see *Questio*, 82.

52

Avicennan faculty psychology, all of which were outlined in Chapter 1. The form of a visible object enters the eye along a rectilinear path and is transmitted through a hollow nerve (filled with visual spirit) to the "imaginativa," the internal sense or faculty which was believed to retain these images. In an important section of Book IV of the *Convivio*, Dante takes up some of these matters again and uses his knowledge of the visual cone to develop an interesting analogy between the process of vision "per modo quasi piramidale" and the way in which human beings apprehend earthly goods without reference to their divine source. During the pilgrimage in this life, the eyes of the soul fail to perceive God at the distant base of the cone because man's mental vision is enticed by ever-increasing objects of desire, and it cannot comprehend the fact that God subsumes all these desirables:

Per che vedere si può che l'uno desiderabile sta dinanzi a l'altro a li occhi de la nostra anima per modo quasi piramidale, che 'l minimo li cuopre prima tutti, ed è quasi punta de l'ultimo desiderabile, che è Dio, quasi base di tutti. (IV, xii, 17)

In Chapter 1, we saw how these ideas were developed by Euclid and Ptolemy and how the visual pyramid was refined by Arabic writers, including Alhazen, who applied the conical figure to an intramission model of vision in which emphasis was given to the role of the central perpendicular ray.[20] From Alhazen, the "perspectivists" incorporated the principle of the unrefracted perpendicular ray in their analyses of vision.[21] On the basis of Dante's use of

---

[20]For the pyramid and perpendicular entry in Alhazen's *De aspectibus*, see Ch. 1, n. 45.

[21]Bacon, *Opus maius*, lib. V, p. iii, dist. 2, c. 1 (Lindberg, p. 288): "dictum est [...] quod non frangitur aliquis radius pyramidis visualis super corneam nec humorem albugineum, nec super anterius glacialis, quoniam tota pyramis cadit perpendiculariter super ista tria corpora" ("it was stated [...] that no ray of the visual pyramid is refracted at [the surfaces of] the cornea, albugineous humour, or anterior glacial humour, since the entire pyramid is incident perpendicularly on these three bodies," trans. Lindberg, p. 289); Pecham, *Perspectiva communis*, lib. I, prop. 38 (Lindberg, pp. 118-120): "Unde sola perpendicularis illa que axis

53

terms such as "per una retta linea" and "co[me] una retta linea" several Dante scholars have argued that Dante is indeed directly dependent upon Alhazen and "perspectivist" works.[22] However, the *Convivio* commentators, Busnelli and now Vasoli, have also cited an important passage from Albert the Great's *De sensu et sensato* which comments on the visual pyramid and the rectilinear ray and discusses the role of the perpendicular visual axis, the relation of optics to geometry, and writings on "perspective":

Ad hoc autem sciendum, notandum est omnem visum fieri sub figura pyramidis, cuius basis est super rem visam, et angulus in centro crystallini: propter quod cum oculus sit sphaericus,[23] omnes lineae a basis pyramidis ad angulum eius protractae, perpendiculares sunt super oculum, et super humorem crystallinem. Cum autem sic fiat visus sub pyramide, fortior tamen est visus ad lineam quae vocatur axis sphaerae oculi, et axis pyramidis, quae quidem linea mobilis est [...] Haec autem omnia supponenda sunt, probanda autem in libris de visu in perspectivis, quae scientia compleri non potest nisi primum consideremus ea quae pertinent ad geometriam.[24]

---

dicitur, que non frangitur, rem efficaciter representat" ("Hence, the visual object is only made properly clear by that perpendicular which is called the axis and is not refracted"); idem, *Tractatus de perspectiva*, c. 5 (Lindberg, p. 44); Witelo, *Perspectiva*, lib. III, prop. 17 (Risner, p. 93): "Fit ergo visio distincta solum secundum perpendiculares lineas a punctis rei visae ad oculi superficiem productas" ("Therefore, distinct vision only takes place along lines extended perpendicularly from points on a visible object to the surface of the eye").

[22]Giovanni Busnelli and G.V. Vandelli, *Il Convivio*, 2 vols. 2nd edn (Florence: Le Monnier, 1967), I, p. 167; Bruno Nardi, "Alla illustrazione del 'Convivio': A Proposito dell'edizione di Giorgio Rossi," *GSLI* 95 (1930), pp. 83 and 91; Parronchi, "La perspettiva dantesca," pp. 27-28, 39-41; Vasoli, *Il Convivio*, p. 191.

[23]Jammy's edition reads *spiritus*, whereas a recent critical edition gives *spicus*, see Cemil Akdogan, *Optics in Albert the Great's "De Sensu et Sensato": A Critical Edition, Translation, and Analysis*, unpublished Ph.D. dissertation, University of Wisconsin, 1978, p. 101. Given Albert's subsequent discussion, *sphaericus* would seem to be the most appropriate reading.

[24]*De sensu et sensato*, tr. I, c. 14 (Jammy, V, pp. 19b-20a), cited by both Busnelli and Vasoli: "In order to know this it should be noted that all vision takes place according to a conical figure the base of which is on the visible object and the apex of which is at the centre of the

54

While these scholars have succeeded in identifying Albert as one of the most probable direct sources for Dante's geometrical model of vision, no one has noted that Dante would have come across the schema of visual ray geometry in a number of commentaries by Albert and Aquinas.[25] Despite the fact that Parronchi and almost all writers on Dante's optics since him have tended to discount their influence, the Aristotelian works of Albert and Aquinas are in fact all that is required to substantiate Dante's optical references and his understanding of visual theory in the *Convivio*. The documentation that will be presented in this chapter and in the remainder of Part One of this study indicates that this is especially true of commentaries on *De caelo, De anima, De sensu,* and *Meteorologica*. It is, however, also well worth noting that in the thirteenth century the geometry of vision was sufficiently well known to be alluded to in works as diverse as astronomical textbooks, treatises on scholastic theology, and handbooks for preachers. It is also noteworthy that the encyclopaedias compiled by Vincent of Beauvais and Bartholomew the Englishman contain many relevant optical doctrines, including detailed sections on the geometrical

---

crystalline humour. Since vision takes place pyramidally in this way, then sight is stronger along the line which is called the axis of the spherical eye, and along the axis of the cone which is a moveable line [...] All of these things are to be taken as suppositions and are to be demonstrated in the writings on perspective, which is a science that cannot be perfected unless consideration is first given to that which pertains to geometry." On Albert's presence in the *Convivio*, see Cesare Vasoli, "Fonti albertine nel *Convivio* di Dante," in *Albertus Magnus und der Albertismus: Deutsche philosophische Kultur des Mittelalters*, ed. J.F.M. Maarten and Alain de Libera (Leiden and New York: E.J. Brill, 1995), pp. 33-49.

[25] Albert, *De anima*, lib. II, tr. 3, c. 14 (Stroick, pp. 119-120); idem, *De caelo et mundo*, lib. II, tr. 3, c. 8 and esp. c. 11, ed. Paul Hossfeld (Münster: Aschendorff, 1971), pp. 159 and 168; idem, *De sensu et sensato*, tr. I, cc. 5, 10, 14 (Jammy, V, pp. 5a-6a, 14a-16a, and 19b-20a); idem, *Quaestiones de animalibus*, lib. I, qq. 29-31, ed. Ephrem Filthaut (Münster: Aschendorff, 1955), pp. 98-100; idem, *Super Dionysium De divinis nominibus*, c. 1, § 21, ad 1 and c. 12, § 15, sol. ed. Paul Simon (Münster: Aschendorff, 1972), pp. 10-11, 440; Aquinas, *In De anima*, lib. II, lect. 15, §§ 434-435, ed. Angeli M. Pirotta, 4th edn (Turin: Marietti, 1959), p. 108; idem, *In De sensu*, lib. un., lect. 4, § 55, ed. Raymund M. Spiazzi, 3rd edn (Turin: Marietti, 1973), p. 19. See also Averroës, *De sensu*, 192vb, in *Compendia librorum Aristotelis qui Parva naturalia vocantur*, ed. A.L. Shields (Cambridge, MA: Mediaeval Academy of America, 1949), p. 28.

model of vision.[26]   While Albert and Aquinas remain Dante's most likely sources, these examples reinforce the point that the optical information used by Dante was so commonplace that it is not necessary to adduce more technical and less widely diffused sources such as the *perspectivae*.

The propagation of light: "luce," "lume," "raggio," and "splendore"

An emphasis on general sources, rather than specific texts, is especially pertinent to Dante's discussion, in Book III, Chapter xiv of the *Convivio*, of a set of technical terms that he uses to categorise the different ways by which light is propagated:

Ma però che qui è fatta menzione di luce e di splendore, a perfetto intendimento mostrerò [la] differenza di questi vocabuli, secondo che Avicenna sente. Dico che l'usanza de' filosofi è di chiamare "luce" lo lume, in quanto esso è nel suo fontale principio; di chiamare "raggio," in quanto esso è per lo mezzo, dal principio al primo corpo dove si termina; di chiamare "splendore," in quanto esso è in altra parte alluminata ripercosso.[27]

As was shown in Chapter 1, the original distinctions were made by Avicenna in Book III of his *Liber de anima*, a work which is particularly rich in optical and visual ideas.[28]   Because Dante does not seem to have known Avicenna's treatise, several scholars have attributed his familiarity with the technical

---

[26]For astronomical works and encyclopaedias, see the references given in Ch. 1, nn. 13, 54-55. On the geometry of vision in theological and homiletic works, see Clark, "Optics for Preachers" and Ch. 3, nn. 56-57.

[27]*Con.* III, xiv, 5. For Dante's application of radial transmission to light from the stars, see further *Con.* II, vi, 9.

[28]The Avicennan distinctions are quoted in Ch. 1, n. 44. On Dante's limited knowledge of Avicenna, see Carlo Giacon, "Avicenna," in *ED* I, pp. 481-482.

56

terminology either to the "perspectivists,"[29] or to the theologian, Bartholomew of Bologna (d. 1292), who drew closely on Alhazen and the *perspectivae*.[30]

Several points of revision need to be made in this connection. First, it should be noted that these terms belong properly to the Avicennan-Aristotelian tradition and were not widely used by the "perspectivists." Only Roger Bacon distinguishes between the light terms by referring to Avicenna as his authority.[31] More significant still is the fact that Avicenna's distinctions had become commonplace by the thirteenth century. Both Busnelli and Vasoli have documented the technical words for light in the *De anima* commentaries of Albert and Aquinas.[32] But another *locus classicus* for discussions of the subject was in commentaries either on Book I (*distinctio* 9) or Book II (*distinctio* 13) of Peter Lombard's *Sentences*;[33] and by 1250, the technical words were

---

[29]Boyde, *Dante Philomythes*, p. 207: "From this applied science [sc. the science of "perspective"] he [Dante] learnt to use some of the more common words for light in their technical acceptations."

[30]Bartholomew of Bologna has attracted attention from Dantists, especially for his *Tractatus de luce*, a theological work which uses light and optical examples to interpret biblical passages. For critics who argue that Dante may well have culled his light terms from this treatise, see Maria Corti, *Percorsi dell'invenzione: Il linguaggio poetico e Dante* (Turin: Einaudi, 1993), p. 162; Maria Simonelli, "Allegoria e simbolo dal 'Convivio' alla 'Commedia' sullo sfondo della cultura bolognese," in *Dante e Bologna nei tempi di Dante* (Bologna: Facoltà di Lettere e Filosofia dell'Università di Bologna, 1967), p. 209 and nn. 5-6. For futher discussion of Bartholomew's possible influence on Dante, see below Ch. 7, pp. 234-238.

[31]Bacon, *De multiplicatione specierum*, pars I, c. 1 (Lindberg, pp. 4-5): "Et Avicenna dicit tertio *De anima* quod lux est qualitas corporis lucentis, ut ignis vel stellae; lumen vero est illud quod est multiplicatum et generatum ab illa luce" ("And Avicenna says in *De anima*, book iii, that *lux* is a quality of a luminous body; but *lumen* is that which is multiplied and generated from that *lux*," trans. Lindberg).

[32]In addition to the examples quoted by Vasoli (*Il Convivio*, pp. 457-458), see Albert, *Super Dionysius De divinis nominibus*, c. 2, § 31, sol. (Simon, p. 63) and the references given in the following two notes.

[33]For relevant discussions in *Sentence* commentary literature, see e.g. Albert, *In I Sent.*, d. 9, a. 8 (Jammy, XVa, pp. 175a-176b); Aquinas, *In II Sent.*, d. 13, q. 1, a. 3, sol. in *Opera omnia*. 25 vols. (Parma, 1852-1873; reprint New York: Musurgia, 1948), VI, p. 500a; Bonaventure, *In I Sent.*, d. 9, dub. 7, in *Opera omnia*, 10 vols. (Florence: Collegium S. Bonaventura, 1882-1902), I, p. 190b: "Et nota quod differunt splendor, radius et lumen, cum omnia dicant influentiam a luminoso, quia radius dicit emissionem secundum diametralem distantiam,

57

sufficiently well known to have been inserted into encyclopaedic works.[34] It is thus not necessary to derive Dante's familiarity with the distinctions from a direct acquaintance either with technical optical sources or with the work of Bartholomew of Bologna. Similarly, one does not need to refer Dante's discussion to texts associated with the so-called "metaphysics of light" such as Grosseteste's *De luce*.[35] One final point to note is that, in the *Comedy*, Dante did not always observe the fine distinctions he reserved for contexts in which "l'usanza de' filosofi" was required.[36]

What is important about Dante's use of this terminology in the *Convivio*, then, is not its derivation but rather how Dante uses the terms for light to make an elaborate analogy between the illumination of the Earth by the sun and the

---

lumen secundum circumferentiam [...] splendor dicit repercussionem ad corpus non trasparens, tersum et limitatum" ("And note that the terms splendour, ray and *lumen* are different, though they are all said of influence from a luminous body; for, ray is said of an emission of light according to a straight-line distance, *lumen* is what surrounds the luminous body [...] splendour is the reflection of light from a polished body which is not transparent and has a boundary").

[34] See Bartholomew the Englishman, *De rerum proprietatibus*, lib. VIII, c. 40 (Richter, p. 426): "lumen est quidam defluxus a luce, sive irradiatio defluens a substantia lucis. Lucem autem dicit ipsam fontalem substantiam, super quam lumen innititur" ("*lumen* is a certain outflow, or irradiation, by the substance of *lux*. *Lux* is said of this source-like substance from which *lumen* is emitted"); Vincent of Beauvais, *Speculum maius*, lib. II, c. 52 (I, cols 112-113): "lux est in propria natura, lumen autem in subiecto recipiente. Porro radius exitus luminis, secundum lineam rectam [...] Splendor autem est ipsa luminis reflexio a reflexione radiorum procedens" ("*lux* is in its own nature, *lumen* is what is received by another body. And I say that ray is the issuing of light in a straight line [...] Splendour is this reflection of light which comes from the reflection of rays").

[35] For this misleading view, see Marta Cristiani, "Platonismo," in *ED* IV, p. 553; James L. Miller, "The Three Mirrors of Dante's *Paradiso*," *University of Toronto Quarterly* 46 (1977), p. 271.

[36] The opposite view in Corti, *Percorsi dell'invenzione*, p. 162; Vasco Ronchi, "De luce et de lumine," *Physis* 8 (1966), p. 15: "per Dante luce e lume non erano sinonimi liberamente interscambiabili." But for examples of non-technical uses of light terms in the *Comedy*, see *Par.* XXXIII, 110 and 116 for "lume" used of God the "source" instead of the more correct "luce" (ll. 67, 100, and 124). For "raggio" used in place of "splendor," see *Con.* II, iv, 17; *Par.* V, 137, and for "splendor" instead of "raggio," see *Par.* XII, 9. These examples can easily be multiplied with references to the light words which Dante uses to describe the radiance of the blessed souls and planets in the *Paradiso*.

58

irradiation of God's goodness into all things.[37] The passage in question will receive more detailed consideration in Chapter 6, but suffice it to say here that that the relevant paragraphs are central to a fuller understanding both of the role of light in Dante's conception of divine action and of the keen interest he displays in intuiting and developing analogies between natural and divine light.

Dante's account of visual theory

In *Convivio*, III, Chapter ix, Dante discusses vision as an act of sensation and perception in considerable detail in order to explain the underlying causes of an apparent contradiction experienced by the human intellect as it contemplates the constant truths of philosophy and seems to observe change in them. The disjuncture between actual and perceived reality allows Dante to digress extensively on visual theory and the problems involved in reaching an accurate judgement of a visual object. The first two paragraphs of this digression deal with the external phases in the transmission of visual images to the eye:

Dove è da sapere che, propriamente, è visibile lo colore e la luce, sì come Aristotile vuole nel secondo de l'Anima, e nel libro del Senso e Sensato. Ben è altra cosa visibile, ma non propriamente, però che altro senso sente quello, sì che non si può dire che sia propriamente visibile, né propriamente tangibile; sì come è la figura, la grandezza, lo numero, lo movimento e lo stare fermo, che sensibili [comuni] si chiamano: le quali cose con più sensi comprendiamo. Ma lo colore e la luce sono propriamente; perché solo col viso comprendiamo ciò, e non con altro senso. Queste cose visibili, sì le proprie come le comuni in quanto sono visibili, vengono dentro a l'occhio – non dico le cose, ma le forme loro – per lo mezzo diafano, non realmente ma intenzionalmente, sì quasi come in vetro transparente. (paras. 6-7)

---

[37]See *Con.* III, xiv, 2-4, quoted in Ch. 6, p. 185.

59

This passage, and indeed the entire chapter, provides a remarkably good illustration of how Dante received a *corpus Aristotelicum* which had been re-elaborated and enriched by thirteenth-century commentators. The terse lecture-note format of Aristotle's text readily lent itself to the kind of exegesis in which commentators added new doctrinal elements and provided explanatory digressions.[38] In line with this exegetical tradition, thirteenth-century commentators such as Albert and Aquinas did not restrict themselves to Aristotle's texts, but reworked Islamic and Jewish sources into their paraphrases and commentaries.

In paragraph 6, for example, Dante maintains that light is visible "propriamente," that is, light is a sensory object which can be perceived directly by its appropriate sense organ, in this case the eye. Yet in the *De anima*, Aristotle argues that it is only colour which sets the transparent medium in motion and becomes the direct object of vision. Light is not itself seen, although luminous bodies dispose the transparent medium, the diaphanous, for the reception of colour by rendering it transparent in act.[39] At times, Aristotle seems to suggest that light and colour are manifestations of the same nature, and he even says that light becomes the "colour of the transparent" when it actualises a potentially transparent atmosphere.[40]

To understand how Dante and other medieval writers came to regard light as inherently visible, some attention must be given to the different conceptions

[38]For a good example of the transformations enacted on Aristotle's theory of light by earlier Neoplatonising commentators, see John J. Finamore, "Iamblichus on Light and the Transparent," in *The Divine Iamblichus: Philosopher and Man of the Gods*, ed. H.J Blumenthal and E.G. Clark (London: Bristol Classical Press, 1993), pp. 55-64; S. Sambursky, "Philoponus' Interpretation of Aristotle's Theory of Light," *Osiris* 13 (1958), 114-126. On Neoplatonic conceptions of light in general, see Part Two of this study.

[39]See *De anima*, II, 7, 418b 4-5, Graeco-Latin trans. in *In De anima*, ed. Pirotta, p. 102: "Diaphanum autem dico, quod est quidem visibile, non autem secundum se visibile, ut simpliciter est dicere, sed propter extraneum colorem" ("By the diaphanous I mean that which is visible, but not visible by itself but on account of extraneous colour").

[40]*De anima*, II, 7, 418b 12; *De sensu*, 3, 439a 17-19.

60

of the phenomenon held by Arabic writers, like Avicenna and Alhazen, who considered light to be visible in its own right and discussed its modes of propagation. Following the lead of Alhazen, the "perspectivists" also classified light amongst the *perceptibilia* proper.[41] But specialist writers on optics were not the only ones to be influenced by such ideas. The detailed consideration which Islamic *physici* paid to the nature and behaviour of light can be gauged from Averroës' Long Commentary on *De anima*. In this commentary, Averroës treats light independently from colour, considers it to be visible in its own right, and discusses the rectilinear and spherical configurations it assumes in the medium.[42] As commentators on *De anima* and *De sensu*, Albert and Aquinas are faithful to Aristotle in treating colour as properly visible (as Dante himself is in an earlier passage from the *Convivio*),[43] but elsewhere they follow Averroës

[41]Alhazen, *De aspectibus*, lib. II, c. 2, sec. 18 (Risner, p. 35): "Lux autem, quae est in corpore illuminato, per se comprehenditur a visu secundum suum esse" ("Light in an illuminated body is itself apprehended by sight according to its being"). For the "perspectivists," see Bacon, *Opus maius*, lib. V, p. i, dist. 8, c. 1 (Lindberg, p. 108): "Lux enim est primo visibile; deinde color et cetera viginti" ("For light is the primary visible thing; then come colour and the other twenty visibles," trans. Lindberg, p. 109); Pecham, *Perspectiva communis*, lib. I, prop. 61 (Lindberg, p. 138): "Nullam intentionem visibilem preter lucem et colorem solo sensu comprehendi" ("No visible intention other than light and colour is perceived by sense alone"); Witelo, *Perspectiva*, lib. III, petitiones (Risner, p. 84): "lux ex se ipsa videtur" ("light is seen by itself"). See also Bartholomew of Bologna, *Tractatus de luce*, pars IV, c. 4, ed. Idrenaeus Squadrani, in *Antonianum* 7 (1932), p. 470: "lux sensibilis est visibilis per seipsam et facit videri alia" ("sensible light is visible by itself and makes other things visible").

[42]Averroës, *Commentarium magnum in Aristotelis de anima libros*, lib. II, comm. 72, ed. F. Stuart Crawford (Cambridge, MA: Mediaeval Academy of America, 1953), p. 240: "Et natura coloris alia est a natura lucis et lucidi; lux enim est visibilis per se, color autem est visibilis mediante lucem" ("The nature of colour is different from the nature of light and of the luminous body; light is inherently visible, whereas colour is visible by means of light"). On the spherical and rectilinear behaviour of light, see further *Commentarium magnum*, lib. II, comm. 80 (Crawford, p. 253): "Lux enim innata est exire a lucido secundum rectas vertificationes ad partem oppositam lucida parti ex corpore luminoso, sicut declaraverunt facientes libros Aspectuum [...] Corpus autem cum fuerit luminosum ex omnibus partibus, non est dubium quod faciet speram lucidam; et hoc declaratum est in Aspectivus" ("By its nature light is apt to leave a luminous body in straight lines to the area opposite, as is demonstrated by those who write books on optics [...] If a body were to be luminous in all its parts then there is no doubt that it would make a luminous sphere; and this is stated in optical works").

[43]See e.g. Albert, *De anima*, lib. II, tr. 3, cc. 7-8 (Stroick, pp. 108-110): "oportet ut color per nihil extrinsecum, sed per suam essentiam sit agens in visum et sit visibilis [...] diaphanum

61

and Avicenna in referring to light as *visibilis per se*.[44] Dante's view that light is visible in its own right should thus be seen as part of a general debt to widely accepted thirteenth-century analyses of the nature of light.

In his subsequent discussion of the distinction between "proper" and "common sensibles" (also para. 6), Dante adheres more faithfully to Aristotelian doctrine. In Aristotle's analysis, a special internal faculty, the "common sense," was required to apprehend the "common sensibles," those sensibles common to two or more senses, which Dante lists as "la figura, la grandezza, lo numero, lo movimento e lo stare fermo."[45] By contrast, Alhazen and his thirteenth-century followers give a greatly expanded version of the "common sensibles" and provide a list of twenty-two *intentiones visibiles*.[46]

---

enim secundum se non est visibile, eo quod nullum habeat colorem" ("it is necessary that colour is active in vision and visible not by anything extrinsic to it but by its essence [...] the transparent medium is not in itself visible, since it has no colour"); Aquinas, *In De anima*, lib. II, lect. 14, §§ 400 and 405 (Pirotta, pp. 103-104): "color in eo quod est color, est visibilis per se [...] lumen est quasi quidem color diaphani" ("colour *qua* colour is inherently visible [...] light is in a sense the colour of the transparent medium"); idem, *In De sensu*, lib. un., lect. 4, § 77 (Spiazzi, p. 26). For these more traditionally Aristotelian ideas on colour visibility in Dante, see *Con.* I, xi, 3; cf. *Par.* X, 42.

[44]Albert, *Summa theologiae, sive De mirabili scientia Dei*, ed. Siedler Dionysius, Wilhem Kübel, and Heinrich Georg Vogels (Münster: Aschendorff, 1978), p. 47: "visibile enim primum, quod est lux, sive sensibilis sive intellectualis, non fit visibile aliquo coadiuvante, sed seipso visibile est" ("the first visible quality, which is light, whether sensible or intellectual, is not visible due to anything assisting it, but is visible by itself"); idem, *Super Dionysius De divinis nominibus*, c. 2, § 29 (Simon, p. 62); Aquinas, *In De anima*, lib. II, lect. 14, § 419 (Pirotta, p. 105); idem, *In II Sent.*, d. 13, q. 1, a. 3, sol. in *Opera omnia*, VI, p. 500b: "Dicendum quod illud quod nos appellamus lucem est illud quod per se videtur"("It should be said that that which we call light is that which is inherently visible"); idem, *ST*, Ia, q. 67, a. 3, resp. (Caramello, I, p. 328): "Lux autem est secundum se visibilis" ("But light is visible by itself"); Ia-IIae, q. 8, a. 2, ad. 2 (Caramello, II, p. 50): "per potentiam visivam sentitur et color, et lux, per quam color videtur" ("colour and light, which allows colour to be seen, are perceived by the visual power"); Ia-IIae, q. 57, a. 2, ad. 2 (Caramello, II, p. 249): "ad eandem potentiam visivam pertinet videre colorem, et lumen, quod est ratio videndi colorem et simul cum ipso videtur" ("the same visual power is responsible for seeing both colour and light, light being both the reason for seeing colour and seen simultaneously with colour").

[45]References to Aristotle's discussion of the "proper" and "common" sensibles are given in Ch. 1, nn. 19-20. Intriguingly, Aristotle himself expands his list of "common sensibles" to include roughness, smoothness, sharpness, and bluntness in *De sensu*, 4, 442b 5-9.

[46]Alhazen, *De aspectibus*, lib. II, c. 2, sec. 15 (Risner, p. 34): "Intentiones particulares quae comprehenduntur sensu visu, sunt multae, sed generaliter dividuntur in 22: et sunt lux, color,

62

In paragraph 7, Dante inquires into the mode of transmission by which the "cose visibili" reach the eye. Once again, he follows the essential outlines of Aristotelian teachings by arguing that it is not the objects of vision which traverse the transparent medium ("il mezzo"), but their representative form, or *species sensibilis.*[47] He nonetheless reveals an indirect reliance on Arabic sources in his use of the adverb "intenzionalmente" to describe this external process. The word *intentio*, which Arabo-Latin translators used to render the Arabic *ma'na*, is one of the most problematic in scholastic vocabulary.[48] Both the Arabic and Latin terms have a number of meanings which vary according to specific usage in technical contexts. In Avicenna's work on logic, for instance, *intentio prima* denotes the type of impression emitted by an object in the external world, whereas *intentiones secundas* are the universal concepts which the mind elicits from these impressions.[49] In the context of Avicenna's faculty-

---

remotio, situs, corporeitas, figura, magnitudo, continuum, discretio et separatio, numerus, motus, quies, asperitas, levitas, diaphanitas, spissitudo, umbra, obscuritas, pulchritudo, turpitudo, consimilitudo et diversitas" ("There are many particular intentions which are apprehended by the visual sense but generally there are twenty-two and these are: light, colour, distance, position, solidity, shape, magnitude, continuity, proximity and separation, number, motion, rest, roughness, smoothness, transparency, opacity, shadow, darkness, beauty, ugliness, similarity and dissimilarity"). All the "perspectivists" and Bartholomew of Bologna give exactly the same list, see Bacon, *Opus maius*, lib. V, p. i, dist. 1, c. 3 (Lindberg, pp. 8-10); Bartholomew, *Tractatus de luce*, pars II, c. 4 (Squadrani, p. 340); Pecham, *Perspectiva communis*, lib. I, props. 55-56 (Lindberg, pp. 134-136); idem, *Tractatus de perspectiva*, c. 9 (Lindberg, p. 54); Witelo, *Perspectiva*, lib. III, petitiones 2 (Risner, p. 84). Bacon states explicitly that this list is an expansion of Aristotle when in his *Opus maius* he writes that the twenty-two *intentiones* "sunt sensibilia communia, de quorum aliquibus exemplificat Aristoteles" ("are the common sensibiles, some of which Aristotle discussed," loc. cit., p. 10).

[47]*De anima*, III, 8, 431b 28-432a 1.

[48]For a valuable introduction to the complexities of the term in a Dantean context, see Tullio Gregory, "Intenzione," in *ED* III, pp. 480-482. For more complete literature on *intentiones*, see Kwame Gyekye, "The Terms 'Prima Intentio' and 'Secunda Intentio' in Arabic Logic," *Speculum* 46 (1971), 32-38; Christian Knudsen, "Intentions and Impositions," in *The Cambridge History of Later Medieval Philosophy*, ed. Kretzmann, pp. 479-495; A.I. Sabra, "Sensation and Inference in Alhazen's Theory of Visual Perception," in *Studies in Perception*, ed. Machamer and Turnbull, pp. 171, 177-178, n. 38, p. 183; idem, *The Optics of Ibn al-Haytham*, II, pp. 70-73; Vescovini, *Studi sulla prospettiva*, pp. 64-69.

[49]Avicenna, *Avicenna latinus: liber de philosophia sive scientia divina*, tr. I, c. 2, ed. S. Van Riet, 3 vols. (Louvain and Leiden: E.J. Brill, 1977-1983), I, p. 10: "Subiectum vero logicae,

63

psychology in his *Liber de anima*, the term has a quite different meaning. An "intention" is here the result of an act of judgement exercised by an interior faculty of the sensitive soul, the *aestimativa*, and not based on information apprehended by external sense organs.[50]

Dante does not use "intenzione" in the second "psychological" sense that Avicenna gives to it, although the notion of a "first intention" (i.e. a non-corporeal replica of a sense object) is crucial to a fuller understanding both of several passages in the *Convivio* and of the processes of visual perception and cognition that he later describes in *Purgatorio* XVIII.[51] In the visual theory expounded in the *Convivio*, the way in which Dante uses the adverb "intenzionalmente" to describe the presence of light and colour in the medium can best be clarified by a passage in Averroës's Long Commentary on the *De anima*. In Book II of this work, Averroës uses *intentio* to denote the type of non-corporeal impression that light makes in the medium.[52] Both Aquinas and

---

sicut scisti, sunt intentiones intellectae secundo, quae apponuntur intentionibus intellectis primo" ("As you know, the subject of logic is the second intelligible intentions, which are dependent upon the first intelligible intentions").

[50]Avicenna, *Liber sextus seu de anima*, pars I, c. 5 (Van Riet, II, p. 86): "intentio autem est id quod apprehendit anima de sensibili, quamvis non prius apprehendat illud sensus exterior, sicut ovis apprehendit intentionem quam habet de lupo" ("an intention is that which the soul apprehends of sensibles even though the external senses did not previously apprehend that, as the sheep apprehends the intention it has of a wolf"); pars IV, c. 1 (Van Riet, I, pp. 6-8): "[intentiones] sunt res quas apprehendit anima sensibilis ita quod sensus non doceat eam aliquid de his; ergo virtus qua haec apprehendantur est alia virtus et vocatur aestimativa" ("intentions are the things which the sensible soul apprehends in such a way that sense does not teach the soul anything about them; hence, the virtue which apprehends them [intentions] is another virtue and it is called the estimative"). To my knowledge, Dante does not use the term *intentio* in this precise technical sense anywhere in his writings.

[51]The notion of a "first intention" as a replica of a sense object, which, as a potential *species cognoscibilis*, permits one to know this object, illustrates the crucial verses in *Purg.* XVIII, 22-23: "Vostra apprensiva da esser verace / tragge intenzione, e dentro a voi la spiega." The idea of *intentio secunda* as a mental image of a real object underlies Dante's use of "intenzione" and "essemplo intenzionale" in *Con.* III, vi, 5-6; IV, x, 11; *Par.* I, 128. For the synonym "intenza," see also *Par.* XXIV, 75, 78.

[52]Averroës, *Commentarium magnum*, lib. II, comm. 70 (Crawford, p. 237): "lux non est corpus, sed est praesentia intentionis in diaphano" ("light is not a body but the presence of an intention in the tranparent medium"). The distinction between "real" and "intentional" *esse* is

64

Albert also describe sense objects in the medium as having *esse intentionale*.[53] Yet once again the "perspectivists" offer a markedly different explanation in arguing that likenesses are transmitted to the eye corporeally.[54]

After describing the external phases of his theory, Dante then moves on to consider the reception of visible forms in the pupil of the eye:

E ne l'acqua ch'è ne la pupilla de l'occhio, questo discorso, che fa la forma visibile per lo mezzo, sì si compie, perché quell'acqua è terminata – quasi come

made quite clear by Bacon, see *De multiplicatione specierum*, pars I, c. 1 (Lindberg, pp. 4-5): "Intentio vocatur in usu vulgi naturalium propter debilitatem sui esse respectu rei, dicentis quod non est vere res sed magis intentio rei, id est similitudo" ("It is called 'intention' by the multitude of naturalists because of the weakness of its being in comparison to that of the thing itself, for they say that it is not truly a thing, but rather the intention, that is, the similitude, of a thing," trans. Lindberg).

[53]For colour as having *esse intentionale* in the medium, see e.g. Aquinas, *In De anima*, lib. II, lect. 14, § 418 (Pirotta, p. 105); idem, *In De sensu*, lib. un., lect. 14, § 62 (Spiazzi, pp. 21-22); cf. idem, *ST*, Ia-IIae, q. 22, a. 2, ad. 3 (Caramello, II, p. 116). Aquinas repeatedly argues that light does not have "intentional" *esse* in the medium, see e.g. *In De anima*, lib. II, lect. 14, § 420; lect. 24, § 553 (Pirotta, pp. 105-106, 138); idem, *ST*, Ia, q. 67, a. 3, resp. (Caramello, I, p. 328). Albert, however, maintains that both light and colour are replicated as *intentiones*, see *De anima*, lib. II, tr. 3, c. 12 (Stroick, p. 116): "Et ideo satis convenienter dicunt quidam, quod lumen est intentio spirituale esse habens in perspicuo, quemdamodum color est habens esse intentionale in medio" ("Therefore some argue sufficiently well that light is an intention having spiritual being in the transparent medium, in a similar way that colour has intentional being in the medium"); idem, *De meteoris*, lib. I, tr. 2, c. 6 (Jammy, II, p. 14b): "lumen [...] est intentio formae corporis luminosi in perspicuo secundum esse spirituale generata" ("light is an intention of the form of a luminous body, which, having spiritual being, is generated in the transparent medium").

[54]On corporeal transmission, see Bacon, *De multiplicatione specierum*, pars III, c. 2 (Lindberg, p. 190): "ideo absolute diffinio quod species rei corporalis est vere corporalis et habet esse vere corporale" ("I therefore state unconditionally that the species of a corporeal thing is truly corporeal and that it has truly corporeal being," trans. Lindberg); idem, *Opus tertium*, c. 35 (Brewer, p. 114). A corollary of this doctrine is the non-Aristotelian view that visible objects are transmitted in time, see Alhazen, *De aspectibus*, lib. II, c. 2, sec. 21 (Risner, p. 37): "Lux ergo non pervenit ex aere, qui est extra foramen, ad aerem, qui est intra foramen, nisi in tempore" ("it is only in time, therefore, that light comes from the air, which is outside the aperture, to the air which is in the aperture"). For the related notion that vision takes place in time, see Bacon, *Opus maius*, lib. V, p. i, dist. 9, c. 3 (Lindberg, p. 134): "Quod species visus et visibilis fiat in tempore" ("That the species of sight and visible objects are produced in time," trans. Lindberg, p. 135); Pecham, *Perspectiva communis*, lib. I, prop. 53 (Lindberg, p. 134): "Omnia que videtur tempore comprehendi" ("All that is seen is understood in time").

specchio, che è vetro terminato con piombo –, sì che passar più non può, ma quivi, a modo d'una palla, percossa si ferma; sì che la forma, che nel mezzo transparente pare, [ne la parte pare][55] lucida e terminata. E questo è quello per che nel vetro piombato la imagine appare, e non in altro. (para. 8)

Dante's account of the reception of the visible "forma" by the watery substance of the eye ("l'acqua," as he calls it) is here in perfect accordance with Aristotle.[56] But one detail in paragraph 8 which appears to run counter to Aristotle is Dante's comparison of the reception of visible forms in the eye to the way in which a mirror receives forms. For Aristotle, seeing involves some sort of reflection, but he argued that his predecessor Democritus had shown a faulty appreciation of the nature of sight in relating vision to "mirroring."[57] Moreover, Aristotle never applied the mechanistic example of a rebounding ball to light, which he regarded as a non-corpuscular phenomenon.[58] In this paragraph of *Convivio*, then, Dante shows a certain degree of independence from Aristotle's

[55]In his *La "Rotta Gonna": Gloses et corrections aux textes mineurs de Dante* (Florence: Sansoni, 1967), pp. 219-220, André Pézard proposes the reading '[ne la parete pare], arguing that the image appears on the retina. Vasoli (*Il Convivio*, pp. 407-408) also finds this view acceptable; however, as Bruce S. Eastwood has demonstrated there is no precedent in medieval sources for referring to image formation as taking place on the retina, see his "Averroes' View of the Retina – A Reappraisal," *Journal of the History of Medicine and Allied Sciences* 24 (1969), 77-82.

[56]*De anima*, III, 1, 425a 4-5; *De sensu*, 2, 438 a 12-16; 438 b 5-15.

[57]*De sensu*, 2, 438a 5-7, Graeco-Latin trans. in *In De sensu*, ed. Spiazzi, p. 17: "Democritus autem quoniam quidem aquam dicit esse, bene dixit. Quia autem putavit ipsum videre esse illam apparitionem, non bene, hoc enim accidit, quoniam oculus lenis est, et est non in illo sed in vidente" ("In saying that it [the eye] is water Democritus spoke well. Insofar as he thought this act of seeing is that reflection, he did not speak well, for, this happens because the eye is smooth and [sight] is not in that [the eye] but in the spectator").

[58]The example is found in Hero of Alexandria (fl. 62 A.D.), *Catoptrica*, c. 3, ed. L. Nix and W. Schimdt, in *Opera quae supersunt omnia*, 5 vols. (Leipzig: Teubner, 1899-1914), II, pp. 322-324, and in Ptolemy, *Optica*, lib. III, c. 19 (Lejeune, p. 98). Systematically employed by Alhazen for reflection, e.g. *De aspectibus*, lib. IV, c. 3, secs. 17-18 (Risner, pp. 112-113), the comparison became standard in "perspectivist" works and elsewhere, e.g. Avicenna, *Liber de anima*, pars III, c. 6 (Van Riet, p. 239).

66

account, and this divergence again illustrates how Arabic sources absorbed by Albert and Aquinas passed from them to Dante.

The concept of "mirroring" is repeatedly used by Avicenna and Averroës to elucidate the reception of forms in the eye,[59] and this tradition clearly influenced Albert and Aquinas in their *De sensu* commentaries. With reference to the eye, both commentators discuss the nature and properties of mirrors at considerable length,[60] and they both adduce the example of the ball rebounding in order to explain how forms are brought to a halt in the eye.[61] In addition, Aquinas

[59]Avicenna, *Compendium de anima* (Venice, 1546; reprint, Farnborough, 1969), pp. 17-18: "[...] et non manifestatur [sc. "species rerum" in the eye] nisi in corpore terso recipiente ipsam, sicut speculum, et quae sunt illi similia" ("the visible species of things is not made visible unless it is received by a smooth body, such as a mirror, or objects that are like mirrors"); Averroës, *Colliget*, lib. II, c. 15 (Apud Iunctas, X, 25rb-va): "Sed oculus recipit colores per sua corpora pervia per modum quem recipit speculum, et quando colores sunt impressi in ipsum, tunc apprehendit eos spiritus visibilis" ("But the eye receives colours through its pervious bodies in the way that a mirror does, and when colours are impressed in it [the eye], then the visible spirit takes them"); idem, *De sensu*, 193rb (Shields, p. 37): "Et ideo recipit [sc. oculus] formas ex aëre, quia est quasi speculum, et reddit eas aque" ("And thus the eye receives forms out of the air because it is like a mirror and passes them to the aqueous humour"). For the significance of this idea, see Lindberg, *Theories of Vision*, p. 49.

[60]For Albert's references to the eye as a *speculum animatum*, see *De anima*, lib. II, tr. 3, cc. 14-15 (Stroick, pp. 122-123); idem, *De intellectu et intelligibili*, lib. II, tr. un., c. 2 (Jammy, V, p. 253b); *De sensu et sensato*, tr. I, c. 5 (Jammy, V, p. 6a). Albert's *De anima* passage is reproduced in Beauvais' *Speculum maius*, lib. II, c. 77 (I, col. 128). Aquinas is typically more measured in his language, see *In De sensu*, lib. un., lect. 4, § 48 (Spiazzi, p. 17): "illa enim passio [sc. reception of forms in the eye] est reverberatio, idest causatur ex refractione sive reverberatione formae ad corpus politum" ("that 'passion,' namely, that the form of a visible object appears in the eye, is a repercussion, that is, it is caused by the reflection or reverberation of forms from a polished body"). The comparison of the eye to a mirror is also found in medical and encyclopaedic traditions, see Alexander Neckham, *De naturis rerum*, c. 154, ed. Thomas Wright (London: Longman, 1863), p. 239; Galen, *De usu partium*, lib. X, c. 8 (May, II, p. 479); Isidore, *Differentiarum*, lib. III, c. 17, § 53 (*PL* 83, col. 78C).

[61]Cf. Albert, *De sensu et sensato*, tr. I, c. 8 (Jammy, V, p. 10b): "reflexio fit ex repercussione sicut repercutitur pila" ("reflection takes place due to repercussion as a ball is rebounded"); Aquinas, *In De sensu*, lib. un., lect. 4, § 48 (Spiazzi, p. 17): "quodam modo reflectitur ad similitudinem pilae, quae repercutitur proiecta ad parietem" ("in a way it is reflected like a ball which is rebounded when thrown against a wall"). In their use of terms such as *repercussio, reverberatio*, and *repulsio*, Albert and Aquinas (and Dante too) followed a long-standing optical tradition dating back to the first and second centuries, but notably reinforced by the scholars of the Arabic world.

observes that it is necessary to add a lead coating to the back of glass in order to make a reflective surface.[62]

In paragraphs 8-9, it is also important to distinguish Dante's treatment of the transmission of forms in the eye from the views of Alhazen, who was one of the first writers to refer to refracted light in order to explain the transmission of light rays within the eye. In the thirteenth century, this principle was reworked by all the "perspectivists" in their analyses of vision.[63] And yet, Dante's eye-mirror comparison shows that, unlike Alhazen and thirteenth-century optical specialists, the principle of refracted rays has no place in his conception of the visual process.

Dante's presentation of the visual act in this section of the *Convivio* also contrasts with another essential feature of the *perspectivae*: the notion of dividing an object of vision into a series of points. All historians of medieval optics agree that this idea was one of the most important foundational principles of "perspectivist" optics. Unlike the Aristotelian theory of vision, which rather unsatisfactorily considered the visible "form" as a coherent whole, Alhazen and

---

[62]Aquinas, *In De sensu*, lib. un., lect. 4, § 49 (Spiazzi, p. 18): "Et ideo videmus, quod nisi in vitro apponatur plumbum vel aliquod huiusmodi, quod impediat penetrationem, ne ulterius procedat apparitio, non fit talis immutatio" ("And thus we see that such a change does not take place unless lead or something of the kind is attached to glass, which prevents the visible object passing through and its image proceeding any further"); cf. idem, *In IV Sent.*, d. 44, q. 2, a. 1, qla. 4, ad 5, in *Opera Omnia*, VII, p. 1087b. For additional references to lead as a backing for mirrors, see e.g. Alexander Neckham, *De naturis rerum*, c. 154 (Wright, p. 239): "subtrahe plumbum suppositum vitro, iam nulla resultabit imago inspicientis" ("take away the lead to which glass is attached, now no image will be visible to the spectator"); Bartholomew of Bologna, *Tractatus de luce*, pars I, c. 4 (Squadrani, p. 339); Bonaventure, *In II Sent.*, d. 13, a. 3, q. 1, ad. 2 (Quaracchi, II, p. 326a); Vincent of Beauvais, *Speculum maius*, lib. II, c. 78 (I, col. 129): "inter omnia melius est speculum ex vitro, et plumbo" ("a mirror made from glass and lead is the best of all"). Dante also refers to lead-backed mirrors in *Inf.* XXIII, 25-27; *Par.* II, 89-90.

[63]On Alhazen's originality in this respect, see *De aspectibus*, lib. VII, c. 6, sec. 37 (Risner, p. 270): "Hoc autem quod quicquid comprehendatur a visu, comprehendatur refracte, a nullo antiquorum dictum est" ("This point, then, that what is apprehended by sight is apprehended by refracted rays has not been said by any of the ancient writers"). For the "perspectivists," see above n. 21.

68

the "perspectivists" divided the visual field up into an array of point forms.[64] In his study of the history of optics, Vasco Ronchi adopts Dante's text (paras. 6-11) as an example of the "medieval failure" to grasp the implications of what he calls Alhazen's "punctiform" analysis of vision.[65] Although this is not a fair assessment of thirteenth-century optics as a whole, Dante's analysis of the visual act in the *Convivio* seems to support Ronchi's view. In the text of the *Convivio*, Dante never refers to the object of vision as anything other than a "forma" or "imagine," and these forms are said to enter the eye as such (i.e. holistically). In short, Dante seems to show no acquaintance with one of the principal premises of Alhazen's work and "perspectivist" optics.[66]

Having described the external phases of his visual theory, Dante considers (para. 9) the transmission of the visual image to the internal senses and the more psychological aspects of vision: seeing, as such, takes place in the brain:

Di questa pupilla lo spirito visivo, che si continua da essa, a la parte del cerebro dinanzi – dov'è la sensibile virtude sì come in principio fontale – subitamente sanza tempo la ripresenta, e così vedemo. (para. 9)

He again reveals his awareness that vision is not merely the reception of forms in the eye by describing how spirits conduct the image to the "sensibile virtude"

---

[64]On the division of the visual field into point forms in Alhazen, see Ch. 1, n. 46. For the "perspectivists," see Bacon, *De multiplicatione specierum*, pars I, c. 2 (Lindberg, pp. 38-40); Pecham, *Perspectiva communis*, lib. I, prop. 20 (Lindberg, pp. 96-98), idem, *Tractatus de perspectiva*, c. 3 (Lindberg, p. 33); Witelo, *Perspectiva*, lib. III, prop. 20 (Risner, p. 94); cf. Bartholomew of Bologna, *Tractatus de luce*, pars I, c. 3 (Squadrani, p. 235). Note also its application to natural force in Grosseteste, *De lineis* (Baur, p. 64): "completa actio est quando ab omnibus punctis agentis [...] veniet virtus agentis ad quemlibet punctum patientis" ("action is complete when virtue comes from every point on an agent to every point of the receiving body"). This principle seems to have originated with Alkindi, see above Ch. 1, n. 33.

[65]Vasco Ronchi, *The Nature of Light: An Historical Survey*, trans. V. Barocas (London: Heinemann, 1970), pp. 57, 62-69, 83, 87.

[66]E.g. *Con.* III, ix, 14: "perché la imagine loro vegna dentro (i.e. inside the eye) più lievemente e più sottile."

located in the anterior part of the brain.[67] All modern commentators agree that the "virtude" mentioned here is the "common sense," the faculty responsible for collating the various sense impressions gathered by the external senses in order to form a composite image. Dante follows this tradition when he describes this faculty as a "fontale principio," although it is not clear whether the seeing he envisages here is conceived in the same manner as the earlier passage in which true vision was in the "imaginativa."[68]

In a chapter that is itself a "disgressione," Dante then intercalates another minor digression (para. 10) in which he discusses the extramission theory of vision advanced by "Plato e altri filosofi" and refuted by Aristotle in his *De sensu*.[69] In this paragraph, Dante endorses Aristotle's refutation of extramission, even though he makes widespread use of eye rays throughout his lyric poetry, in the *Vita Nuova*, and even on occasion in the *Comedy*.[70]

---

[67]In contrast to the "perspectivists," Dante does not include details of the optical transmission of images after sensation, cf. Alhazen, *De aspectibus*, lib. I, c. 4, sec. 16; lib. II, c. 1, sec. 6 (Risner, pp. 8, 25-27); Bacon, *Opus maius*, lib. V, p. i, dist. 7, c. 1 (Lindberg, pp. 96-98); idem, *Opus tertium*, c. 34 (Brewer, pp. 113-114); Pecham, *Perspectiva communis*, lib. I, props. 33 and 40 (Lindberg, pp. 118, 122-124); Witelo, *Perspectiva*, lib. III, prop. 21-22 (Risner, pp. 94-95).

[68]According to Wolfson, "The Internal Senses," pp. 97-98, there was considerable indecision amongst Islamic writers over the relationship between the "common sense" and the "imagination," and this may help to explain the ambiguity in Dante. For relevant Aristotelian commentary texts on the *sensus communis*, see Vasoli, *Il Convivio*, p. 408.

[69]*De sensu*, 2, 437b, 10-24.

[70]For extramitted rays in Dante's earlier poems, see the references given in n. 6 above. Despite his allegiance to Aristotle's intramission theory of vision in the *Convivio*, eye rays are used for dramatic purposes in the *Comedy*, see *Purg.* XXXIII, 18; *Par.* III, 128-29; IV, 139-42; XXVI, 77-78; and for futher discussion, see below Ch. 3, pp. 86-87 and n. 15. The rich heritage left by the Platonic (see Ch. 1, n. 14) and Augustinian (e.g. *De Genesi ad litteram*, lib. I, c. 16, § 31, in *PL* 34, col. 258; *Confessiones*, lib. X, c. 6, § 9, ed. Verheijen, in *CCSL* 26, pp. 159-160) models of vision influenced poets in the lyric tradition from the Sicilians to Petrarch and beyond, and this fact helps in part to resolve the contradiction in Dante. The theme was especially prominent in "stilnovisti," see e.g. Cavalcanti, *Rime* VII, 9-10; IX, 23, in *Poeti del Duecento*, ed. Contini, II, pp. 498, 501. The influence of the extramission theory can still be documented in Grosseteste in the 1230s, see *De iride* (Baur, p. 73): "Perspectiva igitur veridica est in positione radiorum egredientium" ("Perspective is therefore veridical when rays proceed outwards [from the eye]"). Visual rays are still assigned a role by Bacon (who disagrees with Aristotle in this regard) and by Pecham in the 1260s, see Bacon, *Opus*

70

After dealing with the final act of vision in the brain, Dante concludes his treatment of vision with a discussion of visual error (paras. 10-16). In this section, he outlines the potential causes of star scintillation, an optical illusion discussed by Aristotle in his *De caelo*:[71]

[...] per più cagioni puote parere [sc. la stella] non chiara e non lucente. Però puote parere così per lo mezzo che continuamente si transmuta. Transmutasi questo mezzo di molta luce in poca luce, sì come a la presenza del sole e a la sua assenza; e a la presenza lo mezzo, che è diafano, è tanto pieno di lume che è vincente de la stella, e però [non] pare più lucente. Transmutasi anche questo mezzo di sottile in grosso, di secco in umido, per li vapori de la terra che continuamente salgono: lo quale mezzo, così transmutato, transmuta la imagine de la stella che viene per esso, per la grossezza in oscuritade, e per l'umido e per lo secco in colore. Però puote anche parere così per l'organo visivo, cioè l'occhio, lo quale per infertade e per fatica si transmuta in alcuno coloramento e in alcuna debilitade. (paras. 11-13)

In these paragraphs, Dante assigns two causes of error, external ones such as solar and atmospheric changes to the medium, and internal ones such as various forms of eye infirmity. Albert discusses both of these generic causes in a passage from his *De anima* that seems not to have been noticed by Dante scholars:

---

maius, lib. V, p. i, dist. 7, c. 4 (Lindberg, p. 104); Pecham, *Perspectiva communis*, lib. I, prop. 46 (Lindberg, p. 128); idem, *Tractatus de perspectiva*, c. 4 (Lindberg, p. 38). Note, however, that in his exposition of vision in the *Convivio* Dante never assigns such a role to extramitted rays and that as poet of the *Comedy* he also treats the visual process as a reception into the eye, see e.g. *Inf.* X, 69; *Par.* XX, 31-2; XXX, 46-47. The idea of sense passivity is Aristotelian, see *De anima*, II, 12, 424a 26-28.

[71]*De caelo*, II, 8, 290a 18-24. The example of an optical illusion involving star scintillation was widely used elsewhere, see e.g. Neckham, *De naturis rerum* (Wright, pp. 93-94); Robert Grosseteste, *Commentarius in Posteriorum analyticorum libros*, ed. Pietro Rossi (Florence: Olschki, 1981), pp. 190-191.

71

Et multae variationes possunt fieri circa esse suum [sc. the being of the sensory object of sight], quod habet in abstractione, tam in medio quam in organo: quoniam si contingat aerem qui est medium in visu esse humidum multum, forte videbitur album esse perfusum rubare vel croceitate; et si forte pupilla sit infirma, ex humore in oculum defluente alterabit esse color quod habet in abstractione.[72]

The next chapter will examine Dante's poetic treatment of these and other visual errors by comparing his examples with those found in the *perspectivae*, in Aristotelian commentaries, and in other sources available to the late thirteenth century. But one issue raised by this passage which calls for discussion here is the fact that Dante shows a non-Aristotelian familiarity with more complex eye anatomy in referring to the "tunica de la pupilla" (para. 13). Because tunics are repeatedly discussed by Alhazen and the "perspectivists,"[73] it is this detail above all others in Dante's discussion which has been used to argue that it is impossible to reconcile his treatment of vision with sources such as Aristotelian commentary texts.[74] It is, however, important to recognise that details of eye anatomy were more widely available to Dante's time than has often been

---

[72]Albert, *De anima*, lib. II, tr. 3, c. 5 (Stroick, p. 103): "And many variations can take place in the being that it has in abstraction, both in the medium and in the organ; for if it happens that the air, which is the medium of sight, is very damp then perhaps a white object will be seen to be suffused with red or green; and if perhaps the pupil is infirm, due to a humour flowing into the eye, it will change the being that colour has in abstraction."

[73]For Alhazen's three humour and four tunic anatomy, see *De aspectibus*, lib. I, c. 5, sec. 4 (Risner, pp. 3-4). Identical eye anatomies are found in Pecham, *Perspectiva communis*, lib. I, prop. 31 (Lindberg, pp. 112-114); idem, *Tractatus de perspectiva*, c. 5 (Lindberg, p. 42); Witelo, *Perspectiva*, lib. III, prop. 4 (Risner, pp. 85-87). Bacon, *Opus maius*, lib. V, p. i, dist. 2, c. 2 (Lindberg, pp. 26-30), differs from these writers by assigning either three or six tunics to the eye.

[74]Parronchi, "La perspettiva dantesca," p. 39: "nessuno di questi testi [sc. the commentaries of Albert and Aquinas] offre riferimenti con l'intera «disgressione»"; Guidubaldi, *Dante Europeo*, II, pp. 346-348; III, pp. 305-307. Both critics point out the affinities between Roger Bacon's anatomy of the eye and Dante's.

72

assumed.[75] In thirteenth-century sources, *tunica* appears once in the Latin translation of Averroës' *De sensu* paraphrase, twice in Aquinas's commentary on the *De sensu* (c. 1270), and frequently in Albert's earlier paraphrases (c.1254-1257) on Aristotle's psychological works as well as in his own independent treatises.[76] As is the case with the visual concepts and optical terminology contained in Dante's entire digression on visual theory, there is no justification at all for emphasising his use of the term "tunica" as evidence that he was dependent on the *perspectivae*.

Optics and Vision from the Minor Works to the *Comedy*

Before examining Dante's optical knowledge in the *Comedy*, it is worth offering some concluding remarks on Dante's optical and visual thought in his minor works. This chapter, especially the sections on the *Convivio*, has shown that Dante's optics and visual theory lack the technical sophistication found in Alhazen and the "perspectivists" and that Dante tended to draw on ideas which had become commonplace in a wide variety of general sources but were more

[75]On Chalcidius' anatomy of the eye, see Ch. 1, n. 16. The term *tunica* was widely used in Latin translations of Arabic works, and was transmitted to several twelfth-century writers, see e.g. Hugh of St. Victor (d. 1140), *De unione corporis et spiritus* (*PL* 177, col. 287); William of Conches (d. 1154), *De philosophia mundi*, lib. IV, c. 25 (*PL* 172, col. 95); idem, *Glosae super Platonem*, c. 138, ed. Edouard Jeauneau (Paris: Vrin, 1965), p. 238; William of St. Thierry (1085-1148), *De natura corporis et animae*, lib. I (*PL* 180, col. 704).

[76]References in order: Averroës, *De sensu*, 192rb (Shields, p. 13); Aquinas, *In De sensu*, lib. un., lect. 3, § 46; lect. 4, § 58 (Spiazzi, pp. 16, 19); Albert, *De anima*, lib. II, tr. 3, c. 13 (Stroick, p. 118); idem, *De animalibus*, lib. I, tr. 2, c. 7, ed. H.J. Stadler, 2 vols. in *Beiträge* 15-16, I, p. 75; idem, *De meteoris*, lib. III, tr. 4, c. 26 (Jammy, II, p. 139); idem, *De sensu et sensato*, tr. I, c. 13 (Jammy, V, p. 18a); idem, *Quaestiones de animalibus*, lib. I, qq. 30-31 (Filthaut, pp. 99-100). See also Bartholomew the Englishman, *De rerum proprietatibus*, lib. III, c. 17 (Richter, pp. 62-63); Thomas of Cantimpré, *Liber de natura rerum*, ed. H. Boese (Berlin and New York: De Gruyter, 1973), lib. I, c. 6, pp. 18-21; Vincent of Beauvais, *Speculum maius*, lib. XXVIII, c. 49 (I, col. 2025). For the unconvincing view that Aquinas is relying on Alhazen for his eye anatomy, see Adriana Caparello, "Il termine «tunica» e la sua portata scientifico-storica nella dottrina aristotelico-tomista della visione," *Divus thomas* 79 (1976), 369-399.

fully developed and readily available to him in thirteenth-century Aristotelian commentary. In the chapters that now follow I will continue to document and discuss the intellectual context which underlies Dante's optics, drawing further attention to Aristotelian texts and related commentaries. Yet if in the *Convivio* it is possible to trace almost all of Dante's optical and visual references to the Aristotelian tradition, this kind of approach is less valid for a poetic work as richly complex as the *Comedy*.[77] The investigation into Dante's optics in his poem requires a more wide-ranging approach, one which incorporates a sense of other traditions and is willing to put aside the search for "sources" in order to concentrate upon *how* Dante absorbs, re-combines, and transforms optical and visual ideas in his own poetic syntheses.

---

[77]For pertinent observations on the inherent inadequacies of a source-hunting approach to the *Comedy*, see Stephen Botterill, *Dante and the Mystical Tradition: Bernard of Clairvaux in the "Commedia"* (Cambridge: Cambridge University Press, 1994), pp. 177-179; Peter Dronke, *Dante and Medieval Latin Traditions*, p. xii; Etienne Gilson, Review of Nardi's *Dal "Convivio" alla "Commedia,"* in *GSLI* 138 (1961), p. 573; Bruno Nardi, *Saggi di filosofia dantesca*, ed. Paolo Mazzantini, 2nd edn (Florence: La Nuova Italia, 1967), introd. p. viii.

CHAPTER 3

Aspects of Vision in the *Comedy*: Blinding, Optical Illusions, and Visual Error

Dante's intense interest in the "I" that sees, his power to visualise at even the most abstract moments, and his skill at incorporating realistic descriptions of visual phenomena into his poetry are distinctive features of the *Comedy*. Dante's similes have long been admired for their visual potency and pictorial sensibility; and, given the frequency, richness, and dramatic force of his visual imagery, it is hardly surprising that the poem has exercised an enduring influence on the visual arts.[1] As is well known, the poem's narrative is filtered through Dante-*personaggio*'s own visual processes and at least one modern commentator has referred to such techniques as evidence of Dante's cinematographic qualities.[2] The reader's point of view often coincides with that of Dante the protagonist, whose angle of sight is

---

[1]On Dante's presence, for example, in Renaissance artists in the context of light and optical imagery, see Martin Kemp, "In the Light of Dante: Meditations on Natural and Divine Light in Piero della Francesca, Raphael and Michelangelo," in *Sonderdruck aus Ars naturam adiuvans: Festschrift für Matthias Winner* (Mainz-am-Rhein: Philipp von Zabern, 1996), pp. 160-177.

[2]See Guido di Pino, *La figurazione della luce nella "Divina Commedia"* (Florence: Nuova Italia, 1962), p. 34: "una intuizione spaziale che fa già pensare al Brunelleschi e, quindi, all'Alberti e a Leonardo. Gli ambienti delle inquadrature dantesche e le loro proprietà sembrano veramente anticipare, nella poesia, lo studio della «piramide visiva»"; Giovanni Fallani, *Dante e la cultura figurativa medievale* (Bergamo: Minerva Italica, 1971) pp. 17-18; Oliva, *Per altre dimore*, pp. 80-81: "l'esatta percezione della lontananza e dell'avvicinamento"; Tibor Wlassics, "La percezione limitata nella «Commedia»," *Aevum* 47 (1973), esp. p. 507: "una naturalezza che tiene conto (e non sbaglia mai) delle 'proporzioni' del suo palcoscenico." For the notion of "Dante-cinematografo," see *La "Divina Commedia,"* ed. Natalino Sapegno (Milan and Naples: Ricciardi, 1957), p. 408; Tibor Wlassics, *Dante narratore: Saggi sullo stile della Commedia* (Florence: Olschki, 1975), p. 114.

76

always shifting, narrowing its focus and isolating objects, encountering different visual conditions, and actively seeking out new sights.

There are of course many possible approaches to Dante's "visibilità" and recent scholarship has related the optical qualities of his narrative to models as diverse as Virgil's *Aeneid* and Aristotelian perceptual theory.[3] The nature and extent to which the visual arts provided a basis for Dante's imagery continues to fascinate many scholars;[4] and valuable insights into his visual strategies can be gained from comparing Dante's presentation of the visual process with that found in other medieval and Renaissance poets.[5] The approach adopted in this chapter is more limited and aims to take account of Dante's preoccupation with visual perception in passages which either contain technical vocabulary and/or are informed by the optical and visual traditions discussed in previous chapters. My

---

[3] On the *Aeneid* as a model for Dante's visual narrative, see Wlassics, *Dante narratore*, pp. 113-140. For the Aristotelian aspects of Dante's approach to, and representation of, the visual process, see Patrick Boyde, *Perception and Passion in Dante's "Comedy"* (Cambridge: Cambridge University Press, 1993), pp. 93-118; idem, "Perception and the Percipient in *Convivio*, III. ix and the *Purgatory*," *Italian Studies* 35 (1980), 19-24.

[4] For a general overview, see Christopher Kleinhenz, "Dante and the Tradition of the Visual Arts in the Middle Ages," *Thought* 65 (1990), 17-26. See also Fortunato Bellonzi, "Arti figurative," in *ED* I, pp. 400-403 with additional bibliography; Fallani, *Dante e la cultura figurativa*, passim; Christie K. Fengler and William A. Stephany, "The Visual Arts: A Basis for Dante's Imagery in the *Purgatory* and the *Paradiso*," *Michigan Academician* 10 (1977), 127-141; Alison Morgan, *Dante and the Medieval Otherworld* (Cambridge: Cambridge University Press, 1990), pp. 186-192.

[5] On the spatial quality of Dante's narrative and other medieval writings, see Charles Muscatine, "Locus of Action in Medieval Narrative," *Romance Philology* 17 (1963), 115-122. For Chaucer's control over the visual frame, see Linda Tarte Holley, *Chaucer's Measuring Eye* (Houston, Texas: Rice University Press, 1990). More generally on Chaucer and the medieval optical tradition, see Brown, *Chaucer's Visual World*; Carolyn Collette, "Seeing and Believing in the *Franklin's Tale*," *The Chaucer Review* 26 (1992), 395-410. On broader questions raised by the descriptive practices of medieval verse in general, see Sarah Stansbury, *Seeing the Gawain-Poet: Description and the Act of Perception* (Philadelphia: University of Pennsylvania Press, 1991), esp. pp. 116-140. For a reading of Spenser's visual poetry through the categories of modern psychology used by the art historian Ernst Gombrich, see John B. Bender, *Spenser and Literary Pictorialism* (Princeton, NJ: Princeton University Press, 1972). Dante's narrative interest in scenes where a visual object is gradually revealed bears interesting similarities to the "progressive style" found in certain ninteenth-century novelists, see William J. Berg, *The Visual Novel: Emile Zola and the Art of His Times* (University Park, PA: Pennsylvania State University Press), esp. pp. 1-7.

principal concern will be to examine in detail how the poet structures his narrative around Dante-*personaggio*'s own viewpoint, and describes the protagonist being dazzled and blinded by light and confused by a variety of optical illusions. By considering these aspects of the "visibile" in the *Comedy*, we will gain a better understanding of the distinctive ways in which he presents the visual act and the obstacles, changing conditions, and modifications which his sight undergoes throughout the poem.

"Io muovo l'occhio 'ntorno": Dante's Visual Narrative

After passing through Hell's foreboding gate, Dante becomes aware of disparate sighs and cries which reach him through a dark atmosphere and assault his ears in a violent cacophony. In response to his consternation, Virgil explains that the uproar is made by the souls who lived "sanza 'nfamia e sanza lodo" (l. 36). It is, however, only after Virgil has offered a more detailed explanation of the condition of these souls that hearing resolves itself into vision and Dante-*personaggio* has his first sighting of them. The lines in which the poet describes how, as protagonist, he saw these souls provide the first of many good examples in the poem of his concern with depicting a scene exactly as it was registered by his own eyes. In the darkened conditions, Dante-*personaggio* looks hard and his eyes, rather than focusing immediately on the souls themselves, are drawn to a fast-moving flag. Having established the flag as a point of reference, his gaze then moves back and is able to discern a great swathe of souls, some of whom he recognises before he narrows his focus still further in order to isolate one shade in particular:

> E io, che riguardai, vidi una 'nsegna
> che girando correva tanto ratta,
> che d'ogne posa mi parea indegna;
> e dietro le venìa sì lunga tratta
> di gente, ch'i' non averei creduto

78

> che morte tanta n'avesse disfatta.
>   Poscia ch'io v'ebbi alcun riconosciuto,
> vidi e conobbi l'ombra di colui
> che fece per viltade il gran rifiuto.
>   Incontanente intesi e certo fui
> che questa era la setta d'i cattivi
> a Dio spiacenti e a' nemici sui. (III, 52-63)

With its stress on an active involvement with visual data and its desire to penetrate visual surfaces to make sense of perceptions, to order and to understand them, this passage is characteristic of Dante's visual narrative. After observing these souls closely Dante immediately understands that they are the *ignavi* and the poet begins to describe their miserable appearance. He first presents a general view of their naked bodies tormented by flies and wasps and then provides a more visually-focused close-up, which sweeps from top to bottom: the description once again closely replicates the protagonist's eye movements by tracing the trail of blood from their faces to their feet. Finally, Dante demonstrates a visual disdain comparable to Virgil's earlier verbal admonition to "guarda e passa" (l. 51) as he casts his sight out beyond the *ignavi* to a new group of shades amassed at the bank of a great river:

> erano ignudi e stimolati molto
> da mosconi e da vespe ch'eran ivi.
>   Elle rigavan lor di sangue il volto
> che, mischiato di lagrime, a' lor piedi
> da fastidiosi vermi era ricolto.
>   E poi ch'a riguardar oltre mi diedi
> vidi genti a la riva d'un gran fiume. (65-71)

The *Inferno* with its darkened conditions and murky atmosphere has many such passages in which the narrative structure re-enacts the movements of Dante's eyes as they are attracted by sound, rapid movement, or an eerie light source, as he approaches a scene from a distance, or as an object moves towards him, or even as

79

he peers down from an elevated vantage-point.[6] As we shall see, Dante's heightened attention to visual phenomena is often developed in a technical vein in other cantos which deal specifically with optical illusions. But a similar preoccupation with the reactions and discriminations of the eye at varying distances is also present in the series of passages which deal with Dante-*personaggio*'s reactions to light and which reveal further aspects of Dante's imaginative use of the visual process in his narrative.

Dante's Eyes: The "Soverchio" and the "Abbaglio"

The motifs of dazzling and blinding play an important part in the development of the themes of seeing and blindness in the *Comedy*. From the dramatic and mysterious "luce vermiglia" of *Inferno* III to the final "fulgor" of the *Paradiso*, Dante pays close attention to how, as protagonist, he responds to light of differing colours, intensities, and kinds. Naturally, these descriptions are especially prevalent in the *Paradiso*, where the pilgrim is subjected to ever-increasing intensities of light. But it is in the *Purgatorio* that Dante's reaction to intense light becomes a recurrent theme and a pattern is established which is reworked and developed throughout the rest of poem.[7] All of the scenes of dazzling and blinding in the final two *cantiche* can be related to a gradualistic process by which, as Dante ascends, the intensity of natural and supernatural light sources increases and so does his ability to withstand them. The principle that greater intensities of light promote increased visibility is explicitly outlined by Virgil in *Purgatorio* XV (ll.

---

[6]Selected references in order: (sound attracting sight), XXXI, 10-12; (sudden light), *Inf.* VIII, 4-6; (approach from a distance), XXXI, 7-45; XXXIV, 4-7; (high vantage-point), IV, 115-117; XX, 7-12; XXVI, 43-45. For hearing resolving itself into vision, see *Inf.* V, 28-49; IX, 67-72; cf. *Purg.* XXIV, 133-138.

[7]For Dante's descriptions of dazzling and blinding, see *VN*, lxi, 6; *Con.* III, viii, 14; *Inf.* III, 134-135; *Purg.* II, 37-40; VIII, 34-36; IX, 81; XV, 7-11, 22-24; XVII, 52-54; XXVII, 58-60; XXXII, 10-12; *Par.* I, 58-60; III, 128-129; IV, 139-142; V, 3; XIV, 76-78; XXIII, 31-33, 77-78; XXV, 27, 118-123; XXVI, 20; XXVIII, 16-18; XXIX, 8-9; XXX, 11, 46-51.

80

31-33), where Dante is reassured that his eyes will soon find delight in luminosity rather than the discomfort he now experiences. Not all of the scenes in the *Purgatorio* and the *Paradiso* involve blinding or what Dante called the "abbarbaglio," however. The majority are in fact concerned with the overwhelming of the senses by a sensible object that is too strong for them, a phenomenon which Dante referred to as the "soverchio."[8] As far as the question of sources is concerned, it is worth noting that, although optical specialists such as the "perspectivists" discuss blinding and the pain caused by light, a scientific description of sight being destroyed by excessive brightness can be found in Aristotle's *De anima*. According to Aristotle and his medieval commentators, sensory objects and the senses are proportioned to a mean and hence an object which is too strong for a sense organ destroys its proportion:

Et propter id corrumpit unumquodque excellens, et acutum et grave auditum et in humoribus gustum, et in coloribus visum fortiter fulgidum et opacum, et in olfactum fortis odor et dulcis et amaris, tamquam ratio quaedam sit sensus.[9]

[8]For the term "abbarbaglio," see *Par.* XXVI, 20. The "soverchio" is mentioned in *Purg.* XV, 15; cf. *Con.* III, xv, 6.

[9]*De anima*, III, 2, 426a 30-426b 3, Graeco-Latin trans. in *In De anima*, ed. Pirotta, p. 148: "For this reason anything excessively shrill or deep destroys the hearing; and the same in flavours destroys the taste, and in colours, the sight, whether the excessively brilliant or the dark; and in smell, a strong odour, whether sweet or bitter; as if the sense were a certain proportion" (trans. Foster and Humphries, p. 359). See also ibid., III, 4, 429b 2-3; 13, 435b 8-10, 16: "Omnis enim sensibilis superfluitas corrumpit sensum" ("All sensible excess destroys sensation"). For discussions of this idea in Aristotelian commentaries, see Albert, *De anima*, lib. II, tr. 4, c. 9 (Stroick, p. 161); idem, *De animalibus*, lib. XIX, tr. un., c. 3 (Stadler, II, p. 1251); Aquinas, *In De anima*, lib. III, lect. 2, § 149 (Pirotta, p. 149): "Et quia quaelibet proportio corrumpit per superabundantiam, ideo excellens sensibile corrumpit sensum, sicut [...] fortiter fulgidum vel obscurum corrumpit visum" ("And since any proportion is destructive due to overabundance, then a strong sensible quality destroys sensation just as [...] what is intensely bright or dark destroys vision"). See also Beauvais, *Speculum maius*, lib. XXV, c. 40 (I, col. 1801): "excellens color destruit visum" ("intense colour destroys vision"). For the after-effects of dazzling, see Albert, *De anima*, lib. III, tr. 2, c. 15 (Stroick, p. 198): "forma fortis sensibilis reddita organo tenetur ab ipso aliquamdiu, et tunc sub forma illius primi sensati sentitur secundum. Et ideo videns aliquamdiu intensam et claram albedinem et statim convertens se ad virorem vel alium colorem, videt viride sicut tectum tenui panno albo" ("a strong sensible form impressed on the sense organ is held by it for some time and then other sensible objects are perceived under the form of the first

81

The fact that this doctrine can be found in a thirteenth-century *florilegium* of Aristotelian passages seems to indicate that it was very widely known indeed.[10] But to concentrate exclusively on "scientific" sources for the relevant passages in the *Comedy* is to miss the more crucial points raised by Dante's treatment of "soverchi" and "abbagli" and to fail to appreciate how he incorporates other elements of his optical knowledge into his poetry at such moments. For example, the first episode that involves the dazzling light of an angel occurs at the end of a celebrated sequence in *Purgatorio* II which is perhaps the best example of Dante's fascination not only with the optical effects produced by distant objects as they approach the viewer but also with the distortions caused by atmospheric vapours. In this canto, from line 15 onwards Dante describes how he saw the light of the "angelo-nocchiero" appear to him as reddening through the vapours of early morning. His narrative charts the gradual perception of the angel's true form and the whole sequence skilfully exploits Dante-*personaggio*'s (and the reader's) perceptual responses to create an atmosphere of expectation and spiritual trepidation (ll. 13-36). As the angel draws close to the shore, Dante's eyes are unable to withstand its light and he is forced to avert his gaze:

> Poi, come più e più verso noi venne
> l'uccel divino, più chiaro appariva:
> per che l'occhio da presso nol sostenne,
> ma chinail giuso; e quei sen venne a riva. (37-40).[11]

---

one. And hence if you look for some time at a clear and intense whiteness and suddenly turn to green or some other colour you see green as if attached to a thin white cloth"); idem, *De sensu et sensato*, tr. I, c. 14 (Jammy, V, p. 20a). The after-image is also discussed in Augustine, see *De Trinitate*, lib. 11, cap. 2, § 4 (*PL* 42, col. 987).

[10]See Jacqueline Hamesse, *Les "Auctoritates Aristotelis": Un florilège médiéval* (Louvain and Paris: Peeters and E.J. Brill, 1974), p. 182: "Excellens sensibile corrumpit sensum" ("an intense sensory object destroys sensation").

[11]For a fine analysis of the entire passage in terms of the Aristotelian doctrine of "proper" and "common" sensibles, see Boyde, *Perception and Passion*, pp. 93-94.

82

All of the "soverchi" in AntePurgatory and on the "gironi" of Mount Purgatory are caused by the radiant countenances of angels encountered by the protagonist, but in each case the poet strikes subtle variations. In canto VIII, the arrival of two angels is presented as seen from below and amidst a blaze of colour, for Dante is able to watch the angels as they descend from on high and he observes their green garments, their fiery swords, and even their blonde hair. But when he tries to focus more closely on their faces his visual power is confounded by an excess of light: "ma ne la faccia l'occhio si smarria, / come virtù ch'a troppo si confonda" (ll. 35-36). The next overpowering of Dante's sight in canto IX is worthy of detailed examination, since it provides another excellent example of the poet's attentiveness to the visual and optical elements of the scene he is describing. In this canto, Dante-*personaggio* approaches the gate of Purgatory from a considerable distance and it is for this reason that the gate at first appears to him to be no more than a fissure in the rock. Gradually, Dante draws sufficiently close to be able to observe firstly that the fissure is actually a gate, then that it has steps and each is of a different colour, and finally that there is an angelic-gatekeeper. He subsequently perceives the angel's position on the uppermost step but he is unable to raise his eyes any further and fails to see the angel's face, because of its radiance and the light which is reflected from its sword:

> Noi ci appressammo, ed eravamo in parte
> che là dove pareami prima rotto,
> pur come un fesso che muro diparte,
> vidi una porta, e tre gradi di sotto
> per gire ad essa, di color diversi,
> e un portier ch'ancor non facea motto.
> E come l'occhio più e più v'apersi,
> vidil seder sovra 'l grado sovrano,
> tal ne la faccia ch'io non lo soffersi;
> e una spada nuda avëa in mano,
> che reflettëa i raggi sì ver' noi,
> ch'io dirizzava spesso il viso in vano. (73-84)

83

There are three more examples in the *Purgatorio* of "soverchi" due to angelic effulgence, and the first of these in canto XV occasions an elaborate optical simile based on the law of light reflection which will receive closer consideration in the next chapter. In canto XVII, Dante describes the effect of angelic light on his closed eyes as he is awakened from the imaginings of his second purgatorial dream by a "lume" which is said to be "maggior assai che quel ch'è in nostro uso" (l. 45). Despite the assertion of its supernatural intensity, the way in which this angel conceals itself in light (l. 57), the "soverchio," and the overcoming of Dante's visual power ("virtù": l. 52) are likened to a natural phenomenon: the effects of the sun on our sight, "nostra vista" (ll. 52-54).[12]

Having passed through the flames of the final cornice, in which Dante is again temporarily dazzled by the last of the Purgatorial angels,[13] the protagonist enters Earthly Paradise where he will witness the two processions and the central event of Beatrice's coming. The meeting with Beatrice in canto XXX has been anticipated on more than one occasion earlier in the poem and references to her eyes, and their radiance, formed the dominant motif.[14] When Beatrice does appear, though, the poet dramatically - and quite unexpectedly - delays the theme of dazzling by *not* placing emphasis on the effects of her radiance: she is at a distance from him, veiled, and, obscured still further by a cascade of flowers, her appearance is compared to the way in which the sun is made visible by the tempering haze of early morning vapours (ll. 22-30). Dante has still not viewed Beatrice in an unmediated way when he is overwhelmed by her mysterious power ("occulta virtù": l. 38) and, following her later reproaches to him, he lowers his eyes.

---

[12]There are important classical precedents for the effect of light on a sleeping person, see Virgil, *Aeneid*, IX, 109-111; Statius, *Thebaid*, X, 123-124.

[13]*Purg.* XXVII, 58-60: "'*Venite, benedicti Patris mei,*' / sonò dentro a un lume che lì era, / tal che mi vinse e guardar nol potei."

[14]*Inf.* X, 130-131; cf. II, 55; *Purg.* XXVII, 54.

84

It is not until line 76 of canto XXXI that Dante-*personaggio* raises his eyes fully again; the flowers have stopped falling and he now sees Beatrice staring at the griffin, and, even though she is still veiled and at a distance, his senses are overcome to such a degree that he faints (ll. 79-90). Having been carried across Lethe by Matelda, the next canto opens with Dante's eyes feasting on his lady to the exclusion of the other senses until he is called away by three ladies, who are allegorical representations of the theological virtues, with their cry of «Troppo fiso!» (l. 9). Almost immediately, he recovers his visual powers by looking at the lesser brightness of the procession and, rather than a description of blinding, Dante describes how his sight was "reformed":

> e la disposizion ch'a veder èe
> ne li occhi pur testé dal sol percossi,
> sanza la vista alquanto esser mi fée.
> Ma poi ch'al poco il viso riformossi
> (e dico 'al poco' per rispetto al molto
> sensibile onde a forza mi rimossi). (10-15)

The way in which Dante-*personaggio*'s sight is rehabilitated is an important thematic and narrative concern in the *Paradiso* where the successive tempering of Dante's eyes by a series of different lights serves as a preparation for the final vision of God as light. Paradoxically, his capacity to withstand light is heightened by repeated dazzlings and blindings, with such experiences of what might be called visual tempering being especially pronounced in the final cantos of the poem. For much of the *cantica*, the light of Beatrice's eyes and smile play an especially important role in the narration of Dante's physical and intellectual progress. Her eyes not only help to propel him through the heavens, but their radiance also betokens his intellectual and spiritual progress.[15]

[15]In canto III, Dante loses sight of Piccarda as she fades into the distance, and he returns his gaze to Beatrice who fulminates him "sì che da prima il viso non sofferse" (l. 129). At the close of canto IV Dante's "virtute" is overwhelmed by the sparks of divine love which issue from

The Heaven of the Sun is one of the most visually charged in the poem and it is here that Dante first describes in detail the reciprocity between blinding and recovery of greater vision that he undergoes as protagonist. In canto XIV, Dante's eyes are overpowered by the sudden luminosity of a new circle of souls, but he recovers his sight by looking at Beatrice and is immediately transported to the next heaven:

> Oh vero sfavillar del Santo Spiro!
> come si fece sùbito e candente
> a li occhi miei che, vinti, nol soffriro!
> Ma Bëatrice sì bella e ridente
> mi si mostrò, che tra quelle vedute
> si vuol lasciar che non seguir la mente.
> Quindi ripreser li occhi miei virtute
> a rilevarsi; e vidimi translato
> sol con mia donna in più alta salute. (76-84)

The motif of dazzling and visual tempering finds yet another narrative variation in canto XXI when Beatrice refuses to smile for fear that her increasing beauty would dazzle Dante's "mortal podere" (l. 11). And yet later in canto XXIII (ll. 46-48), he is able to behold her smile on having first tried and failed to see Christ's light. Similarly, during his examination on the three theological virtues, Dante undergoes a temporary blinding due to the light of St. James (l. 27) but he soon recovers his sight by following Beatrice's request to mature himself in James's rays (ll. 35-39). The longest period of blindness Dante experiences in the *Paradiso* is during his examination on Charity by St John in Cantos XXV and XXVI. Dante blinds himself by looking for John's body (an experience that is compared to looking at an eclipse of the sun: ll. 118-123), but on this occasion he does not immediately recover his powers of vision through the customary glance back to Beatrice. And his state of prolonged blindness despite his proximity to

---

Beatrice's eyes and he is forced to look down (ll. 139-142). The dazzling is explained at beginning of next canto (ll. 1-12).

86

Beatrice is used to provide the canto's ending with considerable emotional tension and dramatic suspense.

In *Paradiso* XXVI, which opens on the pilgrim's "viso spento," Dante remains blinded until he has completed his examination on Charity, and it is only at this point that Beatrice restores Dante's sight with a ray from her own eyes:

> E come a lume acuto si disonna
> per lo spirto visivo che ricorre
> a lo splendor che va di gonna in gonna,
>   e lo svegliato ciò che vede aborre,
> sì nescïa è la sùbita vigilia
> fin che la stimativa non soccorre;
>   così de li occhi miei ogne quisquilia
> fugò Beatrice col raggio d'i suoi,
> che rifulgea da più di mille milia:
>   onde mei che dinanzi vidi poi. (70-79)

In a limited sense, these lines provide a variation on the idea of the "soverchio," and they show how the poet deploys his knowledge of visual anatomy ("spirito visivo" is a calque of *spiritus visivus*, whereas the tunics of the eye are referred to as "gonne") and the psychology of vision (the "stimativa" is a post-sensory faculty). But several other important points are raised by this passage and call for further comment. Here, for the first time Beatrice's eye ray, which had earlier dazzled Dante in cantos III and IV, rekindles his "viso spento" and heightens his visual powers: "onde mei che dinanzi vidi poi" (l. 79). Even more significantly, these lines fulfil and develop a crucial Pauline allusion that was set up earlier in the canto when Beatrice was said to have "ne lo sguardo / la virtù ch'ebbe la man d'Anania" (ll. 11-12).[16] In addition to its narrative role and thematic implications,

---

[16] Ananias is of course the disciple chosen by God to restore Saul's sight after his blinding on the road to Damascas, see Acts 9:10-22. It may be possible that line 77 contains an echo of Augustine's *Confessiones*, lib. X, c. 27, § 38, in *CCSL* 26, p. 175: "Vocasti et clamasti et rupisti surditatem meam, coruscasti, splenduisti et fugasti caecitatem meam" ("You called and cried out and burst my deafness; you shone and sparkled and chased away my blindness"). But for the

87

the simile also shows Dante the poet at his most varied, resourceful, and resonant. Not only does the passage evince an incredibly broad linguistic and stylistic range with its learned latinisms, technical terms, and unusual rhymes, but the whole effect of sudden awakening is brought out first in the pattern of strong accentuation in line 74 and then in the quickening rhythms of the bisyllables in line 76.

The reference to "Anania" in canto XXVI marks the beginning of a set of Pauline allusions which have direct bearing upon Dante-*personaggio*'s sight and occur at structurally-marked moments in the remainder of the poem. Several of Dante's other descriptions of dazzling in the poem have some foundation in biblical passages that describe bright light,[17] but it is Pauline episodes from Acts of the Apostles that provide the most important sub-text for the later stages of the pilgrim's visual experience of light in the *Paradiso*. Dante's next and most direct allusion to Paul's writings in the context of vision comes in canto XXX. On entering the Empyrean heaven, Dante compares his experience of its light to the reaction of a human eye when its visual spirits are scattered by a sudden effulgence:

> Come sùbito lampo che discetti
> li spiriti visivi, sì che priva
> da l'atto l'occhio di più forti obietti,
> così mi circunfulse luce viva,
> e lasciommi fasciato di tal velo

---

effects of sleepiness on vision, see also Beauvais, *Speculum maius*, lib. XXV, c. 34 (I, col. 1797): "contingit, quod post somnum in quo spiritus omnes ad interiora moventur ad locum digestionis relinquendo sensuum organa exteriora, statim aperiens oculos non bene videt ante reditum spirituum et caloris ab interioribus ad exteriora sensuum" ("after sleep, in which all the spirits leave the external sense organs and move within to the place of digestion, it happens that if someone suddenly opens their eyes they do not see clearly until the spirits and heat have returned to the external senses from within the body").

[17]See e.g. Dan. 10:6: "Et facies eius velut species fulgoris et oculi eius ut lampas ardens" ("And his face was like the appearance of lightning and his eye as fiery lanterns"); Rev. 1:16-17: "Et facies eius sicut sol lucet in virtute sua. Et cum vidissem eum cecidi ad pedes eius tamquam mortuus" ("And his face was as the sun shines in all its power. And when I saw him I fell at his feet as if I were dead"); 22:1: "Et ostendit mihi fluvium aquae vivae splendidum tamquam crystallum procedentem de sede Dei et Agni" ("And he showed me a river of living water resplendent as a crystal, proceeding out of the throne of God and the Lamb"). Cf. *Inf.* III, 98, 133-136; *Purg.* I, 37-39; *Par.* XXX, 61-63.

88

del suo fulgor, che nulla m'appariva. (46-51)

The poet has again allowed anatomical details to penetrate the language of the *Paradiso*, enmeshing them with a direct allusion to the light that surrounded Paul on the Road to Damascus. As is well known, Dante's "circunfulse" translates the *circumfulsit* found in Paul's account of his own blinding in Acts of the Apostles:

Factum est autem eunte me et adpropinquante Damasco, media die subito de caelo circumfulsit me lux copiosa.[18]

The whole sequence is a highly original fusion of visual science and scriptural reminiscence, and it serves to reiterate the crucial series of parallels which Dante draws throughout the *Paradiso* between his experience and that of Paul's *raptus* to the third heaven.[19] The tempering effect on Dante's eyes is such that he acquires miraculous new powers of sight, and, rekindled with "new sight," his eyes are able to withstand any degree of light:

e di novella vista mi raccesi
tale, che nulla luce è tanto mera,
che li occhi miei non si fossero difesi. (58-60)

---

[18]Acts 22:6: "as I made my way and approached Damascus at noon suddenly a light from heaven shined all around me". See also 9:3: "et subito circumfulsit eum lux de caelo" ("and suddenly a light from heaven shined all around"); 26:13: "Die media in via vidi rex de caelo supra splendorem solis circumfulsisse me lumen" ("At noon, King, I saw in the way a light from heaven surpassing the brightness of the sun, shining all around me").

[19]On the Pauline aspects of Dante's visions in the final cantos, see Piero Boitani, "The Sibyl's Leaves: *Paradiso* XXXIII," *Dante Studies* 96 (1978), pp. 93-94; Giuseppe di Scipio, "Dante and St. Paul: The Blinding Light and Water," *Dante Studies* 98 (1980), 151-157; Kenelm Foster, *The Two Dantes and Other Studies* (London: Darton, Longman & Todd, 1977), pp. 60-70; Joseph A. Mazzeo, "Dante and the Pauline Modes of Vision," *Structure and Thought in the Paradiso* (Ithaca: Cornell University Press, 1958), pp. 84-110; Lino Pertile, "*Paradiso* XXXIII: 'L'estremo oltraggio'," *Filologia e critica* 6 (1981), pp. 6-7, 12. For a detailed treatment of Paul's presence in Dante, see Giuseppe di Scipio, *The Presence of Pauline Thought in the Works of Dante* (Lewiston, Queenston, and Lampeter: Edwin Mellen, 1995).

Such radically heightened visual powers are refined still further in the final canto of the *Paradiso*, where Dante sees into the eternal light of God and the poet again reworks the visual paradoxes of greater light inducing greater visual power that were developed earlier in cantos XIV, XXVI, and XXX. Having been repeatedly blinded in order to withstand more light, the pilgrim now sees into the true and exalted light of God. One of the final paradoxes in the poem is that Dante is no longer blinded by this light. Quite the opposite. If he did not see divine light, he would be blinded:

> Io credo, per l'acume ch'io soffersi
> del vivo raggio, ch'i' sarei smarrito,
> se li occhi miei da lui fossero aversi. (76-78)

Optical Illusions in the *Comedy*

In the *Comedy*, Dante reworks many of the optical illusions he had earlier discussed in the closing paragraphs of Book III, Chapter ix of the *Convivio*. The types of deception he puts forward have many causes, both internal and external, but almost always these deceptions, like the "abbarbaglio" and "soperchio," consist in a lack of the requisite *proportio* between sense organ and sensed object. As in the *Convivio*, he describes scintillations of light,[20] the visual distortions caused by a dark atmosphere,[21] and the optical effects resulting from "li vapori de la terra che continuamente salgono."[22] He also outlines a great variety of other optical phenomena: the persistence of light; distortions at distance; illusory movements;

---

[20]*Par.* IX, 113-114; XIV, 67-69; XXVIII, 91.

[21]*Inf.* XV, 17-21; XXXI, 10-11; XXXIV, 5-6; *Purg.* VIII, 49-51; XV, 139-141; *Par.* XIV, 70-72; cf. *Rime* XC, 13-15; *Par.* V, 133-135.

[22]*Con.* III, ix, 12. For the effects caused by vapours in the *Comedy*, see *Inf.* XXXI, 34-39; XXXIV, 4-7; *Purg.* I, 115-117; II, 13-15; XV, 142-145; XVIII, 1-9.

90

variable appearance; the assimilation of form to colour; foreshortening; and apparent reflection.[23]

Many of these striking visual effects are presented as part of similes to provide accommodated images for Dante's supernatural experiences, but not all of these phenomena are incorporated into the poem as metaphorical analogies. As has been seen, it is characteristic of Dante's narrative art that he frequently describes his own faltering, limited perceptions as protagonist and several of the passages that deal with optical illusions develop this theme through technical allusions to visual doctrines. There are many scenes in the *Comedy* which reveal a sophisticated appreciation of vision as an act, which, though prone to error, can nonetheless attain an accurate perception of visual objects in the world. The senses are not an inadequate guide to knowledge, but they can be unreliable. A difficult tercet in *Purgatorio* XVIII illustrates the importance that Dante attributed to a model of visual perception which would allow man to make correct assessments from the mass of potentially confusing sensory data in the world:

> Vostra apprensiva da esser verace
> tragge intenzione, e dentro a voi la spiega,
> sì che l'animo ad essa volger face;
>   e se, rivolto, inver' di lei si piega
> quel piegare è amor, quell' è natura
> che per piacer di novo in voi si lega. (22-27)

In these lines, Dante demonstrates his philosophical concern with ensuring that the apparatus of human perception ("apprensiva")[24] can draw in ("tragge") a replica of

---

[23]References in order: (i) persistence of light, *Purg.* XXIX, 73-75; (ii) distance, *Inf.* XXXI, 20-27; *Purg.* X, 49-54; XXIX, 43-50; (iii) illusory movements, *Inf.* XXXI, 136-139; *Purg.* III, 58-60; VIII, 103-105; *Par.* XXIV, 13-15; (iv) variable appearance, *Par.* XIV, 112-114; (v) assimilation of form to colour, *Purg.* XIII, 43-45; *Par.* III, 13-15; (vi) foreshortening, *Purg.* X, 112-120; (vii) false reflection, *Par.* III, 10-18.

[24]"Apprensiva" refers to both the internal and the external senses, cf. Bartholomew the Englishman, *De rerum proprietatibus*, lib. III, c. 9 (Richter, p. 52): "Apprehensiva vero dividitur

91

a sense-object ("intenzione") that is a faithful representation of true external reality ("esser verace"). The implications of faulty sensory perception require Dante to confront this issue directly before turning to the central questions of free will and love later in the canto.

The *perspectivae* deal with a great variety of *deceptiones visus*, describing them in meticulous detail, establishing their causes, and setting out the preconditions and mental processes required for accurate vision.[25] And it is the presence of similar concerns in the *Comedy* that has led some critics to suggest that Dante was inspired by these optical writings.[26] It is undoubtedly true that, as both philosopher and poet, Dante showed a remarkable sensitivity to the way in which the visual act could sift, adjudicate, and correct the potentially unreliable impressions of the sensory world. Yet it must be said that it is extremely difficult to prove that Dante based this conception of vision on the *perspectivae*. After all, the principal schools of Greek and Roman thought (Platonic, Aristotelian, Epicurean, and Stoic) were passionately concerned with the problem of faulty perception; and almost all ancient writers developed the idea that a reasoning power exercises a corrective function in vision.[27] What is more, Christian writers were also deeply aware of the problems raised by erroneous sense-perception. Faulty sensory impressions or the inability to evaluate and/or correct sense data

---

in sensum communem sive interiorem et in sensum particularem sive exteriorem" ("The apprehensive is divided into the common or interior sense and the particular or exterior sense").

[25] The principal discussions of optical illusions in Alhazen and the "perspectivists" are: Alhazen, *De aspectibus*, lib. III, c. 7, secs. 22-72 (Risner, pp. 91-102); Bacon, *Opus maius*, lib. V, p. ii, d. 1-3 (Lindberg, pp. 160-245); lib. V, p. iii, d. 1, cc. 2-6 (Lindberg, pp. 258-286); lib. V, p. iii, d. 2, c. 4 (Lindberg, pp. 308-320); Witelo, *Perspectiva*, lib. IV, props. 138-159 (Risner, pp. 179-189).

[26] See esp. Parronchi, "La perspettiva dantesca," p. 26; for similar remarks, see also, pp. 13, 25-26, 44.

[27] See e.g. Plato, *Republic*, X, 602D, trans. and comm. Francis M. Cornford (London and New York: Routledge and Kegan & Paul, 1941), p. 334; Pliny, *Historia naturalis*, lib. XI, c. 54, § 146, ed. and trans. H. Rackham et al., 10 vols. (London: Heinemann, 1938-1962), III, p. 522; Lucretius, *De natura rerum*, lib. IV, ll. 469-521, ed. W.H.D. Rouse, 2 vols. (London: Heinemann, 1978), I, pp. 280-284; Seneca, *Naturales quaestiones*, lib. I, c. 3, §§ 9-10, ed. Thomas H. Corcoran, 2 vols. (London: Heinemann, 1971-1972), I, p. 38. On the fallacies of sight, see also Summers, *The Judgment of Sense*, pp. 42-49.

92

would not have allowed man to make the freely-elected choices worthy of recompense or punishment in the afterlife. In Book XII of his *De Genesi ad litteram*, Augustine discussed the fallibility of the senses in his well-known account of three different forms of *visiones*. Intellectual vision is free from errors of any sort, but the two lower modes of seeing, spiritual and corporeal, can be misled in a number of ways.[28] In his account, Augustine gives an interesting, if rather topical, list of the optical illusions that deceive bodily eyes, and his examples were passed on to later writers by Alcher of Clairvaux in his *De spiritu et anima*.[29]

One of the best early medieval examples of this Christian concern with establishing the veracity of the senses is found in Anselm's *De veritate*. In Chapter vi of this work, Anselm mentions several optical illusions such as an object seen through coloured glass, a stick bent in water, and a face viewed in a mirror. Anselm is quite explicit in arguing that the senses are in no way at fault, it is rather a question of misjudgement by the soul.[30] Although both Augustine and Anselm guarantee the veracity of sensation, they do so by having recourse to the judgement of the soul, rather than by positing a corrective function in the faculties

[28]*De Gen. ad litt.*, lib. XII, c. 6, § 15 and c. 7, § 16 (*PL* 34, cols 458-459). On this classical scheme and the *Comedy*, see Pietro Chioccioni, *L'agostinismo nella "Divina Commedia"* (Florence: Olschki, 1952), pp. 43-44; F.X. Newman, "St. Augustine's Three Visions and the Structure of the *Commedia*," *Modern Language Notes* 82 (1967), 56-78.

[29]*De Gen. ad litt.*, lib. XII, c. 25, § 52 (*PL* 34, col. 475): "Fallitur ergo in visione corporali, cum in ipsis corporibus fieri putat quod fit in corporibus sensibus; sicut navigantibus videntur in terra moveri quae stant, et intuentibus coelum stare sidera quae moventur, et divaricatis radiis oculorum duas lucernae species apparere, et in aquam remus infractus, et multa huiusmodi" ("It [sc. the soul] fails therefore in corporeal vision when it thinks that the way we perceive bodies is the same as what is happening to these bodies, as when sailors see things move on the ground which are stationary and people looking at heaven see stars as still when they are moving; and the spreading of eye rays makes one lantern appear as two and an oar in water appear bent and many things of this kind"); the same list is given by Alcher of Clairvaux, *De spiritu et anima*, lib. I, c. 24 (*PL* 40, col. 797).

[30]Anselm (1033-1109), *De veritate*, c. 6 (*PL* 158, col. 474): "cum multa alia nobis aliter videntur visus et alii sensus nuntiare quam sint, non culpa sensuum est, qui renuntiant quod possunt, quoniam ita posse acceperunt; sed iudicio animae imputandum est" ("and when sight and the other senses seem to report to us many things as being other than they really are, the fault is not with the senses, which report what they can in the way that they are able to receive, but must be attributed to the soul's judgement").

93

of the mind or in the senses themselves. Dante's position is somewhat different and can be related quite closely to Aristotelian psychology. Given Aristotle's belief that concepts originated in the senses, it followed that if the latter were to pass misleading sensory impressions to the percipient, the formation of concepts would be undermined. According to Aristotle, a post-sensory faculty, the "imagination," was vital to the process of intellection, since it retained the sensible forms received by the five senses and made these forms available to the intellect. But because of its dependence on sense-perception, the "imagination" was prone to errors either due to excessive variations in the conditions of vision or because of problems with the eye itself.[31] Many thirteenth-century thinkers drew upon this heritage and offered similar solutions to the problem of faulty sensation by attempting to guarantee the truthfulness of the senses by recourse to a system of internal senses.[32] For example, the "perspectivists" followed a broadly Aristotelian tradition in developing a conception of visual perception as a process controlled by mental powers.[33] Of course, these ideas were also developed at considerable length in medieval commentaries upon Aristotle's psychological works, and this seems to be the most likely basis upon which Dante's own views are founded.[34]

---

[31]References to the "imagination": (i) its dependence on sense-perception, *De anima*, III, 3, 427b 14-16; (ii) as a prerequisite of ratiocination, III, 3, 427b 28-30; (iii) retention of *species sensibilis*, III, 1, 425b 25. The imagination could also err due to false compositions of images, see III, 3, 428a 12-15. On this important faculty, see M.-D. Philippe, "Phantasia in the Philosophy of Aristotle," *The Thomist* 35 (1971), 1-42.

[32]For the role of the internal senses in correcting errors in *De oculo morali* by Peter of Limoges (1230-1306), see Clark, "Optics for Preachers," pp. 329-343. The way in which Limoges uses optical illusions to persuade his readers to convert sensory stimuli into "interior sight" is especially noteworthy.

[33]On the adaptation of Aristotelian psychology by the "perspectivists," see Smith, "Getting the Big Picture," pp. 583-584; idem, *Witelonis Perspectivae liber quintus*, introd. p. 31: "the perspectivists chose to construct their perceptual model on Aristotelian foundations"; Vescovini, *Studi sulla prospettiva*, p. 64.

[34]Aquinas makes this point with admirable concision, see *In De sensu*, lib. un., lect. 4, § 50 (Spiazzi, p. 18): "Ipsa autem visio secundum rei veritatem non est passio corporalis, sed principalis eius causa est virtus animae" ("This vision in truth is not a passion of the body but its principal cause is a power of the soul"). For the role of the brain in vision, see also lect. 5, § 64

94

As well as his digression on visual theory in *Convivio* III, ix, Dante shows his familiarity with an Aristotelian conception of the visual act by using ideas and examples from his writings when discussing the deceptions that arise from the "common sensibles" in Book IV.[35]

In fact, a wide variety of the optical illusions mentioned by Dante in the *Comedy* can be found in the *corpus Aristotelicum*, but also – and especially – in the medieval commentaries.[36] On several occasions, Dante describes the illusory effects caused by vision at excessive distance, an example which has some theoretical basis in the *De anima*.[37] It is also conceivable that a passage in *Purgatorio* XXVII, which describes how the stars appeared more clearly to Dante from an enclosed space, was influenced by an identical observation in Aristotle's *De generatione animalium*.[38] The "specchiati sembianti," those faint, ethereal outlines of the blessed souls in *Paradiso* III have a counterpart in Aristotle's description of a man of weak sight seeing his reflection before him in the

---

(Spiazzi, p. 22): "principium visionis est interius iuxta cerebrum" ("the principle of vision is within near to the brain").

[35]*Con.* IV, viii, 6-7: "[...] li sensibili communi, là dove lo senso spesse volte è ingannato. Onde sapemo che a la più gente lo sole pare di larghezza nel diametro d'un piede, e sì è ciò falsissimo" This idea corresponds to Aristotle's observations in *De anima*, III, 3, 428b 18-25; *De sensu*, 4, 442b 7-10. The example of a visual deception given by Dante here and also found in *Ep.* XIII, 7 (the sun appearing to be a foot wide) is borrowed from Aristotelian sources, see *De anima*, III, 3, 428b 2-4; *De somniis*, 2, 460b 18-20.

[36]Descriptions of optical illusions were not of course restricted to optical treatises and commentary texts. For other medieval works, see e.g. the lists of deceptions, often associated with mirrors, discussed by Alain de Lille, *Contra haereticos*, lib. I, c. 58 (*PL* 210, col. 362); Jean de Meun, *Roman de la Rose*, ll. 18123-18166, 18200-18216 (Lecoy, III, pp. 44-47); Vincent of Beauvais, *Speculum maius*, lib. IV, c. 77 (I, col. 280); cf. lib. XXVI, c. 78 (I, col. 1887).

[37]For descriptions of visual illusions in Aristotle's *De anima*, see: (i) falsity of "common sensibles," III, 3, 428b 23-25; (ii) false perception of coloured objects, III, 3, 428b 21-22 and 7, 430b 29-30; (iii) illusion due to distance, III, 3, 428b 29-30; cf. *De caelo*, II, 8, 290a 23-25. See also Irvine Block, "Truth and Error in Aristotle's Theory of Sense Perception," *Philosophical Quarterly* 11 (1961), 1-9.

[38]*De gen. animalium*, V, 1, 780b 18-22. Cf. *Purg.* XXVII, 88-90: "Poco parer potea lì del di fori; / ma, per quel poco, vedea io le stelle / di lor solere e più chiare e maggiori."

95

Meteorologica.[39] Dante's observation of the variable appearance of particles of dust in a ray of sunlight in *Paradiso* XIV is reproduced in Aristotelian commentaries.[40] Similar sources can be adduced for other optical illusions in Dante. The effects of star scintillation,[41] of a more intense light obscuring a lesser

[39]See Aristotle, *Meteorologica*, III, 4, 373b 4-7: "Propter visus autem debilitatem saepe et sine inspissatione facit refractionem [sc. reflection from air], qualis aliquando accidit cuidam passio debiliter et non acute videnti; semper enim idolum videbatur praecedere ambulantem ipsum" Graeco-Latin trans. in *In De meteorologicorum*, ed. Spiazzi, p. 624 ("On account of sight's weakness it [our sight] is often reflected without air being condensed as sometimes happens with the kind of passion that people who are weak-sighted experience. The reflection will always be seen in front of the spectator who is walking"). This example became classical in the literature of optics and elsewhere, see e.g. Alexander Neckham, *De naturis rerum*, c. 153 (Wright, p. 238); Honoré d'Autun, *Philosophia mundi*, lib. IV, c. 27 (*PL* 172, col. 96); Meun, *Roman de la Rose*, ll. 18167-18176 (Lecoy, III, pp. 45-46). But for Dante's simile in *Par*. III, 10-15, see also Albert, *De sensu et sensato*, tr. I, c. 8 (Jammy, V, p. 11a): "aliquando radius penetrat per aquam ad profundum aquae, sicut quando videmus lapides albos in fundo aque iacentes; et aliquando repercutitur ab aqua in aliud, sicut quando videmus simulachrum alicuius apparentis in aqua" ("sometimes a ray penetrates through water to its depths as when we see white stones lying at the bottom of a pool of water; sometimes the ray is bounced back from the water to something else as when we see the likeness of something appearing in the water").

[40]See Albert, *De anima*, lib. I, tr. 2, c. 1 (Stroick, p. 17); idem, *De meteoris*, lib. III, tr. 4, c. 26 (Jammy, II, p. 139a): "[...] sicut pulvis videtur in radio solis non in umbra" ("as dust is seen in a ray of sunlight which is not in the shade"); Aquinas, *In De anima*, lib. I, lect. 34, § 3 (Pirotta, p. 10). See also Averroës, *Commentarium magnum*, lib. I, comm. 20 (Crawford, p. 27). Another possible source of the image (Epicurean atom theory) is Isidore of Seville, see *Etymologiarum libri XX*, lib. XIII, c. 2, § 107 (*PL* 82, cols 472-473): "[atomos] per inane totius mundi irrequietis motibus volitare [...] sicut tenuissimi pulveres, qui infusis per fenestras radiis solis videntur" ("Atoms ... fly about with restless movements through the space of the entire world [...] like the very fine pieces of dust which through windows are seen in incoming rays of sunlight"). The following authors transmitted Isidore's example: Honoré d'Autun, *De imagine mundi*, lib. II, c. 4 (*PL* 172, col. 147); Rhaban Maur, *De universo*, lib. IX, c. 1 (*PL* 111, col. 262); Vincent of Beauvais, *Speculum maius*, lib. II, c. 2 (I, col. 80). The image is also found in the "perspectivists," see Parronchi, "La perspettiva dantesca," pp. 58-60. The *locus classicus* is, of course, Lucretius, *De natura rerum*, lib. II, ll. 114-117 (Rouse, I, p. 92).

[41]For star scintillation, see Albert, *De caelo*, lib. II, tr. 3, c. 8 (Hossfeld, p. 159); idem, *De sensu et sensato*, tr. I, c. 9 (Jammy, V, p. 13a): "res multum longe apparentes tremere videntur, sicut stelle fixe videntur scintillare" ("things which are very far away appear to tremble, as the fixed stars appear to scintillate"); Aquinas, *In De caelo*, lib. II, lect. 12, § 405 (Spiazzi, p. 202): "Tremor autem qui accidit in visu nostro, facit videri quod astrum moveatur, vel secundum scintillationem, sicut stella fixa, vel etiam secundum circumgyrationem, sicut sol" ("the tremor which happens to our sight makes it appear that the star is in motion, either due to scintillation as with a fixed star, or also due to circular motion, as with the sun"). The scintillation that Dante describes in *Purg*. XII, 90 echoes biblical texts (Dan. 12:3; Ecclus. 50:6).

96

one,[42] and of illusions due to movement[43] are all discussed at length by both Albert and Aquinas. The notion that two objects of the same colour can become indistinguishable from one another is found in Albert;[44] and the effects of foreshortening are described by the Pseudo-Aquinas.[45] As will be seen in Chapter 4, passages from commentaries can also be used to provide the intellectual context for Dante's numerous descriptions of atmospheric phenomena involving light. These examples provide further confirmation that it is not necessary to use specialist optical treatises to explain Dante's familiarity with a wide range of optical illusions and related phenomena. In order to investigate how Dante poetically reworks his technical information on visual deceptions in the *Comedy*, let us now look more closely at his representation of the visual process in three important episodes, one from each *cantica*, which involve Dante-*personaggio* correcting the optical distortions and sensory limitations to which his sight is prone.

---

[42]See Albert, *De anima*, lib. II, tr. 3, c. 16 (Stroick, p. 122); Aquinas, *In De anima*, lib. II, lect. xiv, § 430 (Pirotta, p. 108); Pseudo-Aquinas, *In De meteorologicorum*, lib. III, lect. 3, § 265 (Spiazzi, p. 618). See also Avicenna, *Liber de anima*, pars III, c. 3 (Van Riet, I, p. 196).

[43]For illusions caused by movement, see e.g. Albert, *De sensu et sensato*, tr. I, cc. 11, 14 (Jammy, V, pp. 16a-b, 19b-20a); Aquinas, *In De caelo*, lib. II, lect. 12, § 405 (Spiazzi, p. 202); lib. II, lect. 26, § 531 (Spiazzi, p. 265): "non enim videtur moveri quod tardius movetur iuxta corpus velocius motum" ("that which is moving more slowly next to a body in more rapid motion does not appear to move").

[44]E.g. Albert, *De sensu et sensato*, tr. I, c. 7 (Jammy, V, p. 9a): "[...] confusos et incertos videmus colores eorum que eminus apparent" ("we see colours as confused and uncertain in those who appear at a distance"); De *meteoris*, lib. III, tr. 4, c. 9 (Jammy, II, p. 127a): "Adhuc autem et si detur esse pervia, tunc debent colores iridis videri sicut obiecti subtili panno nigro; quod non est verum, cum expresse et clare videantur saepe" ("Thus, were it to be pervious then the colours of the rainbow must be seen as a faint object on a black cloth, which is not the case for they are often seen fully and clearly").

[45]Cf. Pseudo-Aquinas, *In De meteorologicorum*, lib. III, lect. 6, § 294 (Spiazzi, p. 633): "Quando enim aliquod corpus distans videtur per alterum vel iuxta alterum, tunc apparet esse in eadem superficie cum ipso, et propter eandem causam omnia a remotis visa videntur plana" ("When any distant body is seen either through or next to another body then it appears to be in the same plane as it, and for the same reason everything which is seen from a distance appears to be flat").

97

The opening section of *Inferno* XXXI describes the observational difficulties that Dante-*personaggio* experiences when he first sees the biblical and mythological giants who encircle the well that leads down to the icy wastes of Cocytus, the river from which the ninth circle of Hell is formed. Before his misperception of the giants takes place, Dante carefully outlines the conditions which are responsible for the visual distortion; and, as in his encounter with the *ignavi* in canto III, it is the sense of hearing which initially alerts his eyes: the sudden blast of a horn causes him to direct his line of sight back along the path travelled by this sound:

> Quiv' era men che notte e men che giorno,
> sì che 'l viso m'andava innanzi poco;
> ma io senti' sonare un alto corno,
> tanto ch'avrebbe ogne tuon fatto fioco,
> che, contra sé la sua via seguitando,
> dirizzò li occhi miei tutti ad un loco. (10-15)

The dominant motif of canto XXXI is the "torre," a word which as noun and verb punctuates the entire canto and, with the assistance of a pithy historical reference to the towers at "Montereggion" (l. 41), places a prominent emphasis on the stark, monolithic, and threatening presence of the giants.[46] The pilgrim's first sighting of the giants is in fact a mistaken judgement of them as towers: "che me parve veder molte alte torri" (l. 20). The most important section for our purposes (ll. 22-39) is found at this point in the narrative as Dante shows how it is possible for himself to correct the mistaken judgements of the visual sense. In presenting and resolving optical illusions here, and later in *Purgatorio* XXIX, Dante meditates upon crucial questions about how the senses are related to external reality and how they can reach certainty. In both these cantos, moreover, Dante underscores a

---

[46]References from *Inf.* XXXI: ("alte torri"), l. 20; ("terra"), l. 21; ("torri"), l. 31; ("torri"), l. 41; ("torreggiavan"), l. 43; ("una torre così forte"), l. 107; ("la Carisenda"), l. 136.

distinction between a primitive level of sensation and a more complete and sophisticated form of perception.

In *Inferno* XXXI, Dante draws on broadly Aristotelian doctrines in order to describe the processes by which his senses and "imagination" were initially deceived, but then attained a veridical perception:

> Poco portäi in là volta la testa,
> che me parve veder molte alte torri;
> ond' io: «Maestro, dì, che terra è questa?».
> Ed elli a me: «Però che tu trascorri
> per le tenebre troppo da la lungi,
> avvien che poi nel maginare abborri.
> Tu vedrai ben, se tu là ti congiungi,
> quanto 'l senso s'inganna di lontano;
> però alquanto più te stesso pungi». (19-27)

The idea of misperception at a distance (ll. 22-27, 28) is thoroughly Aristotelian: erroneous sense-impressions of a visual object are the result of a disproportion between the eye and its object. This point is made explicitly in a number of Aristotelian commentaries. Aquinas, for example, discusses the "proper" and "common" sensibles, and introduces a series of comments which are quite close to the explanations Virgil offers to Dante:

Cum operamur cum certitudine circa sensibile actu, scilicet ipsum sentiendo, non dicimus quod hoc videtur nobis homo, sed magis hoc dicimus cum non manifeste sentimus, *sicut cum a remotis aliquid videmus, vel cum videmus aliquid in tenebris* [...] Quod enim album sit quod videtur, non mentitur sensus; sed si album sit hoc aut illud, puta vel nix, vel farina, vel aliquid huiusmodi, hic iam contingit mentiri sensum, et maxime a remotis [...] Et circa huiusmodi maxime est deceptio, quia iudicium de his variatur secundum diversitatem distantiae.[47]

Unfavourable conditions lead to false estimates of size, shape, and figure in an inner faculty such as the "imagination," though in normal conditions (i.e. a healthy eye and proper proportions) no visual error should occur. In canto XXXI, reduced visibility and the great distance between the pilgrim and the giants cause him to represent a false mental image, to "maginar" (l. 24) an image of the giants as towers. It is only when Dante draws closer to these illusory towers that his imagination receives and presents an image of these edifices that corresponds to their "esser verace." After the more technical emphases of lines 22-26, Dante offers the reader a simile as a direct illustration of his gradual perception of the giants. As vaporous mists dissipate and allow the observer to see with greater clarity, so Dante comes to see the true form of the giants:

> Come quando la nebbia si dissipa,
> lo sguardo a poco a poco raffigura
> ciò che cela 'l vapor che l'aere stipa,
> così forando l'aura grossa e scura,
> più e più appressando ver' la sponda,

---

[47]Aquinas, *In De anima*, lib. III, lect. 5, § 646; lect. 6, §§ 662-663 (Pirotta, pp. 159, 162; italics mine): "When we are moved by an actual sense-experience to act immediately and without hesitation, we never say e.g., 'That seems to us a man?' We are more likely to speak thus when we are uncertain, as when we see things at a distance or in the dark [...] What seems to be white is indeed white as the sense reports; but whether the white thing is this or that thing, is snow, e.g., or flour, is a question often answered badly by the senses, especially at a distance [...] And these [sensible qualities] are very likely to give rise to error; for in their case our judgement has to adjust itself to differences of distance" (trans. Foster and Humphries, pp. 390, 397). For other relevant discussions, see Albert, *De meteoris*, lib. I, tr. 3, c. 3 (Jammy, II, p. 16a); Averroës, *De sensu*, 192ra (Shields, p. 11): "Et cum in aëre accidit fumus aut vapor, qui prohibeant lucis, debilitatur visio" ("Vision is weakened when smoke or vapour, both of which impede the passage of light, come into the air"); Pseudo-Aquinas, *In De meteorologicum*, lib. III, c. 6, § 294 (Spiazzi, p. 633): "visus propter nimiam distantiam eorum [sc. colours of the rainbow] ab oculo non percipit remotionem unius ab altero" ("due to their excessive distance from the eye sight does not perceive that one is more distant than the other"); Vincent of Beauvais, *Speculum maius*, lib. II, c. 60 (I, cols 117-118).

100

fuggiemi errore e crescémi paura. (34-39)

The whole drama of a beclouded vision resolving itself into a horrific one is encapsulated in the forceful and lapidary line – "fuggiemi errore e crescémi paura" – with its elegantly chiastic arrangement of verbs and nouns.

Canto XXIX of *Purgatorio* provides an even more notable example of Dante's philosophical conception of the visual act as an activity which is prone to error but also capable of more complete and reliable modes of perception. Once again, in using his own visual processes to structure the narrative, Dante displays a keen interest in the difficulties of establishing veridical vision and shows his philosophical command of the technical aspects of perceptual problems. After an invocation to the Muses, the poet describes what he mistakenly believes to be seven golden trees which he sees from a distant point of view on the far bank of the river Lethe:

> Poco più oltre, sette alberi d'oro
> falsava nel parere il lungo tratto
> del mezzo ch'era ancor tra noi e loro;
>   ma quand' i' fui sì presso di lor fatto,
> che l'obietto comun, che 'l senso inganna,
> non perdea per distanza alcun suo atto,
>   la virtù ch'a ragion discorso ammanna,
> sì com' elli eran candelabri apprese. (43-50)

As in *Inferno* XXXI, it is the effect of distance which makes what are in fact candlesticks appear to Dante as "sette alberi d'oro" (l. 43). And it is only when the pilgrim draws nearer to these illusory "alberi" and hence narrows his angle of sight that he attains a clear and accurate perception: their "obietto comun," or "common sensible" (i.e. their true size and shape) can now be adequately judged by a faculty which is periphrastically referred to as "la virtù ch'a ragion discorso ammanna" (l. 49). By charting the appearance of the candlesticks from a series of indistinct,

101

blurred sense impressions to a reliable judgement of their true appearance, Dante's narrative skillfully exploits his own perceptual responses as protagonist to create a heightened sense of expectation, suspense, and drama.

The first of the technical terms that Dante uses ("obietto comun": 1. 47) is clearly Aristotelian, but the poet is studiously guarded about naming the faculty "which provides reason with its raw material." As Nardi suggested, this perceptive faculty may be the "stimativa" mentioned by Dante in *Paradiso* XXVI.[48] In Avicenna's faculty psychology, the *aestimativa* was believed to provide knowledge about the practical value of a sensory object and to dictate a response in the percipient: the classical example was that of a sheep fleeing from a wolf. It is, however, possible to understand "la virtù ch'a ragion discorso ammanna" as either the "imagination" or the "common sense," since both these internal senses were widely believed to act as intermediaries between sensation and ratiocination and to have some discriminative power.[49] The important point is that in this canto Dante highlights how it is possible for man to overcome the limitations imposed on sensation by spatial and atmospheric factors. The human eye is able to correct,

[48]Bruno Nardi, *Dante e la cultura medievale*, ed. Paolo Mazzantini, 2nd edn (Rome and Bari: Laterza, 1983), pp. 139-140; cf. Alessandro Niccoli, "Stimativa," in *ED* V, p. 445. References to the Avicennan doctrine are given in Ch. 2, n. 50. Aquinas introduced greater precision in discussing this faculty and followed Avicenna in arguing that the *aestimativa* is only found in animals, see *ST*, Ia, q. 81, a. 3, resp. (Caramello, I, p. 397). For further details on "estimation," see Pierre Michaud-Quantin, *Études sur le vocabulaire philosophique du moyen âge* (Rome: L'Ateneo, 1970), pp. 18-24; Summers, *The Judgment of Sense*, pp. 206-210.

[49]See Pierre Michaud-Quantin, "La classification des puissances de l'âme au douzième siècle," *Revue du moyen âge latin* 5 (1949), 15-34; Richard W. Southern, *Robert Grosseteste: The Growth of an English Mind in Medieval Europe* (Oxford: Clarendon Press, 1986), pp. 40-45. On the "common sense" in the thirteenth century, see Alain de Libera, "Le sens commun au XIIIe siècle: De Jean de la Rochelle à Albert le Grand," *Revue de métaphysique et de morale* 96 (1991), 476-496. For a history of the common sense in antiquity, the Middle Ages, and the Renaissance, see also Summers, *The Judgment of Sense*, pp. 71-109. For Albert's and Aquinas' discussions of the internal senses, see Pierre Michaud-Quantin, "Albert le Grand et les puissances de l'âme," *Revue du moyen âge latin* 11 (1955), 59-86; Nancy G. Siraisi, *Taddeo Alderotti and His Pupils* (Princeton, NJ: Princeton University Press, 1981), pp. 114-124; Nicholas H. Steneck, "Albert on the Psychology of Sense Perception," in *Albertus Magnus and the Sciences*, ed. Weisheipl, pp. 263-290.

clarify, and enrich visual data, reaching beyond the deceptive surface of appearances to establish a world of depth and veracity. As we have seen, this is far from being the only example in the poem of Dante's close concern with the problems associated with the perceptive process and the internal senses.[50] But in this particular canto, the ideas and terminology underlying Dante's conception of the visual process are expressed in an exquisitely technical form of language. Remaining faithful to the Aristotelian doctrine of "common sensibles" (named in line 47), Dante assigns a central role to the medium and emphasises distance as one of the key obstacles to correct perception (ll. 44-45). In this way, the poet shows that the perceptions of sight may be "accidentally" prone to errors, but they can also be properly assessed; and he also provides the reader with yet another fine illustration of how a set of contemporary ideas on visual perception allowed him to present his narrative with a very high degree of visual verisimilitude.

"Perspectiveless" Vison

Having explored how Dante uses the theme of vision at a distance and the narrative strategies connected with it, the final section of this chapter now turns to examine how this theme is ruptured in *Paradiso* XXX as Dante-*personaggio* moves beyond time and space and enters the Empyrean, the tenth heaven of pure intellectual light. So far, his journey through the physical heavens, though assisted by divine powers (*Par.* I, 75: "col tuo lume mi levasti") and Beatrice's eyes, has nonetheless been a physical one, an ascent through time and space. Despite the deliberate ambiguities

---

[50]For Dante's other references to the internal senses in the *Comedy*, see *Purg.* XVII, 7, 13-18, 25-45; *Par.* I, 53; XXVI, 75. On the "imaginativa" in *Purgatorio* XVII, see Foster, *The Two Dantes*, pp. 108-113. On the history of the imagination, see J.M. Cocking, *Imagination: A Study in the History of Ideas*, ed. Penelope Murray (London and New York: Routledge, 1991), with a section on Dante, pp. 156-167. For the connections between faculty-psychology and Italian literary works, see Robert L. Montgomery, *The Reader's Eye: Studies in Didactic Literary Theory from Dante to Tasso* (Berkeley, Los Angeles, and London: University of California Press, 1979), with a section on Dante, pp. 50-92.

103

evoked in canto I of the *Paradiso* concerning Dante's bodily status (ll. 73-75, 98-99), it seems almost certain that his senses are still essentially human and still rely on bodily organs. Throughout the first thirty cantos of the *cantica*, Dante continues to encounter difficulties in seeing at a distance, and it is not until the final cantos that such factors no longer affect his human – if "transhumanised" – eyes.

As was seen earlier, Dante's temporary blinding by the "luce viva" of the Empyrean is followed by the pilgrim's recovering of heightened powers of vision, "novella vista" (l. 58). He now experiences a series of visions: a fulgent light in the form of a river, its banks adorned with flowers, gives off bright sparks in a scene of mobility and transformation. Beatrice tells Dante to drink from this river with his eyes (ll. 73-74), and as he does so, the "fiume" becomes a circle (ll. 88-90). She then explains that his earlier visions were "umbriferi prefazi" (l. 78), unfolding representations of the two celestial courts (the blessed and the angels) that constantly redefine themselves in Dante's eyes. These "shadowy prefaces," she maintains, do not change as such; the mutability is entirely Dante's: "... è difetto da la parte tua, / che non hai viste ancor tanto superbe»" (ll. 80-81).[51]

As Dante's powers of sight strengthen in canto XXX, the souls "unmask" (ll. 91-96) to reveal "lor vero" (l. 78): they now appear to Dante as angels and blessed souls seated in a Celestial Rose (miraculously, they appear with their resurrected bodies as they will be at the time of the Final Judgement).[52] In the

---

[51]Cf. *Par.* XXXIII, 109-114. On the mutability of sight in beatific vision, see especially Albert, *Super Matthaeum*, cap. 5, § 8, ed. Bernhard Schmidt, 2 vols. (Münster: Aschendorff, 1987), I, p. 113: "cum deus infinitus sit, a nemine comprehenditur, et ideo per eum transiens visio per infinita vadit et semper nova invenit [...] visio angelis et homini decurrit super pelagus substantiae eius infinitum et ideo semper nova invenit" ("since God is infinite, He is not comprehended by anyone, and hence vision on passing through Him goes through infinite things and always finds new things [...] the vision of angels and men flows over the infinite depths of His substance and thus always encounters new things"). Cf. Os. 12:10: "Ego visiones multiplicavi eis et in manibus prophetarum assimilatus sum" ("I have multiplied visions to him and have been taken up into the hands of prophets").

[52]Dante's vision of the souls with their ressurected bodies is foretold by Peter Damian in *Par.* XXII, 60-62.

104

final visual encounters of the canto, as Dante's eyes explore the Rose, his sight acquires a radically different form of stability and surety. Mutability gives way to stable seeing:

> La vista mia ne l'ampio e ne l'altezza
> non si smarriva, ma tutto prendeva
> il quanto e 'l quale di quella allegrezza.
> Presso e lontano, lì, né pon né leva:
> ché dove Dio sanza mezzo governa,
> la legge natural nulla rileva. (118-123)

Dante is quite explicit about the miraculous quality of this new visual power, and in the following canto, he reworks this idea at the very moment when he looks up to see Beatrice reflecting divine light from her seat in the Rose:

> Sanza risponder, li occhi sù levai,
> e vidi lei che si facea corona
> reflettendo da sé li etterni rai.
> Da quella regïon che più sù tona
> occhio mortale alcun tanto non dista,
> qualunque in mare più giù s'abbandona,
> quanto lì da Beatrice la mia vista;
> ma nulla mi facea, ché süa effige
> non discendëa a me per mezzo mista. (70-78)

Although these two sequences have received some comment, they have not been given the attention that they deserve. First of all, it has not been noted that these lines contrast sharply with Dante-*personaggio*'s many earlier difficulties in seeing at a distance in the *Paradiso*.[53] What is more, these passages represent Dante's most adventurous application of the geometry of vision, the same geometry of vision which he discussed earlier in the *Convivio* and which had previously helped him as poet of the *Comedy* to structure his narrative through the dramatic

[53]*Par.* III, 124-126; VII, 8-9; XXI, 29-30; XXIII, 115-120; XXVII, 73-75.

possibilities of his own limited eye. In the preceding ninety-seven cantos, Dante has described himself seeing "secondo le leggi della visibilità": objects appeared less clearly either because they subtended finer angles in the eye or were altered by interposing media. In the Empyrean, as Dante nears the final vision, the geometry of vision ceases to operate: it fails precisely because "[...] dove Dio sanza mezzo governa, / la legge natural nulla rileva" (ll. 122-123). And yet, the end of visual "perspective" brings with it a far greater clarity of sight; Dante is free from the confines of space and untroubled by the distorting effects of the medium.[54]

This is not the first time in the poem that the poet has insisted upon the need to avoid measuring and classifying the divine by human standards.[55] But there is an especial propriety about the fact that, in a heaven outside space, vision takes place independently from spatial factors. In other words, Dante points out the irrelevance of the spatial distinctions provided by the geometry of vision, because his journey is no longer organised from an optical point of view but *sub specie aeternitatis*. It may even be possible that Dante's abandonment of the geometry of vision in this canto is in part related to a specific system of ideas which was developed by some later thirteenth-century commentators on Peter Lombard's *Sentences*. In his commentary, for example, Aquinas discusses whether the senses of glorified bodies are active, and puts forward a possible objection by noting that if their vision were affected by the visual pyramid this would imply that there is no difference between seeing on Earth and the visual powers of souls in heaven. Aquinas goes on to argue that, since this would be absurd, the sight of glorified bodies is not in fact subject to the same limitations as man's vision on Earth:

[54]The medium is, of course, a central element in Aristotle's theory of physical vision, see Aquinas, *In De anima*, lib. II, lect. 15, § 433 (Spiazzi, p. 108): "necesse est ergo esse aliquod medium inter visibile et visum" ("it is thus necessary for there to be a medium between the visible object and the eye").

[55]See *Purg.* XXVIII, 91-129; *Par.* XXVIII, 73-78; and see also Boyde, *Dante Philomythes*, p. 78.

106

[...] quanto sensus est perfectior, tanto ex minori immutatione facta potest obiectum suum percipere. Quanto autem sub minori angulo visus a visibili immutatur, tanto minor immutatio est; et inde est quod visus perfectior magis a remotis aliquid videre potest quam visus debilior; quia quanto a remotiori videtur, sub minori angulo videtur. Et quia visus corporis gloriosi erit perfectissimus, ex parvissima immutatione poterit videre; unde sub angulo multo minori videre poterit quam modo possit, et per consequens multo magis a remoto.[56]

Albert the Great also puts forward a similar line of argument in his earlier commentary (c. 1243) and in other independent treatises.[57] The principle upon which these commentary discussions rest is somewhat similar to Dante's treatment of his own new powers of sight in cantos XXX and XXXI, even though Dante goes beyond these texts in describing his visual powers *in patria* as being entirely unaffected by the spatial implications of seeing that human beings experience *in*

---

[56]Aquinas, *In IV Sent.*, d. 44, q. 2, a. 1, qla. 4, ad 6, in *Opera omnia*, VII, pp. 1087b-1088a: "the more perfect sensation is, the less alteration is required to allow the sense organ to perceive its object. The smaller the angle under which a visible object alters the eye, the smaller the alteration; and hence it is that someone with better vision can see a distant object more clearly than someone with weaker vision, because what is seen from a distance is seen under a smaller angle. And since the vision of the glorified body will be most perfect, it will be able to see due to the smallest alteration. Thus, a glorified body will be able to see under a much smaller angle than it presently can and as a result will see much more from a distance."

[57]Albert, *In IV Sent.*, d. 13, q. 2, ob. 5, ed. F.M. Henquinet, "Une pièce inédite du commentaire d'Albert le Grand sur le IVe livre des sentences," *RThAM* 7 (1935), on pp. 280-281. See also Albert, *De resurrectione*, tr. 2, q. 8, a. 3, § 13, ed. Wilhelm Kübel (Münster: Aschendorff, 1958), p. 276; idem, *Quaestio de sensibus corporis gloriosi*, art. 2, p. 1, ad 3-4, in *Summa theologiae*, ed. Wilhelm Kübel and Henrico Georgio Vogels (Münster: Aschendorff, 1978), p. 116. The earliest precedent for this idea that I have found is William of Auvergne, *De universo*, lib. I, pars II, cap. 32 (Paris, 1674; reprint, Frankfurt-am-Main: Minerva, 1963), p. 737a: "necesse est virtutem sensibilem in omnibus glorificandis hominibus glorificandam esse. Quae igitur erit gloria illius, si per eam non vident homines tunc aeque longe ut propre, et aequaliter remota ut propinqua?" ("it is necessary for sensible virtue in all who are to be glorified to be itself glorified. What will their glory be if through it men do not see equally well what is distant and what is close, and also what is far and what is near"). For the elaboration of similar doctrines by later writers, see Michael Baxandall, *Painting and Experience in Fifteenth-Century Italy: A Primer in the Social History of Pictorial Style*, 2nd edn (Oxford and New York: Oxford University Press, 1988), pp. 172-173.

via. This would not be the first time in the poem that Dante has dealt with the problem of visual sensation in the afterlife in a way which distantly echoes medieval discussions of related topics in *Sentence* commentaries.[58] In canto XIV, he asks Solomon whether the light that encircles the souls will remain with them when their bodies are rehabilitated, and, if so, how their eyes will be able to withstand the new intensity of light (ll. 13-18). If Dante is dealing with related issues in canto XXX, he is doing so in the most oblique and original fashion possible. Having corrected his own visual errors throughout the poem and tempered his vision with increasing intensities of light, he now seems to describe himself as seeing in the manner of a glorified body.[59]

---

[58]Albert, *In IV Sent.*, d. 44, aa. 27-31 (Jammy, XVI, pp. 850b-853b); Aquinas, *In IV Sent.*, d. 44, q. 2, a. 4, qla 2-3, in *Opera omnia*, VI, pp. 1098a-1099b; Bonaventure, *In IV Sent.*, d. 49, a. 2 (Quaracchi, IV, pp. 1024a-1026b). In *Paradiso* XIV, Dante marks his independence from these writings with his exceptional sense of the value of the resurrected body: the light of the body overcomes the light of the soul "sì come carbon che fiamma rende, / e per vivo candor quella soverchia" (ll. 52-53). A similar comparison is found in the above commentaries, but it is the light of the soul, not that of the body, which has the greater intensity; for relevant texts and further discussion, see Anna Maria Chiavacci Leonardi, "«Le bianche stole»: Il tema della resurrezione nel *Paradiso*," in *Dante e la Bibbia*, ed. Giovanni Barblan (Florence: Olschki, 1988), p. 261, nn. 17-18.

[59]Further support for this view is provided by the text from Albert quoted in n. 57 and the final section of Ch. 7, pp. 250-256.

CHAPTER 4

Light Reflection, Mirrors, and Meteorological Optics
in the *Comedy*

Throughout the *Comedy*, one of the most prominent features of Dante's optical knowledge is his scientific interest in light reflection and the properties of mirrors. Dante was clearly fascinated by the behaviour of light as it strikes different surfaces, and, as poet, he uses images drawn from reflected light in almost all phases of his development.[1] In the *Comedy*, his most technical applications of the science of mirrors, or catoptrics, are two similes based on the optical law of light reflection and the mirror experiment described by Beatrice in *Paradiso* II. But Dante's fascination with light and optics also finds its expression in a variety of imagery based on the rainbow and moon-halo, phenomena which he believed to be caused by the reflection of light in the atmosphere. All these important clusters of images are the subject of the present chapter in which the primary concerns will again be to investigate the intellectual context of Dante's ideas, to examine how he assimilates contemporary optical doctrines, and to consider the contexts and ends to which he puts them in his poetry.

---

[1]For the lyric *topos* of light reflected from the God of Love to the lady and from her to the lover, see *Rime* XC, 28-30; CII, 5. For the main categories of imagery drawn from reflected light in the *Paradiso*, see n. 27.

110

The Law of Light Reflection

Purgatorio XV begins, like several other cantos in the second *cantica*, with a densely technical astronomical periphrasis (ll. 1-6). The sun, playing with its light like a child, has just entered the vespertinal hours and is positioned at three o' clock. As Dante and Virgil proceed north-westwards along the terrace of the envious, the rays of the sun strike them (l. 7) and the intensity of light increases to such an extent that Dante is forced to shield his eyes. At this point in the narrative, Dante makes a sort of parasol with his hands over his eyebrows (ll. 12-13), and no explanation is offered for the increased brightness until line 30, when Virgil announces the reason for the greater luminosity: a new light source, the angel of humility, is approaching Dante and Virgil.

As we saw in Chapter 3, encounters with the radiant countenances of angels are not new by this stage in the journey up Mount Purgatory. The simile of canto XV is, however, singular in introducing a greater amount of scientific detail and technical vocabulary than before:

> Come quando da l'acqua o da lo specchio
> salta lo raggio a l'opposita parte,
> salendo su per lo modo parecchio
>   a quel che scende, e tanto si diparte
> dal cader de la pietra in igual tratta,
> sì come mostra esperïenza e arte;
>   così mi parve da luce rifratta
> quivi dinanzi a me esser percosso;
> per che a fuggir la mia vista fu ratta. (16-24)

The comparison takes its starting point from the reflected ray, "lo raggio" (l. 17) that "springs" upwards from either a body of water or a mirror.[2] Dante

---

[2]In his commentary on these lines, Porena proposes the reading "come quando de l'acqua da lo specchio" i.e. water is the only reflective surface, see *La "Divina Commedia,"* ed. Manfredi Porena, 3 vols. (Bologna: Zanichelli, 1953), II, p. 138. Petrocchi forcefully argues against this

uses the verbs "salta" and "salendo" in prominent position to express the upward movement, and, as we shall see, this is a verbal energy provided by the poet – it is not in any of his scientific source-material. The reflected ray emerges on the opposite side, "l'opposita parte" (l. 17) from where the incident ray, the effectively delayed "quel che scende" (l. 19), had first struck the reflective surface. Clearly, this reflection must have taken place at an oblique angle: perpendicular reflection would deflect the ray straight back towards its source. In the second tercet of the vehicle, Dante elaborates on the idea of "per lo modo parecchio" (l. 18) and demonstrates his knowledge of the fundamental catoptrical law which states the equality of the angles of incidence and reflection. Dante expresses the constant relationship between the angles of the incident and the reflected ray in terms of the distance each straight-line ray travels from the perpendicular. He uses the phrase "il cader de la pietra" (l. 20) to refer to this perpendicular, since until Galileo proved otherwise, it was commonly believed that any heavy object, when dropped to the Earth from a given height, described a straight line.[3]

In the final line of the vehicle, Dante refers to an unnamed "arte," a term which can mean productive activity (e.g. *Inf.* XI, 103), although in this technical context, "arte" is obviously related to the mathematical disciplines of the *quadrivium*. "Esperïenza" also has a specific, technical connotation and stands for

---

view, maintaining that the accepted reading is a typical scholastic expression, see *La "Commedia" secondo l'antica vulgata*, ed. Giorgio Petrocchi, 4 vols. (Milan: Mondadori, 1966-1967), III, pp. 246-247. Dante does use the metaphor "specchio" for water in *Inf.* XXX, 128, and elsewhere he points out its mirror-like properties (*Purg.* XXIX, 67-69; *Par.* III, 10-12; IX, 112-114; XXX, 109-111). But there are important examples of the poet's desire to use a balanced series of alternatives in the first line(s) of analogous similes (e.g. *Par.* XXIX, 25 and esp. III, 10-12, where water and a mirror are the reflective surfaces). In terms of poetic balance and syntactic structure, then, Petrocchi's reading is also preferable.

[3]The "perspectivists" termed this perpendicular line the *cathetus*, see e.g. Bacon, *Opus maius*, lib. V, p. iii, dist. 1, c. 2 (Lindberg, p. 262); Pecham, *Perspectiva communis*, lib. II, prop. 20 (Lindberg, p. 170): "Cathetus est linea perpendicularis ducta a re visa super superficiem speculi" ("The cathetus is the perpendicular line drawn from the visible object to the mirror's surface"); Witelo, *Perspectiva*, lib. V, diff. 8 (Risner, p. 214).

112

an "experiment" of some kind.[4] An analogous usage of "arte" and "esperïenza" during a scientific-scholastic passage in *Paradiso* II provides confirmation of this interpretation.[5]

Critics and commentators who have attempted to find a possible source for Dante's knowledge of the law of light reflection are presented with a major obstacle. No one credits Dante with a direct knowledge of either Euclid or Ptolemy, and Aristotle does not mention the equality of angles in his treatment of geometrical optics in the *Meteorologica*.[6] This central and fundamental optical proposition was known in Dante's time in Latin translations of Greek and Arabic optical works and in the "perspectivist" treatises.[7] Bruno Nardi, one of the first twentieth-century scholars to propose a direct "perspectivist" influence on Dante, drew attention to most of the relevant sources in a celebrated reading of this canto in 1953:

---

[4]The terms *experimentum* and *experientia* were often used in the Middle Ages to refer to carefully generalised observations drawn from experience, not the artificially controlled experiments performed by scientists today.

[5]*Par*. II, 94-96: "Da questa instanza può deliberarti / esperïenza, se già mai la provi, / ch'esser suol fonte ai rivi di vostr' arti." For the *locus communis* that all science comes from experience, see Aristotle, *Metaphysica*, I, 1, 980b 28-981a 3.

[6]Aristotle does, however, speak of a ratio between the angles at *Meteorologica*, III, 5, 376a 1-5; for his catoptrical knowledge, see *Meteorologica*, III, 2-5, 371b 18-377a 28. For further details, see Carl B. Boyer, "Aristotelian References to the Law of Reflection," *Isis* 36 (1945-1946), 92-105.

[7]The law of reflection was first elaborated in Euclid, see *Optica*, lib. I, prop. 19 (Thiesen, p. 72); *Catoptrica*, lib. I, prop. 1, ed. J.L. Heiberg and H. Menge, in *Opera omnia*, 9 vols. (Leipzig: Teubner, 1883-1916), VII, pp. 286-289. For its presence in Alexandrian and Islamic opticians, see Alhazen, *De aspectibus*, lib. IV, c. 3, sec. 10 (Risner, p. 108); Alkindi, *De aspectibus*, prop. 16 (Vogl, p. 28); Hero of Alexander, *Catoptrica*, prop. 4, in *Opera quae supersunt omnia*, ed. L. Nix and W. Schmidt, 5 vols. (Leipzig: Teubner, 1899-1914), II, pp. 325-326; Pseudo-Euclid, *De speculis*, prop. 5 (Björnbo and Vogl, p. 100); Ptolemy, *Optica*, lib. III, prop. 15 (Lejeune, p. 91). For the "perspectivists," see Bacon, *De multiplicatione specierum*, pars II, c. 6 (Lindberg, pp. 137-147); idem, *Opus maius*, lib. V, p. iii, dist. 1, c. 1 (Lindberg, p. 254); Pecham, *Perspectiva communis*, lib. II, prop. 6 (Lindberg, p. 160); Witelo, *Perspectiva*, lib. V, prop. 10 (Risner, p. 195). See also Grosseteste, *De lineis, angulis et figuris* (Baur, p. 62).

113

Il fenomeno è ben noto a Dante, non tanto per l'esperienza comune che tutti n'abbiamo fatta, quanto perché egli ha studiato il fenomeno della riflessione della luce sulla scorta di quel 'arte che si chiama perspettiva' [...] della quale arte erano maestri ai contemporanei di Dante l'arabo Alhazen e Witelo e altri, nei cui trattati si trova appunto dimostrata coll'esperienza e col ragionamento la legge elementare dell'ottica, che cioè l'angolo d'incidenza è uguale all'angolo di riflessione, calcolati qui l'uno e l'altro in rapporto alla verticale del piano sul quale il poeta cammina.[8]

However, on a closer consideration of primary sources, Dante's use of the circumlocution "il cader de la pietra" proves to be highly significant. Benvenuto da Imola was the first commentator to indicate the presence of the Latin term *casus lapidis* in Albert the Great's *De causis proprietatum elementorum*, a work which Dante indubitably studied because he cites both it and its accompanying treatise, the *De natura locorum*, in the *Convivio*.[9] In Tractatus I, Chapter 6 of the *De*

[8]Bruno Nardi, "Il Canto XV del *Purgatorio*," reprinted in *«Lecturae» e altri studi danteschi*, ed. Rudi Abardo (Florence: Le Lettere, 1990), p. 129. Parronchi repeatedly draws attention to Nardi's remarks and calls his reference to Alhazen "la citazione d'obbligo" for this simile, see "La perspettiva dantesca," p. 68. Similarly, Guidubaldi argues that the simile is "palesemente modellato su [...] Alhazen," see *Dante Europeo*, II, n. 2, p. 251; III, n. 4, pp. 311-312. Many recent commentators also refer to Nardi's comments, see e.g. Umberto Bosco and Giovanni Reggio, *"La Divina Commedia,"* 3 vols. (Florence: Le Monnier, 1979), II, p. 258; Giuseppe Giacalone, *La "Divina Commedia,"* 3 vols. new edn (Rome: Signorelli, 1988), II, p. 355; Charles S. Singleton, *The "Divine Comedy,"* 3 vols. (Princeton, NJ: Princeton University Press, 1971-1975), III, commentary, pp. 318-319.

[9]Benvenuto da Imola, *Comentum super Dantis Aldigherij Comoediam*, ed. Giacomo Filippo Lacaita, 5 vols. (Florence: Barbèra, 1887), III, p. 406: "Albertus libro de proprietatibus elementorum." See Albert, *De causis et proprietatum elementorum*, lib. 1, tr. 1, c. 5, ed. Paul Hossfeld (Münster: Aschendorff, 1980), p. 57: "Propter quod duas habent [sc. inhabitants of the equatorial regions] aestates caldissimas, quia, sicut diximus bis pertransit [sc. the sun] casus lapidis super capita eorum; hoc est, lineam perpendiculariter ductam a centro solis super capita eorum; haec enim vocatur casus lapidis, eo quod lapis et quodlibet simpliciter grave cadit inferius perpendiculariter" ("On this account they have two very hot summers, because, as we said, the sun passes through the *casus lapidis* twice directly over their heads; this is the line drawn perpendicularly from the centre of the sun onto their heads; it is called the *casus lapidis* because a stone and anything naturally heavy falls perpendicularly"). For Dante's reference to this treatise, see *Con.* III, v, 12: "[...] secondo ch'io comprendo per le sentenze de li astrologi, e per quella d'Alberto de la Magna nel libro de la Natura de' luoghi e de le Proprietadi de li elementi." For details of the manuscript tradition of Albert's treatises, see Nardi, *Saggi di filosofia dantesca*, p.

114

natura locorum, Dante would have come across a short optical excursus in which Albert discusses the view that the equatorial regions are uninhabitable. As part of his discussion, Albert describes how the rays of the sun are concentrated upon this area and notes the equality of angles between the incident and the reflected rays:

Et quia via solis directe est super spatium illud bis in anno, dicunt, quod ibi reflectitur radius eius in seipsum, propter quod combustivus est, et facit partem illam ex nimio calore inhabitabilem [...] cum enim omnis radius incidens reflectatur ad parem sive aequalem angulum, oportet necessario, quod radius incidens alicui terrae perpendiculariter in seipsum reflectatur. In seipsum reflexus radius causat adustionem, sicut probatur in Perspectivis.[10]

In another section of one of Albert's works indubitably known to Dante, Book I of his *De meteoris,* the law is not explicitly mentioned but Albert does

---

64.   In addition to *De causis proprietatum elementorum* and *De natura loci,* medieval manuscripts also contain *De natura et origine animae* and *De intellectu et de intelligibili.* These Albertine works, which were all known to Dante, contain detailed discussions of the role of light in the cosmos, the formation of the soul, and human knowing.

[10]Albert, *De natura loci,* tr. I, c. 6 (Hossfeld, pp. 9-10): "And since the path of the sun is above that space twice a year, they maintain that the ray is reflected upon itself here; on account of which it is combustive and makes that part inhabitable due to excessive heat [...] since every incident ray is reflected at an even or equal angle, it must necessarily be that a ray incident on some part of the Earth is reflected upon itself. As will be proved in the writings on perspective, a ray reflected upon itself causes heat." The equality of angles is also recorded in the following Albertine works: *De meteoris,* lib. I, tr, 4, c. 10 (Jammy, II, p. 25b): "quia non est omnino simile radio repercusso ab aqua: quia ille ad aequalem angulum cum radio incidente in aquam repercutit ad parietem oppositum" ("for it is not completely similar to the ray propelled back by water; for that ray is repelled to the opposite wall at an angle equal to that of the ray incident on the water"); lib. III, tr. 4, c. 13 (Jammy, II, p. 130a): "reflectitur radius a quodlibet istorum [sc. polished mirrors] ad angulum aequalem cum angulo radii incidentis in speculum" ("a ray is reflected from any of these at an angle which is equal to that of the ray incident on the mirror"); *De caelo et mundo,* lib. II, tr. 3, c. 3 (Hossfeld, p. 146): "quando sol diametraliter superponitur capitibus [...] tunc reflectitur radius in seipsum, eo quod ad pares angulos semper fit reflexio" ("when the sun is placed directly above the heads [...] then the ray is reflected upon itself, since reflection always takes place at equal angles"); *De sensu et sensato,* tr. I, c. 8 (Jammy, V, p. 11a); *In II Sent.,* dist. 9, a. 8 (Jammy, XIV, p. 175b).

115

comment in detail on reflection from both a mirror and water and he also describes the oblique behaviour of the reflected ray:

[...] nos videmus quod omne lumen repercussum sive a speculo vel a superficie corporis humidi super quod incidit radius corporis luminosi, semper repercutitur ad oppositum situm corporis luminosi, a quo incidit radius, sicut apparet in radio incidente in vas aquae.[11]

These passages confirm how Dante was able to use Albert's writings in order to acquire a fairly detailed knowledge of optical science, to gain familiarity with its main principles, and even to appreciate certain "experimental" procedures.

In addition to Albert's works, Averroës, Aquinas, and the Pseudo-Aquinas also provide detailed discussions of light reflection in their commentaries on *De caelo* and *Meteorologica*.[12] Indeed, it would seem that the relevant information

[11]*De meteoris*, lib. I, tr. 2, c. 4 (Jammy, II, p. 12b), punctuation slightly amended: "we see that all light which is repelled, either from a mirror or from the surface of a wet body upon which the ray of a luminous body is incident, is always propelled back to the opposite part of the luminous body, from where the incident ray came, as is apparent with a ray incident on a vessel of water." In other commentaries, Albert discusses reflection and the properties of mirrors, including burning mirrors, see *De anima*, lib. II, tr. 3, c. 15 (Stroick, pp. 121-122); idem, *De caelo et mundo*, lib. II, tr. 3, c. 1 (Hossfeld, p. 143); idem, *Quaestiones de animalibus*, lib. I, qq. 29-31 (Filthaut, pp. 98-100); idem, *De meteoris*, lib. III, tr. 4, cc. 12-15 (Jammy, II, pp. 128b-132b); idem, *De sensu et sensato*, tr. I, c. 10 (Jammy, V, p. 15a-b): "Proiectio autem luminis in oppositum est ideo quia reddit [sc. mirror] lumen ex sua tersione. Si enim tantum retineret, tunc non fieret quod dictum est: sed quia abundat lumine ad copiam, ideo fundit ipsum in oppositum radii incidentis" ("Light is sent out in the opposite direction because the mirror's polished surface returns light. If it only retained it then what we have said would not happen, but because the mirror abounds in an excess of light it therefore propels it in the opposite direction to that of the incident ray").

[12]For references to the law of light reflection in these commentators, see Averroës, *Meteorologica*, lib. III, summa secunda, c. 2 (Apud Iunctas, V, pp. 451va, 456rb and 458ra); Aquinas, *In De caelo*, lib. II, lect. 10, § 392 (Spiazzi, p. 193); Pseudo-Aquinas, *In De meteorologicorum*, lib. III, lect. 4, § 276 (Spiazzi, p. 622): "sed radii quibus videtur halo, aequaliter distant a perpendiculari, et refranguntur in aequali distantia ad perpendicularem ad angulos aequales; ergo faciunt circulum" ("but the rays under which a halo is seen are equidistant from the perpendicular and are reflected at an equal distance from the perpendicular and at equal angles; they therefore make a circle"). The idea of distance from the perpendicular in the Pseudo-Aquinas is very close to Dante's own presentation of the phenomenon in lines 19-20.

116

was more widely known than has been recognised, for, outside the Aristotelian tradition, Vincent of Beauvais gives a succinct exposition of the law in his *Speculum maius*, and Bonaventure also provides the relevant information in one of his sermons.[13]

Having clarified the scientific content of, and background to, Dante's comparison, we can now turn to the interpretative difficulties it raises in both a local and a wider context. From the general context of the episode, it is apparent that Dante-*personaggio* is subjected to an intensity of light similar to that experienced when rays are reflected from a highly polished surface. But does the simile function merely as an analogy, or is Dante establishing precise points of contact between the details of the vehicle and those of the tenor? In response to the indeterminacy of line 21 ("quivi dinanzi a me esser percosso"), three principal interpretations have been advanced:

1. no reflection occurs, since the angel's light strikes Dante directly;
2. God's light reflects from the angel's face and thence onto Dante;
3. the angel's light reflects from the surface of the terrace to Dante.

Benvenuto of Imola was the first commentator to argue that rays emanate from the angel and strike the pilgrim directly, and this view has found numerous supporters this century.[14] For these commentators, reflected light is described in the vehicle merely in order to convey the experience of direct light by way of analogy.[15] The general dynamics of Dante's similes in the *Comedy* seem to

[13]References in order: Beauvais, *Speculum maius*, lib. II, c. 72 (I, cols 128-129): "Ad pares quidem angulos fit, si radius ex obliquo veniens est ad superficiem speculi. In seipsum autem si perpendiculariter" ("[Reflection] takes place at equal angles if a ray is coming to the surface of the mirror at an oblique angle. It is reflected back upon itself if [it comes] perpendicularly"); Bonaventure, *Dominica tertia in Quadragesima* (Quaracchi, IX, p. 228a-b): "in radiorum reflexione radius incidens et radius resiliens constituunt aequales angulos" ("in the reflection of light rays the incident ray and the reflected ray form equal angles").

[14]Benvenuto, *Comentum*, III, p. 406.

[15]See André Pézard, *Oeuvres complètes* (Paris: Gallimard, 1968), p. 1221. Most recent commentators have favoured this interpretation, see Tommaso Di Salvo, *La "Divina Commedia"* (Bologna: Zanichelli, 1987), pp. 258-259; Giacalone, *La "Divina Commedia,"* II, p. 355; Emilio

support this claim, since there are many examples of similes in which the poet accumulates details in the vehicle which do not have a precise and equal number of counterparts in the tenor.[16]

There is an equally long history to the second principal interpretation which considers the light as coming from a divine source and being reflected from the angel before it strikes Dante. This point of view begins with Buti's assertion that Dante describes "luce rifratta [...] volendo dare ad intendere che la luce eterna; cioè Iddio ferisce nella faccia dell'angiulo, et inde rinfrangesse nel suo volto."[17] Several modern commentators have favoured such an interpretation, because it establishes a consistent set of relations between tenor and vehicle, and some have also used this reading to attribute a variety of allegorical connotations to the simile, such as the importance of the light of grace in the penitential process and/or the absolute necessity of submission to God. Without necessarily balking at possible symbolic suggestions, one has to wonder why such resonances cannot be applied to the angel's direct light. The angel is, after all, God's creation and so infused with his light in the creative act.[18]

The hazards involved in the tendency to treat Dante's similes as completely consistent can also be seen in objections levelled by many modern commentators against the third interpretation, namely, that the light described in the tenor is reflected from the surface of the terrace of the envious.[19] Many recent

---

Pasquini and Antonio Quaglio, *"Commedia"* (Milan: Garzanti, 1987), p. 539; Luigi Scorrano and Aldo Vallone, *La "Divina Commedia,"* 3 vols. (Naples: Ferraro, 1986), II, p. 229.

[16]E.g. *Inf.* XII, 4-12; XV, 4-12; XXI, 7-18; XXIV, 1-18; XXX, 1-27; *Purg.* XVII, 1-9; *Par.* XII, 10-21; XIII, 1-21; XXIII, 1-12.

[17]*Commento di Francesco da Buti sopra la "Divina Commedia" di Dante Allighieri*, ed. Giannini di Crescentino, 3 vols. (Pisa: Fratelli Nistri, 1858-1862), II, p. 350. Modern advocates of this view include: Bosco-Reggio, *La "Divina Commedia,"* II, pp. 254-255; Attilio Momigliano, *La "Divina Commedia,"* 3 vols. (Florence: Sansoni, 1950), II, p. 373; Sapegno, *La "Divina Commedia,"* p. 563.

[18]*Par.* VII, 64-66. For the "sanza mezzo" irradiative creation of angels, see XXIX, 28-30.

[19]Supporters of this interpretation include: Tommaso Casini, *La "Divina Commedia,"* revised by S.A. Barbi, with a new presentation by Francesco Mazzoni, 3 vols. (Florence: Sansoni, 1965), II,

118

commentators have dismissed reflection from the terrace in the belief that its opaque surface, which is twice said to be of "livido color" (*Purg.* XIII, 9 and 48), will absorb rather than reflect light.[20] And yet, for a number of reasons it seems quite absurd to interpret the relations between the vehicle and tenor of a simile solely on the basis of such a criterion. First of all, it is possible to adduce several medieval passages that describe reflection from coloured and opaque surfaces; these passages indicate that polish seems to have been the most important factor.[21] Secondly, to assess Dante's simile on the basis of consistency between tenor and vehicle is to ignore his technique of deliberately refusing to yoke both sides of his similes together to attain direct correspondences. Dante will often creatively pair vehicles with tenors in such a way as to produce symbolic resonances that reverberate beyond the local context of the simile.[22] Finally, it must be said that evidence internal to the poem supports the view that the simile describes light being reflected from the surface of the terrace. In related scenes of dazzling, many of which we examined in the previous chapter, the pilgrim frequently looks down

---

p. 363; Pietro Fraticelli, La *"Divina Commedia,"* (Florence: Barbèra, 1887), p. 346; Giacomo Poletto, La *"Divina Commedia,"* 3 vols. (Rome: Desclée Lefebvre, 1894), II, p. 346; Porena, *La "Divina Commedia,"* II, pp. 138-139; Nardi, "Il Canto XV del *Purgatorio*," p. 129.

[20]See n. 15.

[21]Albert, *De sensu et sensato*, tr. I, c. 8 (Jammy, V, p. 11a): "a qualibet leni superficie fit reflexio" ("reflection takes place from any smooth surface"); Bonaventure, *In II Sent.*, d. 13, a. 2, q. 2 (Quaracchi, II, p. 321a): "corpus opacum [...] per multam tersionem et politionem possit affici luminosum, sicut patet, cum de cinere fit vitrum, et de terra carbunculum" ("an opaque body with much smoothing and polishing can become luminous, as is seen when glass is made from ashes and the carbuncle from earth); Vincent of Beauvais, *Speculum maius*, lib. II, c. 78 (I, col. 129): "Nam si lapis niger valde politus sub sole ponantur [...] videbitur ille lapis albus vel splendidum, propter diffusionem radiorum in superficie" ("For if a black, well-polished stone is placed in the sun [...] that stone will be seen as splendid and white, because of the diffusion of light rays on its surface").

[22]For these aspects of Dantean similes, see Ignazio Baldelli, in *ED* VI, pp. 96-97; Richard H. Lansing, *From Image to Idea: A Study of Simile in Dante's Commedia* (Ravenna: Longo, 1977), passim, but esp. p. 13; Antonino Pagliaro, *Ulisse: Ricerche semantiche sopra la "Divina Commedia,"* 2 vols. (Milan: D'Anna, 1967), II, pp. 647-683; idem, "Similitudine," in *ED* V, pp. 253-259.

119

in order to avoid bright light.[23] In canto XV, Dante places his hands above his brow to shield himself from the light of the angel (ll. 13-14), but even with his eyes covered, the light from the approaching angel continues to strike him with such force that he has to turn away (l. 24). The entire sequence is quite consistent with light being reflected from below, and this reading is sufficiently congruent with medieval views on reflection, Dante's art of simile, and the narrative context to merit far closer attention than it has recently received.[24]

Dante's use of optics in *Purgatorio* XV is not restricted solely to the simile of lines 16-24. Virgil's discourse on charity in the central section of the canto contains some interesting adaptations of images drawn from a technical knowledge of light reflection. In order to clarify a rather elliptical comment on human rapacity made by Guido del Duca in the previous canto (ll. 86-87), Virgil explains how the relentless pursuit of finite, secondary goods creates envy and strife, whereas the possession of spiritual goods entails the very opposite: the more people who seek to attain them, the greater the share for everyone.[25] To illustrate this paradox, Dante uses images of light and reflected light to show how a single spiritual good may, like a single light source, be shared between innumerable objects without detriment to its essential unity:

> Quello infinito e ineffabil bene
> che là sù è, così corre ad amore

[23]For the averted gaze downwards, see *Purg*. II, 39-40; *Par*. IV, 139-142; XIV, 82-83. Elsewhere the pilgrim either turns around (*Purg*. XV, 24), or takes cover behind Virgil and Statius (*Purg*. XXIV, 142-144).

[24]For a detailed critique of previous interpretations of the optical simile in favour of reflection from the surface of the terrace, see Michio Fujitani, "Dalla legge ottica alla poesia: la metamorfosi di «*Purgatorio*» XV, 1-27," *Studi danteschi* 61 (1989) [published in 1994], 153-186. Fujitani's article is also valuable for its reinterpretation of the significance of the simile.

[25]Cf. Aquinas, *In I Sent*. d. 17, q. 2, a. 4, sol. in *Opera omnia*, VI, p. 147b: "ipsa anima, quantum plus recipit de bonitate divina et lumine gratiae ipsius, tanto capacior efficitur ad recipiendum; et ideo quanto plus recipit, tanto plus potest recipere" ("the more this soul receives of divine goodness and the light of its grace, the more it is made able to receive; and hence the more it receives, the more it can receive").

120

> com' a lucido corpo raggio vene.
> Tanto si dà quanto trova d'ardore;
> sì che, quantunque carità si stende,
> cresce sovr' essa l'etterno valore.
> E quanta gente più là sù s'intende,
> più v'è da bene amare, e più vi s'ama,
> e come specchio l'uno a l'altro rende. (67-75)

With its alliterative effects, its patterns of phonic repetition and assonance, its balance of comparatives, and its prominent rhymes on the key words, "amore," "ardore," and "valore," this passage offer a fine example of Dante's doctrinal poetry. The solemn language and syntactic balance provide a sustained and suggestive anticipation of the style and construction of several passages in the *Paradiso*,[26] while the short similes *per brevitatem* at the end of the first and last tercets give a visual illustration of the ideas which Dante has put forward. The felicitous sequence of argument and imagery also shows Dante pressing the symbolism of mirrors and light reflection to intellectual ends, thereby anticipating some of his finest "specular" poetry in the final *cantica* where Dante's use of mirror imagery and the mirror metaphor is especially rich and varied.[27]

The unobtrusive optical examples (a ray moving towards a bright body and the multiplication of light between mirrors) probably derive from Dante's study of

---

[26]For lexical and metrical parallels between these lines and the *Paradiso*, see: (i) rhymes on "ardore" and "valore," XIV, 40-42; (ii) "etterno valor," XXIX, 142-143; (iii) epithets "primo" and "ineffabil," X, 3. For the idea of "sommo bene" developed with light imagery, see *Par.* VII, 80-81; XIV, 47-51. Dante develops the concept of God as limitless good and expresses it through light imagery in *Par.* XIX, 50-51.

[27]The main categories of images drawn from reflected light in the *Paradiso* are: (i) souls as light reflectors, IX, 69, 113-114; XVII, 123; XIX, 4-6; (ii) planets and stars, XX, 6; XXI, 16-18; XXIII, 29-30; (iii) angels, XIII, 52-60; XXIX, 138, 142-145; (iv) Dante's eye-mirrors, XXI, 18; XXX, 85-86; (v) Beatrice reflecting divine light, XVIII, 16-18; XXVIII, 4-12; XXXI, 70-72. Note also that the Empyrean is formed from a ray of reflected light, see XXX, 106-108. On the use of the mirror metaphor with reference to God, see also *Par.* XV, 61-63; XXVI, 106-108; cf. *Dve*, I, ii, 3. On mirrors and mirror imagery in Dante, see Herbert D. Austin, "Dante and Mirrors," *Italica* 21 (1944), 13-17; E. Crivelli, "Il vetro, gli specchi e gli occhiali ai tempi di Dante," *Giornale dantesco* n.s. 12 (1939), pp. 84-89; Simon A. Gilson, "Light Reflection, Mirror Metaphors, and Optical Framing in Dante's *Comedy*: Precedents and Transformations," *Neophilologus* 83 (1999), 241-252; Emilio Pasquini, "Specchio," in *ED* V, pp. 366-367.

121

the action of light in the cosmos. In antiquity, the Middle Ages, and the Renaissance, it was commonly believed that the planets and stars borrowed their splendour from the sun, the primary source of celestial light in the cosmos. This view is advanced in a wide range of ancient and medieval writings,[28] but there is a passage in Restoro d'Arezzo's vernacular work, *La composizione del mondo colle sue cascioni* (c. 1282), which formulates the doctrine by using imagery that affords particularly close parallels with Virgil's "e come specchio l'uno a l'altro rende":

E le stelle, stando pulite e forbite, ragionevolmente dee essere lo corpo sodo; ricevendo la luce dal sole, ripercoterà questa luce l'una all'altra, come l'uno specchio all'altro.[29]

In a quite different context, Macrobius (fl. 400 A.D.) also provides an interesting example of multiple reflection in discussing the triadic emanation of the universe in his *Somnium Scipionis*. In Book I of this work, which Dante seems to have known

---

[28]See e.g. Plato, *Timaeus*, 39B (Waszink, p. 31); Chalcidius, *Comm. in Timaeus*, c. 117 (Waszink, pp. 161-162); Isidore, *Etymologiarum libri XX*, lib. III, c. 61, § 3 (*PL* 82, col. 177). One of the more detailed scholastic discussions is found in Albert, *De caelo*, lib. II, tr. 3, c. 6 (Hossfeld, pp. 153-155): "Et est digressio declarans, qualiter stellae omnes illuminantur a sole"; cf. idem, *De causis proprietatum elementorum*, lib. II, tr. 1, c. 1 (Hossfeld, p. 89); idem, *Metaphysica*, lib. XI, tr. 2, c. 26 (Geyer, II, p. 516). On celestial light, see Edward Grant, *Planets, Stars and Orbs: The Medieval Cosmos, 1200-1687* (Cambridge: Cambridge University Press, 1994), pp. 390-402.

[29]*La composizione del mondo colle sue cascioni*, lib. II, dist. 8, c. 18, ed. A. Morino (Florence: Accademia della Crusca, 1976), p. 227. On the power of mirrors to multiply light, see Albert, *De sensu et sensato*, tr. I, c. 10 (Jammy, V, p. 15a-b): "tersa spissa sive dura sint [sc. mirrors], sive non, reflectunt simul et divaricant lumen; et ideo in oppositum multiplicant ipsum [...] et politum est huius naturae propter convenientiam quam habet ad lumen, quod multiplicat lumen super ipsum: propter quod etiam tegitur color omnium huiusmodi tersorum, et videntur radiantia quando sunt in lumine" ("whether they are smooth and dense, or hard or not, they immediately reflect and project light; and thus they multiply it in the opposite direction [...] and what is polished is of this nature because of the propensity it has to light, which multiplies light onto it; on this account also the colour of all such smooth things is hidden and they are seen as radiant bodies when they are in light").

122

at firsthand,[30] Macrobius uses the mirror image to describe the way in which Soul and Mind derive from the One:

[...] cum ex summo deo mens, ex mente anima fit, anima vero et condat et vita compleat omnia quae sequuntur cunctaque hic unus fulgor illuminet et in universis appareat, ut in multis speculis per ordinem positis vultus unus [...][31]

Whatever the "source" of the image, the important point is that Dante is not simply transcribing a scientific or philosophic notion into his text. Rather, he borrows images and transfers them to quite different contexts to express his thought about charity in terms of light. As we shall see in Chapter 7, it is precisely this type of creative adaptation of familiar materials which underlies much of Dante's extraordinarily rich light imagery in the *Paradiso*.

Dante's only other simile based on the law of light reflection is found in the opening canto of the *Paradiso*, and it offers a similar fusion of scientific ideas with Christian themes to that found in Virgil's discourse in *Purgatorio* XV. Located in Earthly Paradise, the pilgrim has been purged of the last stains of sin and purified in water. He is about to be "transhumanised," to pass beyond the limits of human nature in this life so that he may fly with utmost velocity through the last sublunary

---

[30]For Dante's knowledge of Macrobius's work, see Georg Rabuse, "Macrobio," in *ED* III, pp. 757-759. On Macrobius' influence in the early Middle Ages, see Edouard Jeauneau, "Macrobe, source du platonisme chartrain," *Studi medievali* ser. 3/1 (1960), 3-24.

[31]Macrobius, *Commentarii in Somnium Scipionis*, lib. I, c. 14, § 15, ed. Mario Regali, 2 vols. (Pisa: Giardini, 1983-1990), I, p. 138: "since Mind emanates from the Supreme God and Soul from Mind, and Mind, indeed, forms and suffuses all below with life, and since this is the one splendor lighting up everything and visible in all, like a countenance reflected in many mirrors arranged in a row," trans. William Harris Stahl, in *Commentary on the Dream of Scipio*, 2nd edn (New York and London: Columbia University Press, 1966), p. 145. Tullio Gregory had earlier noted the similarity between this passage and Dante's mirror imagery in his *Anima Mundi*, n. 3, p. 91. The notion of multiple reflection was widely used by Neoplatonic writers to express the relationship between different grades in the hierarchy of being, see Herbert Grabes, *The Mutable Glass: Mirror-Imagery in Titles and Texts of the Middle Ages and the English Renaissance*, trans. Gordon Collier (Cambridge: Cambridge University Press, 1982), pp. 112-115; Gregory, *Anima Mundi*, pp. 91-92.

sphere, the sphere of fire, to the first celestial one, that of the moon. Within this context, the simile describes how Dante mirrors, in an act of almost immediate fidelity, Beatrice's action of gazing at the sun, just as a reflected ray is instantaneously generated upwards from an incident ray:

> quando Beatrice in sul sinistro fianco
> vidi rivolta e riguardar nel sole:
> aguglia sì non li s'affisse unquanco.
> E sì come secondo raggio suole
> uscir del primo e risalire in suso,
> pur come pelegrin che tornar vuole,
> così de l'atto suo, per li occhi infuso
> ne l'imagine mia, il mio si fece,
> e fissi li occhi al sole oltre nostr' uso. (46-54)

Unlike the simile in *Purgatorio* XV, there is no explicit mention of a reflective surface in the vehicle of this simile, and attention is thus immediately drawn to the ascendant reflected ray, the "secondo raggio" (l. 49). There is also a greater syntactic and overall balance to the simile: indeed, it is a remarkably fine example of Dante's art of poetic compression. In the first tercet (ll. 49-51), the one verb "suole" placed emphatically in rhyme suggests the presence of the law-governed, physical phenomenon, which, in the *Purgatorio*, Dante had developed across two tercets. There is no mention of the *casus lapidis*, nor of the angles subtended with it by a ray incident on a plane surface.

The power to compress meaning through suggestion and modification of the syntactic structure is evident throughout. Dante suppresses the article for the reflected ray (l. 49), the noun itself for the incident ray, referred to as "del primo" (l. 50); and he then allows the preposition "del" to carry the idea of generation. This generative action is further refined in the following line where the replication of the reflected ray and its movement upwards are conveyed by a pair of syntactically prominent verbs: the vigorous "uscir" and "risalir." The whole effect

124

is heightened by the strong *enjambement* between verses 49 and 50, and the system of sound repetition within the line: "us*cir* del *pri*mo e *risalire* in suso" (l. 50). Dante then introduces a clause which has no counterpart in the vehicle to explain how his spontaneous action in looking at the sun required him first to register Beatrice's sunward-gaze into his imaginative faculty, "l'imagine" (l. 53). While this interpolated reference detracts from the immediacy of the act, it nonetheless reconfirms Dante's keen interest in the psychological faculties involved in the visual process.

Very few commentators pause to examine such features, or even to consider the character of the light reflection described in the first tercet. They restrict themselves to rather generic comments about the law of reflection and provide a cross-reference to the earlier simile in *Purgatorio* XV.[32] On closer inspection, it seems clear that the reflection is perpendicular: there is no mention of an angle; and the reflected ray is generated upwards. Given the established view that the sun is positioned directly above Dante and Beatrice, as well as Dante's own unending search for variety in the scientific elements he blends into the "sacro poema," it is justifiable to see the similes in *Purgatorio* XV and *Paradiso* I as examples of oblique and perpendicular light reflection respectively.

In the lines which immediately follow this simile, Dante-*personaggio* turns to the sun and is able to withstand its intensity for a brief moment, before he is forced to mediate his vision by returning his gaze to Beatrice (ll. 64-66). It is significant that, throughout this sequence (49-66), Beatrice acts as a mediator for Dante's eyes.[33] The mirroring function of Beatrice's eyes recalls one of the

---

[32]For a particularly generic comment, see Casini, *La "Divina Commedia,"* III, p. 692: "La paragone è la stessa del *Purg.* XV 16-21."

[33]For Beatrice's eyes as mirrors which allow Dante to gaze indirectly on the divine, see esp. *Purg.* XXXI, 121-123; *Par.* XXVIII, 4-12. On the central importance of this emphasis on mirrors and "la funzione rivelante della Beatrice dantesca" at such moments, see Corti, *Percorsi dell'invenzione*, pp. 58-62; Francesco Mazzoni, "Il Canto XXXI del *Purgatorio*," in *Lectura Dantis Scaligera*, ed. Giovanni Getto (Florence: Le Monnier, 1965), esp. pp. 40-56.

principal applications of the mirror image in Western literature, namely, its use as a symbol for the indirect manifestation of the deity.[34] The simile's emphasis on reflection, mediated sight, and indirect vision can be related to the celebrated passage from I Corinthians 13 in which Paul states that, *in statu viatoris*, God is seen indirectly or *per speculum et in enigmate*.[35] In this way, the optical simile takes on a frame of reference beyond its immediate context and suggests that Dante-*personaggio*'s approach to God is an indirect, mediated ascent.[36] If this is the case, then this simile needs to be related closely to a narrative sequence involving the mirroring function of Beatrice's eyes that Dante places at the beginning of canto XXVIII, where the notion of seeing *per speculum* is especially prominent. In this passage, which is notable for its polypton (*vista-veder-vede*), Dante-*personaggio* first views God and the angelic hierarchies as a point of light surrounded by nine concentric circles which are seen as reflections in Beatrice's eyes, before he then turns around to gaze upon the *apparitio dei* without any mediation:

> come in lo specchio fiamma di doppiero
> vede colui che se n'alluma retro,
> prima che l'abbia in vista o in pensiero,
> e sé rivolge per veder se 'l vetro
> li dice il vero, e vede ch'el s'accorda
> con esso come nota con suo metro;
> così la mia memoria si ricorda

---

[34]One of the earliest examples of this kind of metaphorical application is found in Plato's *Phaedo*, 99D-100A, cited by Grabes, *The Mutable Image*, pp. 109-111.

[35]The celebrated passage from I Cor. 13:12 reads: "Videmus nunc per speculum in enigmate, tunc autem facie ad faciem: nunc cognosco ex parte, tunc autem cognoscam sicut et cognitus sum" ("We see now through a mirror in an obscure way, but then face to face. Now I know in part, but then I shall know even as I have been known"). On this highly influential mirror metaphor, and its classical antecedents, see Norbert Hugedé, *La métaphore du miroir dans les Épîtres de saint Paul aux Corinthiens* (Neuchâtel: Delachaux and Niestlé, 1957).

[36]The possibility of relating the optical simile to the Pauline text is strengthened by Dante's explicit reference to II Cor. 12:3 in *Par.* I, 73-75. See also di Scipio, *The Presence of Pauline Thought in the Works of Dante*, pp. 31, 49, 58, 263-264.

126

> ch'io feci riguardando ne' belli occhi
> onde a pigliarmi fece Amor la corda. (4-12)

There are several features common to both the optical simile from canto I and this later sequence. Both passages make use of imagery drawn from reflected light, apply this imagery to Dante's sight as he gazes upon some manifestation of the divine (the sun; a point of light), and emphasise Beatrice's role in mediating his vision. Both passages in short seem to use Dante's visual encounters to enact the oscillations between vision *per speculum* (in the mirrors of Beatrice's eyes) and a more direct apprehension of the divine.[37]

The primary emphasis of the optical simile in canto I is, of course, on the correspondence between two rays and two sun-directed looks, rather than on the wider frame of Pauline resonance. As such, the essence of this simile resides in the completely spontaneous and natural character of Dante's imitative act. It is for this reason that the comparison, and indeed much of the imagery of canto I as a whole, is couched in the language of scientific necessity: as a physical law of nature, under the required conditions and at the propitious moment, cannot fail to occur, so too Dante, like the reflected ray, cannot avoid fixing his gaze on the sun. Seeing becomes both the first stage in his "transhumanisation" and also an ascending movement. This is as it should be. After all, Dante is about to embark on the most truly natural course of action for a human being who is reconciled with his Maker: he is about to return to God. It is a mark of Dante's powers of synthesis, suggestion, and tonal variance that, contrary to Siro Chimenz's reading of the "pelegrin" of line 51 as a peregrine falcon, this term may also introduce the motif of the pilgrim's return.[38] To use an optical example in this context in order to

---

[37]For the view that Dante uses optics as a structuring device for his visions in *Paradiso* XXIII, see Robert Hollander, "The Invocations of the *Commedia*," in his *Studies in Dante* (Ravenna: Longo, 1980), pp. 31-38. See also Fujitani, "Dalla legge ottica alla poesia."

[38]S.A. Chimenz, "Per il testo e la chiosa della *Divina Commedia*," *GSLI* 133 (1956), pp. 180-185. Chimenz excludes the pilgrim reading as "un vago motivo sentimentale di discutibile

anticipate the theme of the soul's return might seem to be a poetic synthesis original to Dante. However, it is interesting to note that Saint Bonaventure, in his *Collationes in Hexaëmeron* (c. 1273), and in other treatises and sermons, frequently adopts similar examples to illustrate the way in which the light of grace initiates the soul's return:

Hae virtutes fluunt a luce aeterna in hemisphaerium nostrae mentis et reducunt animam in suam originem, sicut radius perpendicularis sive directus eadem via revertitur, qua incessit [...] Sapientes in perspectiva dicunt, quod si radius perpendiculariter cadat super corpus tersum et politum; necesse est, quod per eandem viam repercutiatur. Influxus gratiae est sicut radius perpendicularis; dico de gratia gratum faciente, quia gratia gratis data est sicut radius incidentiae; necesse est igitur, quod qui gratiam Dei vere suscipit, quod gloriam Deo reddat [...] Radius perpendicularis, quando cadit recte super obiectum, revertitur in suum originale principium; sic gratia, quae a Deo descendit in beatum Stephanum, ipsum reduxit in suum principium quia recognovit se habere gratiam a Deo.[39]

---

convenienza" (p. 185), without mentioning that the theme of *homo viator* (II Cor. 5:6; Hebr. 11:13) informs the *Comedy*, and Dante's *oeuvre* as a whole, and that pilgrim similes are used in *Par*. XXXI, 43-48, 103-108.

[39]Bonaventure, *Collationes in Hexaëmeron*, coll. VI, § 24 (Quaracchi, V, p. 363b); idem, *De septem donis spiritus sancti*, coll. I, § 9 (Quaracchi, V, p. 459a); idem, *De sancto stephano martyre*, ser. I (Quaracchi, IX, p. 480a): "These virtues [sc. cardinal ones] flow from the eternal light into the hemisphere of our minds and lead the soul back to its origin, as the perpendicular or direct ray returns back along the same path from which it came [...] Those who are well versed in perspective maintain that if a ray falls perpendicularly on a smooth and polished body it must be sent back along the same path. The influx of grace can be compared to the incident ray; I speak of sanctifying grace, for freely bestowed grace is like the incident ray; it thus must be that whoever receives God's grace returns the glory to God [...] The perpendicular ray when it falls directly on an object is returned to its point of origin; so it is with grace which descended from God into blessed Stephen, and returned him to his origin because he recognised that he had grace from God." For an interesting analogy between natural and spiritual reflection in another Bonaventurian sermon, see *Dominica tertia in Quadragesima*, ser. II (Quaracchi, IX, p. 228a-b).

128

My concern is not to suggest these texts as direct sources for Dante, even though the parallels are close and interesting. The important point is that these examples show that optical examples were widely used in theological works and were not simply restricted to thirteenth-century scientific writings, be they the works of the "perspectivists" or even those of commentators on Aristotle.[40] As my earlier discussion of mirrors in Restoro d'Arezzo, Macrobius, and St. Paul indicates, Dante would have found images associated with mirrors and reflected light in a wide range of writings, not only philosophical and scientific, but also scriptural and theological. To conclude the more strictly catoptrical material in the *Paradiso*, however, let us now consider the well-known mirror experiment in canto II.

Beatrice's Three-Mirror Experiment: *Paradiso* II, 97-105

The sources and significance of the three-mirror experiment described by Beatrice in *Paradiso* II have been discussed in some detail by Dante scholars. As is well known, the experiment is used by Beatrice to refute Dante's view that the dark and bright patches on the moon's surface are caused by variations in the density of the aether which makes up the planetary body.[41] By showing that light is not any less intense when seen at different distances, Beatrice uses the experiment to rule out the possibility that the moon is composed of material strata arranged at varying depths:

---

[40]On mirrors in the Middle Ages, see the studies by Hugudé and Grabes (nn. 31 and 35); and see further Ritamary Bradley, "Backgrounds of the Title *Speculum* in Mediaeval Literature," *Speculum* 29 (1954), pp. 100-115; Hans Leisegang, "La connaissance de Dieu au miroir de l'âme ou de la nature," *Revue d'histoire et de philosophie religieuses* 17 (1937), 145-171; Margot Schmidt, "Miroir," *DS* 10 (1974), pp. 1290-1303; Bruno Schweig, "Mirrors," *Antiquity* 15 (1941), 257-268.

[41]*Con.* II, xiii, 9: "non è [sc. the shadow] altro che raritade del suo corpo, a la quale non possono terminare li raggi del sole e ripercuotersi così come ne l'altre parti." This is the view shared by Averroës, Albert the Great, and Jean de Meun, see Vasoli, *Il Convivio*, p. 223. For Dante's revision of the cause of the "macchie lunari," see below Ch. 6, pp. 200-205.

129

Tre specchi prenderai; e i due rimovi
da te d'un modo, e l'altro, più rimosso,
tr'ambo li primi li occhi tuoi ritrovi.
  Rivolto ad essi, fa che dopo il dosso
ti stea un lume che i tre specchi accenda
e torni a te da tutti ripercosso.
  Ben che nel quanto tanto non si stenda
la vista più lontana, lì vedrai
come convien ch'igualmente risplenda. (97-105)

Perhaps the most direct and suggestive "source" for this experiment is found in the Pseudo-Aquinas' commentary on Book III of the *Meteorologica*. The anonymous author of this work uses three mirrors at different distances from one another to show that the area between the bows of a double rainbow is coloured:

[...] si accipiantur tria specula, in tali distantia ad se invicem et ad corpus obiectum secundum aliquam proportionem, qualis est distantia duarum iridum et partis intermediae ad se invicem et ad solem et ad visum, si fit refractio corporis obiecti a duobus speculis extremis ad visum, et apparitio idoli, tunc etiam fit a speculo intermedio, ut patet ad sensum.[42]

Despite the similarities in phraseology, there are nonetheless obvious differences: the light source is not placed behind the observer's back; and no reference is made to the way in which image brightness remains constant at different distances.

---

[42]Pseudo-Aquinas, *In De meteorologicorum*, lib. III, lect. 6, § 296 (Spiazzi, p. 634): "If three mirrors are set up in such a way that the distances between each of them and a facing object are the same as the distances between, on the one hand, both of the rainbows and the area separating the bows, and, on the other, the sun and the spectator, then if the facing object is reflected in the two most distant mirrors, the image of this object is also reflected in the middle mirror, as is evident to the senses." For this experiment as a source for *Par.* II, 97-105, see Giovanni Busnelli, *Cosmogonia e antropogenesi secondo Dante Alighieri e le sue fonti* (Rome, 1922), p. 87; Heinrich Gmelin, *Die göttliche Komodie*, 3 vols. (Stuttgart: Klett, 1954-1957), III, p. 271. For the suggestion that Jean de Meun's *Roman de la Rose* (ll. 16833-16880) is a possible source for Dante's experiment, see also André Pézard, in *Dans le sillage de Dante* (Paris: Société d'études italiennes, 1975), pp. 470-473.

130

Interestingly, Chalcidius' commentary on the *Timaeus* contains a discussion of a mirror-experiment in which one mirror is placed behind the observer:

Etenim si duo specula sic collocentur, ut unum sit ante vultum, alterum a tergo, obliquatum, ne impediatur obiectu corporis visus radii commeantis, tunc occipitium nostrum videmus in speculo quod habetur a tergo.[43]

But the parallels with Beatrice's mirrors are still far from complete: there are only two mirrors, and light is provided by eye rays not any external sources.

It is not surprising that some parallels between Dante and "perspectivist" writings have been pointed out, and as early as the 1880s Ernest Plumptre indicated an experiment with mirrors and candles in Roger Bacon's *Opus tertium*. Unfortunately, he did not observe that Bacon had derived his details from the Pseudo-Aristotelian work, *De causis proprietatibus elementorum*.[44] Albert's commentary on this treatise also contains a similar experiment which uses candles and mirrors to show how the moon receives light from the sun. Given Dante's indebtedness to this Albertine work, it is well worth noting that a common principle underlies Beatrice's "esperïenza" and Albert's experiment: lamps and a mirror are used to model the action of the sun's light on the moon:

---

[43]*Comm. in Timaeus*, c. 241 (Waszink, p. 253): "If two mirrors are arranged in such a way that one is in front of the spectator and the other behind him, at an oblique angle, then unless the rays which issue from sight are impeded by an object, we see the back of our head in the mirror behind us." Chalcidius uses the experiment to support the validity of extramitted visual rays.

[44]See Bacon, *Opus tertium*, c. 37 (Brewer, p. 118): "dicunt omnes quod lux, quae a luna venit, sit lux solis reflexa a superficie lunae, propter hoc quod dicitur in libro De Proprietatibus Elementorum, quod sol est sicut candela et luna sicut speculum" ("everyone says that the light which comes from the moon is the light of the sun reflected from its surface on account of what is said in *De proprietatibus elementorum*, namely, that the sun is like a candle and the moon like a mirror"), cited by E.H. Plumptre, "Two Studies in Dante I: Dante and Roger Bacon," *The Contemporary Review* 40 (December, 1881), pp. 850-852. A similar example was later reproduced by Guidubaldi, see "Il Canto II del *Paradiso*," *Nuove Letture Dantesche*, ed. Enzo Esposito, 6 vols. (Florence: Le Monnier, 1972), V, esp. pp. 288-292; idem, *Dante Europeo*, II, pp. 350-351; III, pp. 312-313.

[...] sicut enim candela speculum illuminat per substantiam luminis, quod est in ea, et hoc recipit speculum per substantiam et formam perspicui terminati, quod est in eo, et est alia substantia candelae et speculi, sic oportet esse substantiam solis illuminantis et lunae et stellarum, quae illuminantur.[45]

It was such apparently "experimental" details in the *Comedy* that led a generation of critics to regard Dante as a precursor of scientific discoveries. This type of anachronistic approach to Dante's science is now generally discredited, but the popular image of Dante as a pioneer of medieval scientific inquiry still flourishes.[46] It is, then, salutary to note that the evidence presented so far in Part One suggests a rather different view: Dante was not fully aware of the latest developments in optical science.

Reflection, Refraction, and the Rainbow

The optical principles used by Aristotle in several sections of his *Meteorologica* provided an important stimulus to later writers and helped to establish a special tradition of meteorological optics. In chapters ii-iv of Book III of the

[45] Albert, *De causis proprietatum elementorum*, lib. II, tr. 1, c. 1 (Hossfeld, p. 90): "As a candle illuminates a mirror due to the luminous substance within it and the mirror receives this because it has the form and substance of a bounded transparent body (the substance of the candle and that of the mirror not being the same) it must be likewise with the substance of the sun which illuminates and that of the moon and the stars which are illuminated."

[46] For a particulary striking demonstration of precursoritis in the context of Dante's visual theory, see Gino Ricchi, "Il meccanismo della visione secondo Dante Alighieri," *Giornale dantesco* 10 (1902), 177-179. For other more recent examples, see Mattioli, *Dante e la medicina*, p. 133; Francesco Pannaria, "Dante e la scienza," *Nuova Antologia* 519 (1973), 247-261, esp. p. 255: "il genio divinatorio di Dante in campo scientifico"; Mark Peterson, "Dante's Physics," in *The Divine Comedy and the Encyclopedia of the Arts*, ed. di Scipio and Scaglione, pp. 163-180. On the absurdities of this kind of approach, see Bortolo Martinelli, "Poesia e scienza in Dante," *Critica letteraria* 9 (1981), pp. 651-652. On the dangers of reading modern concepts into ancient contexts, see further Desmond Lee, "Science, Philosophy, and Technology in the Greco-Roman World," *Greece and Rome* ser. 2/19-20 (1972-1973), 65-78, 180-193; David C. Lindberg, in *Science in the Middle Ages*, ed. Lindberg, preface, p. viii; idem, *The Beginnings of Western Science*, passim.

132

Meteorologica, he discussed the moon-halo and the rainbow, explaining the conditions in which these optical arrays occur, establishing the causes of their colours, and outlining the geometry of their formation. According to Aristotle, both phenomena were caused by the reflection of light in a dense, wet atmosphere. In the thirteenth century, the tradition of meteorological optics was developed and refined by several writers. In the 1230s, Grosseteste made a significant innovation to Aristotelian rainbow theory, by introducing the principle of refraction (the bending of light as it passes through media of differing densities) into his pithy treatise on the bow. Later in the century, Pecham and Witelo both employed refraction in their expositions of the rainbow; and at the beginning of the fourteenth century, Theodoric of Freiburg made the most significant step towards a rational explanation of the bow since antiquity, by positing a series of reflections and refractions within individual raindrops.[47]

It is significant that Dante remains faithful to Aristotle by only ever mentioning reflected light when referring to the rainbow, since this presents a marked contrast with the emphasis placed on refraction by thirteenth-century optical specialists.[48] Further and compelling evidence that Dante did indeed have a

---

[47]*Circa* 1235 Grosseteste had developed a refraction theory of the rainbow in his *De iride* (Baur, pp. 75-78). For Pecham's and Witelo's combined refraction-reflection theories, see *Perspectiva communis,* lib. III, prop. 18 (Lindberg, p. 232) and *Perspectiva,* lib. X, props. 65-79 (Risner, pp. 457-471) respectively. For detailed accounts of rainbow theory in these writers and Theodoric of Freiberg, see Carl B. Boyer, *The Rainbow: From Myth to Mathematics,* with new colour illustrations and commentary by Robert Greenler (Princeton, NJ: Princeton University Press, 1987), pp. 88-97; idem, "The Theory of the Rainbow: Medieval Triumph and Failure," *Isis* 49 (1958), 378-390; Crombie, *Robert Grosseteste,* pp. 124-127, 167, 226-232; Samuel Devons, "Optics through the Eyes of Medieval Churchmen," in *Science and Technology in Medieval Society,* ed. Pamela O. Long (New York: New York Academy of Sciences, 1985), pp. 205-224. The principal discussions of meteorological phenomena in the *perspectivae* are: Bacon, *Opus maius,* lib. VI, cc. 2-12 (Bridges, II, pp. 172-202); Pecham, *Perspectiva communis,* lib. III, props. 18-22 (Lindberg, pp. 232-236); idem, *Tractatus de perspectiva,* c. 12 (pp. 64-68); Witelo, *Perspectiva,* lib. X, props. 65-84 (Risner, pp. 457-474).

[48]The key text is *Purg.* XXV, 92: "per l'altrui raggio che 'n sé si reflette." Contrast Witelo, *Perspectiva,* lib. X, prop. 65 (Risner, pp. 457-458): "Quod vero iris per refractionem [here the term denotes a refraction] etiam radiorum corporis luminosi fiat, patet per hoc quia non generatur iris, nisi in aliqua diaphana materia existente in medio, et prohibente transitum luminis" ("That

133

limited knowledge of refraction is shown by his use of the verb "refrangere" in the *Comedy*. This verb and its cognates would seem to denote refraction, but close scrutiny of the contexts in which Dante employs it reveals that he only ever uses this word of *reflected* light.[49] Dante was not alone in conflating the terms for reflection and refraction in this way – Albert, Aquinas, and many other medieval writers did likewise.[50] By contrast, however, the "perspectivists" did distinguish between the two phenomena, using *refractio* and *refrangere* for refraction and *reflectio* and *reflectere* for reflection. Dante clearly did not make this discrimination,[51] and he seems to have had little theoretical understanding of an optical phenomenon which occupied a full third of the *perspectivae*.[52] And yet, despite his lack of technical knowledge, in both his prose and poetry Dante

---

the formation of the rainbow also takes place through the refraction of rays from a luminous body is clear from the fact that the rainbow is only generated in some diaphanous matter in the medium which impedes the passage of light").

[49]*Purg.* XV, 16-17, 22: "Come quando da l'acqua o da lo specchio / salta lo raggio a l'opposita parte /[...]/ così mi parve da *luce rifratta*"; *Par.* II, 93: "per esser lì *refratto* [sc. lo raggio] più a retro"; XIX, 4-6: "[...] rubinetto in cui / raggio di sole ardesse sì acceso, / che ne' miei occhi *rifrangesse* lui" (italics mine).

[50]See Aquinas, *In De sensu*, lib. un., lect. 4, § 48 (Spiazzi, p. 17): "[...] ex refractione sive reverberatione formae ad corpus politum" ("due to the reflection or rebounding of forms from a polished body"); idem, *In De meteorologicum*, lib. I, lect. 4, § 26 (Spiazzi, p. 405): "nubes visae per refractionem in aque tamquam in quodam speculis nigriores videntur quam visae in seipsis" ("clouds which are seen by reflection in water as in some mirror are seen to be darker than those that are seen in themselves"); Pseudo-Aquinas, *In De meteorologicum*, lib. III, lect. 6, § 285 (Spiazzi, p. 629): "[...] radios refractos, idest reverberatos" ("reflected, that is, reverberated rays"). On Albert's indiscriminate use of *reflexio* and *refractio* in his *De meteoris*, see Boyer, *The Rainbow*, pp. 95-96. Bacon attributes such indiscriminacy to the "vitio translationis" in his *De multiplicatione specierum*, pars II, c. 2 (Lindberg, p. 98).

[51]See the example from Witelo given in n. 48 and Bacon, *De multiplicatione specierum*, pars II, c. 5 (Lindberg, pp. 130-131): "Sunt etiam aliqua corpora mediocris densitas a quibus fit reflexio simul et fractio, ut est aqua" ("There are also certain bodies of moderate density, such as water, in which reflection and refraction occur simultaneously," trans. Lindberg). On the study of refraction in the "perspectivists," see Crombie, *Robert Grosseteste*, pp. 120-124, 219-225; David C. Lindberg, "The Cause of Refraction in Medieval Optics," *British Journal for the History of Science* 4 (1968), 23-38; A.I. Sabra, *Theories of Light: from Descartes to Newton*, new edn (Cambridge: Cambridge University Press, 1981), pp. 93-98.

[52]Note also that Dante's treatment of vision in *Convivio*, III, ix provides no evidence that he was familiar with the refraction doctrine which the "perspectivists" used to explain image transmission within the eye (see above Ch. 2, nn. 21 and 63).

134

repeatedly describes the effects produced by light rays as they pass through glass, transparent media, and atmospheric vapours.[53] The "perspectivists" provide many examples of light being absorbed, concentrated, and coloured,[54] but it is possible to adduce similar details from more general medieval sources, and especially from the medieval Aristotelian commentaries, to which I have referred in previous chapters.[55]

In the *Comedy* Dante frequently deals with the observational difficulties associated with the effects of atmospheric refraction and several of the relevant passages have already been examined in Chapter 3. Dante's recurrent fascination with the optical effects caused by vapours in blurring the outlines of distant objects appears to have no precedent in medieval art and therefore seems to provide an

---

[53]On the optical effects caused by vapours, see Ch. 3, n. 22. For an extended comparison between opacity/transparency and light absorption/transmission, see *Con.* III, vii, 3-5.

[54]For the passage of rays of light through water-filled glass bottles, see Bacon, *De multiplicatione specierum*, pars II, c. 3 (Lindberg, p. 116); idem, *Opus maius*, lib. IV, dist. 2, c. 2 (Bridges, I, pp. 113-114); idem, *Opus tertium*, c. 32 (Brewer, p. 111); Grosseteste, *De natura locorum* (Baur, p. 71). On coloration of light rays, see Bacon, *De multiplicatione specierum*, pars I, c. 1 (Lindberg, p. 11); Pecham, *Perspectiva communis*, lib. I, prop. 14 (Lindberg, p. 88).

[55]The concentration of light by water and mirrors is mentioned in many Albertine works, see e.g. *De caelo et mundo*, lib. II, tr. 3, c. 1 and esp. c. 3 (Hossfeld, pp. 143, 147): "Reflexio autem est causa caloris, eo quod in reflexione multi radii diriguntur ad punctum unum [...] sicut apparet in beryllo vel crystallo vel forte vitro bene rotundo et impleto aqua frigida, quae opposita soli fortissima illuminatione illuminant unum locum post ipsa, et ad illum fit reflexio radiorum" ("reflection is the cause of heat on account of the fact that in a reflection many rays are directed to one point [...] as is seen in the beryl stone or crystal or perhaps a well-rounded glass vessel filled with cold water, all of which when placed opposite the strong illumination of the sun illuminate one place beyond and in that place a reflection of light rays takes place"); idem, *Quaestiones de animalibus*, lib. IV, q. 3 (Filthaut, p. 207): "Quare per crystallum radiis solis oppositum carbo extinctus accenditur" ("Why a crystal when opposite the rays of the sun lights a charcoal"). On the coloration of light, see Albert, *De anima*, lib. II, tr. 3, c. 7 (Stroick, p. 109): "[...] quando radius transit per vitrum coloratum quocumque colore, ille color fit in aëre et in pariete opposito" ("when a ray passes through a glass coloured with some colour, that colour is produced in the air and the wall opposite"); idem, *De meteoris*, lib. III, tr. 4, c. 12 (Jammy, II, p. 129a); Pseudo-Aquinas, *In De meteorologicum*, lib. III, lect. 6, § 292 (Spiazzi, p. 632). For similar comparisons in theological writings, see Anselm, *De veritate*, c. 6 (*PL* 158, col. 473); Bonaventure, *Collationes in Hexaëmeron*, coll. XII, § 14 (Quaracchi, V, 386b); idem, *De sancto marco evangelium*, ser. 1 (Quaracchi, IX, p. 519a); William of Auvergne, *De fide*, in *Opera omnia*, I, p. 6a. See also Meun, *Roman de la Rose*, ll. 18137-18142 (Lecoy, III, pp. 44-45). Cf. *Con.* III, ix, 10.

early literary anticipation of Leonardo da Vinci's "prospettiva aerea."[56] If there is the echo of a contemporary scientific text in the brilliantly worked out details of a scene like the opening of *Purgatorio* II, it is to be found not in an optical work but in a passage such as the one from Book I of Albert's *De meteoris*, which Dante cites when he describes a similar effect in the *Convivio*:

esso [sc. Mars] pare affocato di colore, quando più e quando meno, secondo la spessezza e raritade de li vapori che 'l seguono: li quali per loro medesimi molte volte s'accendono, sì come nel primo de la Metaura è diterminato.[57]

Aristotelian commentaries provide other examples of coloured effects caused by the presence of light in the atmosphere, and there is also a description of optical reddening in Aristotle's *De sensu*; the Thomist commentary *ad locum* reads as follows:

Sol enim secundum se videtur albus propter luminis claritatem; sed quando videtur a nobis mediante caligine sive fumo resoluto a corporibus, fit tunc puniceus, idest rubicundus. Et sic patet quod id quod secundum se est unius coloris, quando videtur per alium colorem, facit apparentiam tertii coloris.[58]

[56]For Leonardo's own scientific interest in the atmosphere and its effects, see *Il Paragone*, ch. 38, in Claire J. Farago, *Leonardo da Vinci's Paragone: A Critical Interpretation with a New Edition of the Text in the Codex Urbinas* (Leiden and New York: E.J. Brill, 1992), p. 266. See also Moshe Barasch, *Light and Colour in the Italian Renaissance Theory of Art* (New York: New York University Press, 1978), pp. 46-47, 50-52.

[57]*Con.* II, xiii, 21. In Book I of his *De meteoris*, Albert repeatedly describes the optical effects caused by vapours, see e.g. lib. I, tr. 4, c. 9 (Jammy, II, p. 25b), quoted by Vasoli, *Il Convivio*, p. 235. But see also lib. I, tr. 3, c. 2 (Jammy, II, p. 15b): "mutatur stella in colore suo secundum mutationem illius vaporis, quia in vapore puro videbitur coma alba, et in humido appareat rubea" ("the colour of a star is changed following alteration in its vapour, for, in a pure vapour its white head will be seen and in a wet vapour it appears ruddy"). For further descriptions of the optical effect of reddening in Dante, see *Purg.* XXX, 22-24; *Ep.* V, 2. For additional examples of atmospheric refraction, see *Purg.* I, 116-117; *Mon.* II, i, 5.

[58]Aquinas, *In De sensu*, lib. un., lect. 7, § 103 (Spiazzi, p. 33): "The sun in itself is seen as white because of the clarity of light, but when it is seen by us through a vapour or smoke given off by

136

This kind of Aristotelian text is all that is required to provide a scientific context for Dante's own references to refracted phenomena. As far as his presentations, in the *Comedy*, of the rainbow and the moon-halo are concerned, it is relatively easy to show that these were directly influenced by Aristotle's *Meteorologica*, a work which Dante knew well, because he cites it in the *Convivio* and the *Questio*, and draws widely upon its doctrines in his poetry.[59] The main technical details that can be extracted from his rainbow and moon-halo comparisons in the *Comedy* reveal four points:

1. he understands the rainbow to be the result of a *reflection*;

2. he believes that this reflection occurs in a moist atmosphere, either a rain-droplet laden sky, or a tenuous cloud, or a dense vapour;

3. the reflected light ray becomes coloured as it enters the dense and humid atmosphere of the cloud;

4. a further reflection can take place to produce a secondary bow.[60]

---

bodies, then it becomes purple, that is red. And so it is clear that when an object which by itself is of one colour is seen through another colour then this object takes on the appearance of a third colour"; cf. Aristotle, *De sensu*, 3, 440a 11-13: "quemadmodum sol secundum se quidem albus videtur, per caliginem vero et fumum puniceus," Graeco-Latin trans. in Spiazzi, p. 31: ("in a similar way the sun by itself is seen as white but through an obscuring mist or smoke it is seen as red"). For atmospheric refraction producing crimson and purple colours, see also Aristotle, *Meteorologica*, I, 5, 342b 6-8; III, 4, 373a 3-4. See further Averroës, *De sensu*, 192b (Shields, p. 14): "diaphanum autem innatum est recipere lucem et perfici ab illa [...] Et hoc manifestum est ex coloribus diversis factis quando lux solis adunantur cum nubibus" ("by its very nature the transparent medium is apt to receive light and be perfected by that [...] and this is evident from the different colours that are made when the light of the sun is concentrated in clouds").

[59]See *Con.* II, xiv, 6; *Questio*, 14 and 83. On Dante's meteorology, see Boyde, *Dante Philomythes*, pp. 74-95, and the notes with references to the Pseudo-Thomist commentary on pp. 317-325.

[60]References in order: (i) reflection, *Purg.* XXV, 92: "per l'altrui raggio che 'n sé si reflette"; (ii) moist atmosphere, *Purg.* XXV, 91: "E come l'aere, quand'è ben pïorno"; *Par.* X, 68: "[...] quando l'aere è pregno"; *Par.* XII, 10: "[...] tenera nube"; *Par.* XXVIII, 24: "quando 'l vapor che 'l porta più è spesso"; (iii) coloured light ray, *Purg.* XXV, 93: "di diversi color diventa addorno"; *Par.* XII, 11: "due archi paralelli e concolori"; (iv) secondary bow as reflected from the primary one, *Par.* XII, 11-13: "due archi paralelli e concolori / [...] / nascendo di quel d'entro quel di fori"; *Par.* XXXIII, 118-119: "e l'un da l'altro come iri da iri / parea reflesso [...]."

It should be noted that all these details, some of which had been condensed into encyclopaedic works,[61] are present in the medieval Latin translations of Aristotle's *Meteorologica*.[62]

Of course, it is essential not only to establish whether Dante was dependent upon one category of medieval optical writing or another, but also to assess the ways in which his optical learning serves his poetry. For this reason, the analysis of his meteorological-optical images that now follows has a two-fold aim: to provide both general and specific references to the Aristotelian framework; and, more significantly, to explore how Dante assimilates, recombines, and transforms these optical ideas for his own purposes and at varying levels of complexity. As in previous chapters, then, consideration will also be given to poetic context, layers of symbolic resonance, intertextual echoes, narrative practices, and stylistic factors.

Dante's Meteorological Optics in the *Comedy*

The first simile in which Dante draws on meteorological optics is found in *Purgatorio* XXV and provides one of the best examples of how he incorporates

[61]For discussions of related phenomena in the encyclopaedic tradition from late antiquity to the tenth century, see Seneca, *Naturales quaestiones*, lib. I, cc. ii-ix, (Corcoran, I, pp. 22-64); Pliny, *Historia naturalis*, lib. II, c. 60 (Rackham, I, pp. 286-287); Isidore, *Etymologiarum libri XX*, lib. XIII, c. 10, § 1 (*PL* 82, col. 744); Bede, *De natura rerum*, c. 31 (*PL* 90, col. 252); Pseudo-Bede, *De mundi constitutione* (*PL* 90, col. 888); Maur, *De universo*, lib. IX, c. 20 (*PL* 111, cols 277-278); Honoré d'Autun, *De imagine mundi*, lib. I, c. 58 (*PL* 172, col. 137). For thirteenth-century encyclopaedists, see Bartholomew the Englishman, *De rerum proprietatibus*, lib. VIII, c. 43 (Richter, p. 433); Thomas of Cantimpré, *De natura rerum*, lib. XVIII, c. 12 (Boese, pp. 402-403); Vincent of Beauvais, *Speculum maius*, lib. II, c. 70; lib. IV, c. 74 (I, cols 125, 278).

[62]References from Book III of Aristotle's *Meteorologica*: (i) rainbow as reflection, 2, 372a 17-21; 4, 373a 33-35; 4, 373b 33; (ii) moist, uniform atmosphere, or vapour, 3, 372b 15-20; (iii) double rainbow, 2, 372a 1-5; (iv) secondary bow as a reflection of the first with inverted colours, 4, 375a 30-375b 15; (iv) coloration of the light ray, 4, 374b 10-35. For a detailed analysis of Aristotle's rainbow theory, see Boyer, *The Rainbow*, pp. 38-55; Aydin M. Sayili, "The Aristotelian Explanation of the Rainbow," *Isis* 30 (1939), 65-83. The first Latin translation of Books I-III was made by Gerard of Cremona from the Arabic before 1187. In the 1220s, Michael Scot made a full translation from the Arabic; and *circa* 1250-1260 William of Moerbeke translated Books I-IV from the Greek.

138

images drawn from Aristotelian literature into his closely reasoned doctrinal poetry.[63] In his important discourse on embryology and the human soul, Statius invokes the comparison of a light ray becoming coloured as it passes through a dense and wet atmosphere in order to illustrate how the aerial bodies of souls are produced by the irradiating effect of the soul's own formative "virtue." Just as the light of the sun takes on a new form (colour) when its rays enter a dense atmosphere, so too this "virtù formativa" (l. 89) shines out and makes the circumambient air into a new body:

> Tosto che loco lì la circunscrive,
> la virtù formativa raggia intorno
> così e quanto ne le membra vive.
>   E come l'aere, quand' è ben pïorno,
> per l'altrui raggio che 'n sé si reflette,
> di diversi color diventa addorno;
>   così l'aere vicin quivi si mette
> e in quella forma ch'è in lui suggella
> virtüalmente l'alma che ristette. (88-96)

The example is an apposite one. Although the light ray does not strictly confer organs on the air ("organa": l. 101) in the same way as the soul's virtue does, there is nonetheless a precise range of correspondences. Both the natural phenomenon and its supernatural counterpart involve a series of relations between a lucent agent (the ray and the "virtù" of the separated soul: ll. 92 and 96), a tenuous but quasi-material structuring effect (colours and the new form of the aerial body: ll. 93 and 95), and a common material substrate (the surrounding air: ll. 91 and 94).[64]

---

[63]Dante's first, but least interesting, reference to the rainbow is the classicising periphrasis in *Purg.* XXI, 50-51: "[...] figlia di Taumante, / che di là cangia sovente contrade." On the context of this passage and its stylistic features, see Boyde, *Dante Philomythes*, p. 85 and n. 29. Because the rainbow always appears opposite the sun, it frequently changes its position ("cangia sovente contrade") in relation to the observer, cf. Aristotle, *Meteorologica*, III, 4, 373b 34-35.

[64]For convincing evidence that the "'n sé" of line 92 is a complement of place and refers to the air, rather than to the ray reflecting upon itself, see Franca Brambilla Ageno, "Interpretazione e punteggiatura in passi danteschi," *Studi danteschi* 52 (1979-1980), pp. 179-182.

139

In fact, Dante shows a precise understanding of the theoretical principles which underlie ancient and medieval colour theory in this technically accurate simile. According to Aristotle, light was essentially white but its colour underwent changes as it passed through dark media. The appearance of colours in a rainbow was thus considered to be the result of light rays being reflected at varying strengths from the darkness of dense and wet clouds.[65] In lines 92-93, Dante adheres to the essentials of this Aristotelian doctrine (one which is discussed in greater detail by his medieval commentators)[66] by describing how the presence of a bright ray of light in the darker dense air causes this ray to become coloured.

Canto XXV of the *Purgatorio* is resonant with many such vivid and direct examples – it is characteristic of Dante's most intellectually inspired discourses that he supplements (and even constructs) his arguments with a densely figurative language, rich in simile and metaphor. Statius constantly exemplifies his discussion of the soul's embryological development, which is itself dense in metaphor (e.g. the

---

[65]*Meteorologica*, III, 4, 374a 3-5; 374b 10-15; see also *Timaeus*, 67D-E. On Aristotle's colour theory, see Paul Kucharski, "Sur la théorie des couleurs et des saveurs dans le "De Sensu" aristotélicien," *Revue des études grecques* 67 (1954), 355-390. For the role of white in Dante's conception of colour formation, see *Dve*, I, xvi, 2: "[...] in coloribus omnes albo mensurantur – nam visibles magis et minus dicuntur secundum quod accedunt vel recedunt ab albo" ("as for colours, all are measured in relation to white – for they are said to be more or less visible according to whether they are nearer to or farther from white"). On colour as a scale from white to black, see *Con.* IV, xx, 2; *Inf.* XXV, 65-66.

[66]Albert, *De meteoris*, lib. III, tr. 4, c. 9 (Jammy, II, 127): "in veritate colores iridis non sunt aliud nisi radii solis humore vaporis vel nubes diversimodi tincti" ("in truth the colours of the rainbow are nothing other than the rays of the sun coloured in different ways by the wet of vapour or clouds"); Aquinas, *In De meteorologicum*, lib. III, lectio 6, § 292 (Spiazzi, p. 632): "color causatur ex praesentia luminis in perspicuo terminato per opacum, et secundum diversam proportionem luminis ad opacum in perspicuo diversificantur colores, quia ex multo lumine et pauco opaco causatur color albus, et e converso color niger" ("colour is caused by the presence of light in a transparent medium which is terminated by an opaque body, and colours are made different according to the different proportion of light to opaque in the medium, for, a white colour is caused by much light and little opaque and a black colour in the opposite conditions"). A background to medieval ideas on colour will be found in Samuel Y. Edgerton, Jr., "Alberti's Colour Theory: A Medieval Bottle without Renaissance Wine," *Journal of the Warburg and Courtauld Institutes* 22 (1969), 109-134. More generally on colour in Dante, see Antonio Lepschy, "Osservazioni sul vocabolario cromatico della *Commedia* di Dante," *Atti dell'istituto veneto di scienze ed arti: Classe di scienze fisiche, matematiche e naturali* 152 (1993-1994), 1-14.

140

commonplace food and drink metaphors of lines 37-39), with visually-satisfying examples. The vegetative soul is, as it develops, first plant- and then sponge-like (ll. 53 and 56); the divine "inspiration" of the rational soul is compared to the heat of the sun making wine (ll. 77-78); and the soul's new body, once formed, pursues the animating spirit responsible for its composition as flames follow a fire (ll. 97-99).

This rainbow simile, then, functions as a clarifying figure in the sense that ideas expressed by Dante in the fine distinctions and abstract terms of scholastic language are contemporaneously illustrated and corroborated by limpid visual images.[67] The simile also refines, extends, and develops a less sophisticated optical analogy that Virgil had suggested earlier in the canto. In lines 25-26, Virgil compared the seemingly material effects suffered by the immaterial bodies of the souls to the way in which a mirror can reproduce (by reflection) even the slightest bodily movement.[68] What is more, the formal elements of the simile provide an excellent illustration of Dante's sense of rhythmical variation. The relative slowness of the first line with its strong caesura and the dieresis of "pïorno" quickens into the alliterative effects of the "div"s in line 91. The sensuous transposed adjective "adorno" (referred to "raggio": l. 90) is effectively delayed and brought into emphasis by its position in rhyme. For technical accuracy, correspondence to the subject illustrated, and metrical features, the rainbow simile of *Purgatorio* XXV is one of the finest examples of a comparative device in a technical discourse.

Dante's last meteorological-optical comparison in the *Purgatorio* is used to describe the appearance of the candles in the first procession of Earthly Paradise.

---

[67]The following terms in Statius' discourse have been rendered directly from scholastic Latin: "virtute informativa" (l. 41); "disposto a patire" (l. 47); "lo motor primo" (l. 70); "forma" (l. 95); and "virtüalmente" (l. 96).

[68]This is not the first time that Virgil has used comparisons drawn from mirrors and the behaviour of light, see *Inf.* XXIII, 25-27; *Purg.* XV, 67-75.

As we have seen in Chapter 3, Dante-*personaggio*'s gradual perception of the light of these candles dominates the first half of canto XXIX during which his own visual processes are used to structure and lend coherence to the narrative. After having clarified and corrected the optical illusions caused by his distance from the candlesticks, he inspects himself in the mirror-like waters of the river Lethe (ll. 67-69).[69] His eyes then focus on the flames of the candles and the pilgrim is party to another optical distortion: the flames appear to leave coloured trails behind them:

> e vidi le fiammelle andar davante,
> lasciando dietro a sé l'aere dipinto,
> e di tratti pennelli avean sembiante;
> sì che lì sopra rimanea distinto
> di sette liste, tutte in quei colori
> onde fa l'arco il Sole e Delia il cinto. (73-78)

The periphrastic comparisons to the colours of the rainbow's arc and the moon-halo's circle (l. 78) offer a rich chromatic backcloth to the scene of the procession as Dante views it from the banks of Lethe. The notion that flames give off a coloured rainbow-like effect in the atmosphere might seem to be an original transposition of meteorological optics to another subject area (fire). Yet it is noteworthy that Aristotle gives the example of flaming lanterns appearing coloured due to atmospheric distortion in his *Meteorologica*. According to Aristotle, colour

---

[69]The notion of the mirror as a means of moral self-examination (cf. Jam. 1:18-19) was extremely commonplace, see Bradley, "Backgrounds of the Title *Speculum*," pp. 100-115. Dante's own action seems to suggest a similar degree of self-scrutiny. Cf. Pseudo-Bernard, *Tractatus de interiori domo*, c. 11, § 18 (*PL* 184, col. 517): "Non immerito conscientiam speculo comparavimus, quoniam in ea tanquam in speculo rationis oculus, tam quod decens, quam quod indecens est in se, claru aspectu apprehendere potest" ("Not without cause did we compare conscience to a mirror, for in it, as in a mirror, the eye of reason is able to apprehend with clear sight both what is decent and what is indecent in it").

142

is produced when the white light of the lamp mixes with the dark sooty smoke of the flames.[70]

Dante's images are, however, also directly motivated by a scriptural source, the Book of Revelations. Descriptive detail from Revelations (as well as from Ezekiel) guides much of the imagery in this canto, including the light and colour effects. As is well known, several passages in these books present the rainbow as a symbol of divine power. In Chapter 4 of Revelations, a rainbow simile precedes, as it does in Dante, the appearance of twenty-four elders clothed in white raiment.[71] It would be representative both of Dante's learning and his capacity for multiple allusion if his comparison was initially drawn from Aristotle, and then recombined and condensed with scriptural details and mythological reminiscences.

Some commentators have suggested that Dante shows considerable independence from the science of his time by referring to "sette liste, tutte in quei colori" (l. 76), i.e. each one of the candlesticks has its own colour. To the modern reader, this observation might not appear extraordinary, but the notion of attributing seven colours to the rainbow was unprecedented both in Aristotle and in thirteenth-century natural philosophy: at the very most, the rainbow was

---

[70]*Meteorologica*, III, 4, 374a 17-20: "Apparet utique iris tota, sicut quae circa lucernas," Graeco-Latin trans. in *In De meteorologicorum*, ed. Spiazzi, pp. 624-625: ("A full rainbow appears as that [which is seen] next to lanterns").

[71]References in order: (i) candlesticks, Rev. 1:13: "et in medio septem candelabrorum similem Filio hominis" ("and in the middle of the seven candlesticks similar to the Son of Man"); (ii) visions of rainbows and appearance of elders, esp. Rev. 4:3-4: "et iris erat in circuitu sedis similis visionis zmargdinae [...] et super thronos vigintiquatuor seniores sedentes, circumamicti vestimentis albis et in capitibus eorum coronae aureae" ("and there was a rainbow around the seat similar to the sight of an emerald [...] and upon the thrones were twenty-four seated elders, clothed in white garments and with crowns of gold upon their heads"); 10:1: "et vidi alium angelum fortem descendentem de caelo amictum nube et iris in capite eius" ("and I saw another angel descending powerfully from heaven wrapped in a cloud and with a rainbow upon its head"). In line 105 of canto XXIX, Dante acknowledges his debt to St. John the Divine's apocalyptic vision. For the rainbow as a manifestation of the deity, see also Ezek. 1:28: "velut aspectum arcus cum fuerit in nube in die pluviae" ("as the appearance of the bow that were in the cloud on a rainy day"); cf. Ecclus 50:8: "quasi arcus effulgens in nebulam gloriae" ("as the bow shining in a cloud of glory").

assigned four colours.[72] Yet it is questionable how far any such assertions of Dante's scientific originality are valid. In this instance, it is not completely clear whether he is saying that every "lista" has a different colour or whether each one is coloured in the rainbow's three or four colours. Dante's allusion to the rainbow's colours is after all a tersely poetic one and, as such, it carries a suggestion of undefined iridescence that has literary precedents in descriptions of Iris' descent by his Latin poets.[73]

In the *Paradiso*, Dante refers to the rainbow on three occasions and makes two allusions to the moon-halo. All five references are used as part of similes in order to illustrate the shape, size, and configuration of groups of blessed souls, the angelic hierarchies, and the Trinity. In the Heaven of the Sun, Dante uses two similes to give a visual impression of the arrangement of the "spiriti sapienti," once these souls have formed a circle of lights around Dante and Beatrice. In canto X, the poet reveals his understanding of the optical process by which a light ray ("il fil": l. 69) from the spherical moon forms a circular figure, "la zona" (l. 69), in a rain-saturated atmosphere. His predilection for combining Aristotelian detail with mythological reference is as pronounced as it had been in the *Purgatorio*:

> Io vidi più folgór vivi e vincenti
> far di noi centro e di sé far corona,

---

[72] Aristotle allowed for three primary colours (red, green, and blue; *puniceum*, *viridem*, and *halurgus* in Moerbeke's Graeco-Latin translation) in *Meteorologica*, III, 4, 374b 30-35; he conceded that it was possible for a fourth colour (yellow; *xanthos* or *citrinus*) to be formed by mixing of the middle region of the arc (see 375a 7-9). The Pseudo-Aquinas, Albert, and Averroës all attribute a maximum of four colours to the rainbow. Dante's "originality" is noted by Boyer, *The Rainbow*, pp. 108-109.

[73] See Ovid, *Metamorphoses*, VI, 65-67; XI, 589 "velamina mille colorum" ("a covering of a thousand colours"); Virgil, *Aeneid*, IV, 701: "mille trahens varios adverso sole colores" ("spreading through the sky a thousand changing colours before the sun"); V, 88-89: "ceu nubibus arcus / mille iacit varios adverso sole colores" ("as in the clouds the bow casts a thousand colours in front of the sun"); 609: "illam viam celerans per mille coloribus arcum" ("making haste along that path with the thousand colours of its bow"). For a critical assessment of the problem in Dante, see Herbert D. Austin, "Dante Notes XI: The Rainbow Colors," *Modern Language Notes* 44 (1929), 315-318.

144

> più dolci in voce che in vista lucenti:
> così cinger la figlia di Latona
> vedem talvolta, quando l'aere è pregno,
> sì che ritenga il fil che fa la zona. (64-69)

At the opening of canto XII, Dante describes the way in which two circles of lights rotate around one another by referring to the formation of a secondary bow. This simile is characterised by its combination with other similes, its expansive use of different sources, and the relatively autonomous development that Dante gives to the vehicle. Similes within similes, like the earlier example from canto I, are an important device in the *Paradiso* as the poet offers successive, redefining images in attempting to adumbrate the often tenuous after-impressions of his experiences. Hence, in this triple simile the direct point of analogy is between the movement of two circles of lights and that of two rainbows; but the *tertium comparationis* is almost lost as Dante allows his learned references to accumulate and interlock:

> Come si volgon per tenera nube
> due archi paralelli e concolori,
> quando Iunone a sua ancella iube,
>   nascendo di quel d'entro quel di fori,
> a guisa del parlar di quella vaga
> ch'amor consunse come sol vapori,
>   e fanno qui la gente esser presaga,
> per lo patto che Dio con Noè puose,
> del mondo che già mai più non s'allaga:
>   così di quelle sempiterne rose
> volgiensi circa noi le due ghirlande,
> e sì l'estrema a l'intima rispuose. (10-21)

In the first tercet, the technically precise and scientific language of lines 10-11 with its "tenera nube" and "archi paralelli e concolori" combines with the first

145

mythological reference to Juno.[74] The procedure adopted in the second tercet is similar, even though the dominant sounds are now plosives rather than liquids. The Aristotelian doctrine of the secondary rainbow is here developed in one line, and the idea of repercussion common to the secondary bow and the echo leads Dante to make an imaginative connection with Ovid's Echo (ll. 14-15).[75] After the splendid sound repetitions of "di quel d'entro quel di fori" (l. 13), the last line of this tercet condenses another meteorological detail into a terse one-line simile (l. 16). Finally, in the third tercet of the vehicle the pattern of allusion becomes biblical (Genesis 9:8-17). This decorative balance of rainbow doctrine, myth, and Scripture, the intricate interweaving of similes and the repetition of phonic units are signs of the poet's virtuosity and intellectualism. And yet, despite its openly decorative appeal, the similes also present a sense of order and harmony, intimating the ideal parallelism between the Orders in Paradise, which is in itself a reproach to the degenerate Dominicans on Earth. It may be significant, then, that these lines present a strong stylistic contrast with the consciously disordered explosion of metaphor (and the accompanying imagery of decay and putrescence) with which Bonaventure vents his diatribe against the Franciscans towards the end of the canto (ll. 112-123).

[74]The classical references provided by most commentators are to Virgil, *Aeneid*, IV, 694; V, 606; IX, 2; Ovid, *Metamorphoses*, I, 270.

[75]For Aristotle, the echo, light reflection, and the rainbow were all related as forms of repercussion or reflection, and, though specifically different, he believed them to share a common genus, see *Posteriora analytica*, II, 15, 98a 28-30. Dante might have known this important work (cf. *Mon.* I, ii, 4, and see Nardi's commentary, in *Opere minori*, II, p. 287); he definitely knew of a similar analogy between sound and light in Aristotle's *De anima*, II, 8, 419b 25-33. The succinct transposition of this theoretical principle into a mythical example is, it seems, original to Dante. For his other creative reworkings of the Narcissus episode (*Metamorphoses*, III, 395-401), see *Inf.* XXX, 128; *Par.* III, 14. For Dante's allusions to Narcissus and their wider implications, see Kevin Brownlee, "Dante and Narcissus (*Pg.* XXX, 76-99)," *Dante Studies* 96 (1978), 201-206; Roger Dragonetti, "Dante et Narcisse, ou les faux-monnayeurs de l'image," *Revue des études italiennes* 11 (1965), 85-146; Michelangelo Picone, "Dante e il mito di Narcisso dal *Roman de la Rose* alla *Commedia*," *Romanische Forschungen* 89 (1977), 382-397; R.A. Shoaf, *Dante, Chaucer, and the Currency of the Word* (Norman, OK: Pilgrim Books, 1983), pp. 21-100.

146

In canto XXVIII, Dante adopts two further optical comparisons to suggest the dimensions of his vision of the deity and the angelic hierarchies as a point of light surrounded by nine concentric circles of fire. In a passage notable for its alliterations of *-anto, -gne, -or*, and *-i*, the poet again shows his awareness of how atmospheric variations alter the types of optical array formed when he describes the proximity of the first circle to the point by analogy to a moon-halo and its light source, the undetermined "luce" (l. 23):

> Forse cotanto quanto pare appresso
> alo cigner la luce che 'l dipigne
> quando 'l vapor che 'l porta più è spesso,
>   distante intorno al punto un cerchio d'igne
> si girava sì ratto, ch'avria vinto
> quel moto che più tosto il mondo cigne. (22-27)

The following simile in this canto is based on the rainbow and describes the relative size of the seventh circle in relation to those that precede it in order to convey the size and spatial configuration of the fiery circles:

> Sopra seguiva il settimo sì sparto
> già di larghezza, che 'l messo di Iuno
> intero a contenerlo sarebbe arto. (31-33)

Like the prominent "Forse" of the previous comparison, Dante relies on hypothetical suggestion (a "se fosse" is implicit in line 33) and in so doing draws our attention to a significant and somewhat paradoxical feature of his treatment of the *apparitio dei* in the final cantos of the poem. The poet represents such "showings" with visually-clear images, whilst often suggesting the impossibility of a completely accurate representation. Thematically, the notion that spatial terms are not adequate designations for God and the angels helps to prepare the central lesson of canto XXVIII: the need to transcend earthly estimates of magnitude and distance in order to readjust to the new laws which govern heaven. The doubt

147

which assails Dante later in this canto is caused by his failure to understand why the largest circle in the configuration of flaming "cerchi" does not stand for the highest angelic rank, the Order of Seraphims. And his perplexity is only resolved when Beatrice tells him to "circumscribe" his judgement, his own "measure," not to the physical appearance of the fiery circles but to their spiritual power, their "virtù" (ll. 73-78).

The final rainbow simile in the canto XXXIII, which is also one of the last comparative devices in the poem, is used by Dante to describe his vision of God in His Triune nature. In canto XII, the poet relied on the Aristotelian idea that a secondary rainbow is formed by a reflection from the primary bow to refer to the revolving circles of the wise souls. Here, the same doctrine allow him to indicate how the Son is generated from the Father, and it also helps to give shape and colour to his description. The colours of the first rainbow give rise to those of the second, and likewise Dante suggests that the Second Person is generated from the First, thus conveying with theological exactitude the doctrine of the Son as proceeding from the Father:

> Ne la profonda e chiara sussistenza
> de l'alto lume parvermi tre giri
> di tre colori e d'una contenenza;
> e l'un da l'altro come iri da iri
> parea reflesso, e 'l terzo parea foco
> che quinci e quindi igualmente si spiri. (115-120)

The theological concepts expressed are relatively clear: one coloured circle of light (the Father) generates another such circle (representing the Son) by reflection; and a third circle of different colour (the Holy Spirit), is breathed out from the first two circles (the First and Second Persons).

The idea of using light imagery to represent the Word-Logos has, of course, biblical precedents; and, as is well known, the Fathers of Nicaea established

148

the Son as *lumen de lumine* in an attempt to assert the non-inferiority of the Second Person.[76] Several scholars have raised the question of a direct source for this image by advancing parallels in Hildegard of Bingen's *Scivias* and in Joachim of Flora's *Liber figurarum*.[77] The rainbow and coloured circles are, however, pre-eminently biblical motifs,[78] and it is quite possible that Hildegard, Joachim, and Dante drew independently on scriptural precedents such as Ezekiel and Daniel in developing their own Trinitarian rainbow symbolism.[79]

From the analyses given in this chapter, it can be seen that Dante uses optical doctrines related to light reflection, mirrors, the rainbow, and the moon-

---

[76]On the biblical precedents for expressing the idea of the generation of the Word-Logos with light imagery, see Wis. 7:26: "candor est enim lucis aeternae et speculum sine macula Dei maiestatis et imago bonitatis illius" ("she [sc. Wisdom] is the brightness of eternal light and the immaculate mirror of God's majesty, and the image of His goodness"); Hebr. 1:3: "splendor gloriae." To assert the non-inferiority of the Second Person, the Fathers of Nicaea established the Son as *lumen de lumine*, drawing on Neoplatonic terminology derived from Plotinus, see below Ch. 6, n. 21. For Trinitarian doctrines couched in light imagery, see also Ch. 7, nn. 11, 18-19.

[77]For the putative parallels with Hildegard, see Heinrich Ostlender, "Dante und Hildegard von Bingen," *Deutsches Dante Jahrbuch* 27 (1948), pp. 161-164. The text cited is *Scivias*, lib. III, visio 1 (*PL* 197, col. 565). The discovery (1946) of Joachim's *Liber figurarum* led to its publication as *Il "Libro delle Figure" dell'Abate Gioachino da Fiore*, ed. Luigi Tondelli, Margaret Reeves, and Barbara Hirsch-Reich, 2 vols. 2nd edn (Turin: Società editrice internazionale, 1961). For the possible connection between Dante's circles and Joachim's, see vol. I, pp. 221-224 and the illustrations in vol. II, tables xi a and xi b. See also Barbara Hirsch-Reich and Margaret Reeves, *The "Figurae" of Joachim of Flora* (Oxford: Clarendon Press, 1972), pp. 317-329. For critical estimates of these attributions, see Arsenio Frugoni, "Gioachino da Fiore," in *ED* III, pp. 165-167; Etienne Gilson, Review of Nardi's *Dal "Convivio,"* in *GSLI* 138 (1961), p. 573; Bruno Nardi, *Dal "Convivio" alla "Commedia" (Sei saggi danteschi)* (Rome: Istituto storico italiano per il Medio Evo, 1960), pp. 360-369.

[78]References in order: (i) rainbow as a symbol of divine power, Ezek. 1:28 and Rev. 4:3; (ii) circles of light and colour, Dan. 3:24-25, 7:9. For the bow as a symbol of the Trinity, see John Gage, *Colour and Culture: Practice and Meaning from Antiquity to Abstraction* (London: Thames and Hudson, 1993), pp. 93 and n. 13, p. 281. See also Philip McNair, "Dante's Vision of God: An Exposition of *Paradiso* XXXIII," in *Essays in Honour of John Humphreys Whitfield*, ed. H.G. Davis and others (London: St. George's Press, 1975), pp. 25-26, citing Ezek. 1:27-28.

[79]For a judicious reconsideration of the coloured circles in Hildegard, Joachim, and Dante, see Dronke, "Tradition and Innovation in Medieval Western Colour-Imagery," pp. 99-106. Dronke also maintains that Dante deliberately "flouts" Aristotelian physics in describing the rainbow as circular. It should, however, be noted that Aristotle was aware that the rainbow could form a circle; its semi-circular shape was due to the interposition of the Earth (see the example of a circular rainbow quoted in n. 70).

halo extensively but with a considerable degree of selectivity and discrimination. Dante develops these doctrines to help to visualise a variety of supernatural phenomena: the effect of angelic light on his eyes; the diffusion and reception of divine goodness; the rapidity with which one gaze elicits another; the formation of soul-bodies; the configuration of souls and angels; and even the arcane geometry of the Trinity. But, in almost all cases, the result is far more than either a slavish imitation of traditional materials or an ostentatious display of the poet's scientific culture. The scientific text often provides a starting-point for a more elaborate pattern of imagery, one which is often fused with mythological and scriptural suggestions and may on occasion have wider implications for the themes pursued in individual cantos. In short, Dante presents us with a wide variety of images drawn from his understanding of reflected light, images that are highly individual poetic syntheses of intellectual inspiration and literary craft.

PART TWO

Theories of Light in Dante

CHAPTER 5

Dante and the "Metaphysics of Light": A Reassessment

In evaluating ideas related to light in thirteenth-century thought one of the most widely used critical categories is the notion of a "light-metaphysics" tradition. The term was coined early this century by a German scholar, Clemens Baeumker, in order to account for a propensity shared by several thirteenth-century philosophers to elaborate physical and metaphysical doctrines by recourse to light.[1] It is as a result of his pioneering work that the notion of a "light-metaphysics" tradition has become commonplace in histories of medieval philosophy and widespread in Dante studies.[2]

In Part Two my aim is to show that as a critical category "light metaphysics" is not sufficiently precise to be of value in assessing Dante's use of light images to describe the operations of the physical universe, to express God's act of creation, and to put forward his changing beliefs on the Empyrean heaven. Many Dante scholars who refer to "light metaphysics" have tended to oversimplify the complexity and heterogeneity of medieval ideas about light, and this has led to

---

[1]See Clemens Baeumker, *Witelo, ein Philosoph und Naturforscher des XIII. Jahrhunderts*, in *Beiträge* 3/2 (1908), p. 360. Baeumker's historical study of the "Lichtmetaphysik" (pp. 372-422) still provides an important source of documentation on the subject.

[2]See e.g. Frederick C. Copleston, *A History of Medieval Philosophy* (London: Menthuen, 1972), pp. 35, 52-53, 171; David Knowles, *The Evolution of Medieval Thought* (London: Longman, 1962), pp. 49, 240, 282, 332; Gordon Leff, *Medieval Thought from Saint Augustine to Ockham* (London: Penguin, 1958), pp. 188-189. For critics who apply the idea of a "light-metaphysics" tradition to Dante, see below n. 16.

152

imprecise estimates of Dante's indebtedness to key figures associated with the putative tradition. To counteract this tendency, I intend to place theories of light in their historical context and to pay close attention to individual thinkers and the writings in which their ideas are formulated. Dante remains of course the ultimate point of reference, and close consideration will be given to three broad contexts in which light has an important role to play in his own ideas and in his poetry: the act of creation; the operations of the cosmos; and theological doctrine. But before turning to these contexts in the next two chapters, it is necessary to justify and substantiate more fully the reasons for my critique of "light metaphysics."

There have been many historical studies of ancient and medieval "light metaphysics,"[3] but almost all these treatments follow a similar pattern. The philosophical development of the light image is traced back from the thirteenth century across Arabic and Jewish philosophy, the Latin and Greek Fathers, and Neoplatonic traditions of thought. This type of reconstruction has its merits, but one less felicitous consequence is that historical complexities are often avoided and leading ideas tend to become wrested from their proper relation to context. Of course, this is true of all such designations. Neoplatonism, for example, is merely

[3]For secondary literature on the "metaphysics of light," see Crombie, *Robert Grosseteste*, pp. 104-116, 128-131; C.K. Mckeon, *A Study of the Summa philosophiae of the Pseudo-Grosseteste* (New York: Columbia University Press, 1948), pp. 156-166; Otto von Simson, *The Gothic Cathedral: Origins of Gothic Architecture and the Medieval Concept of Order*, 2nd edn (Princeton, NJ: Princeton University Press, 1962), pp. 50-55, 119-123; Vescovini, *Studi sulla prospettiva*, pp. 7-32. More recent critical essays include: Barbara de Mottoni Faes, "Il problema della luce nel commento di Bertoldo Moosburg all'«Elementatio Theologica» di Proclo," *Studi medievali* ser. 3/16 (1975), 325-352; David C. Lindberg, "The Genesis of Kepler's Theory of Light: Light Metaphysics from Plotinus to Kepler," *Osiris* n.s. 2 (1986), 5-42; idem, *Roger Bacon's Philosophy of Nature*, introd. pp. xxv-liii; idem, *Theories of Vision*, pp. 95-99; James McEvoy, "The Metaphysics of Light in the Middle Ages," *Philosophical Studies* 26 (1979), 124-140; idem, *The Philosophy of Robert Grosseteste*, pp. 136, 141-162, 182-185, 294, 336, 414 and 449-451; Giuseppe Battisti Saccaro, "Il Grossatesta e la luce," *Medioevo* 2 (1976), 21-76. The most complete modern treatment of the subject is Klaus Hedwig, *Sphaera Lucis: Studien zur Intelligibilität der Seienden im Kontext der mittelalterlichen Lichtspekulation*, in *Beiträge* n.s. 18 (1980). Bibliographical material until 1973 can be found in Dieter Bremer, "Licht als universales Darstellungsmedium: Materialien und Bibliographie," *Archiv für Begriffsgeschichte* 18 (1974), 185-206. My understanding of "light-metaphysics" is indebted to these secondary sources.

153

the modern critical term that is used to describe the work of Plotinus and his followers; and the advantages that such a category may bring in terms of convenience are often outweighed by its tendency to mask a many-sided richness.[4] However, while a greater understanding of the various strands within each tradition of medieval Neoplatonism has emerged due to new critical editions of individual authors and thorough scholarship,[5] this is less true of historians' attempts to reconstruct the tradition of "light metaphysics" by using speculations about light as a single common denominator. It is true that many ancient and medieval philosophers posit a close relationship between light and concepts that are traditionally defined as metaphysical (i.e. concepts such as being, substance, and causation). Yet none of these writers establishes these connections with the directness shown by the anonymous author of an early thirteenth-century treatise, the *De intelligentiis*.[6] This short work shows how light could be used as the basis for a conception of being, and it thus provides a good point of comparison from which to assess the metaphysical function of light in other thinkers.

In the *De intelligentiis*, light is the principle which unifies the universe and ensures continuity between different grades of reality. Divine light is the first substance in which all other beings participate; and the nature, essence, and activity of every entity are linked to this one divine light. Any body which exercises

---

[4] For related problems with the term Neoplatonism, see Raymund Klibansky, *The Continuity of the Platonic Tradition during the Middle Ages* (London: Warburg Institute, 1939), pp. 36-37.

[5] See Barbara de Mottoni Faes, *Il platonismo medioevale* (Turin: Einaudi, 1979); Tullio Gregory, *Anima Mundi*; idem, *Platonismo medievale: Studi e ricerche* (Rome: Istituto italiano per il Medio Evo, 1958); idem, "The Platonic Inheritance," in *A History of Twelfth-Century Western Philosophy*, ed. Peter Dronke (Cambridge: Cambridge University Press, 1988), pp. 54-80. There is now an extensive critical literature on twelfth-century Platonism, see the essays collected in *A History of Twelfth-Century Western Philosophy* and the extensive bibliography on pp. 459-486.

[6] The *De intelligentiis*, which dates from 1220-1230, was edited by Baeumker in *Witelo*, pp. 1-71. On its authorship, see Aleksander Birkenmajer, "Witelo est-il l'auteur de l'opuscule 'De intelligentiis'?," in *Études d'histoire des sciences en Pologne* (Wroclaw: Ossolineum, 1972), pp. 259-339.

154

influence over another does so by virtue of light; and light is identified with the principle of knowledge and of life:

Prima substantiarum est lux. Ex quo sequitur naturam lucis participare alia [...] Omnis substantia influens in aliam est lux in essentia vel naturam lucis habens [...] Unumquodque quantum habet de luce, tantum retinet de esse divini, si lux est ens divinum per essentiam, sicut ostensum est, participatio lucis est participatio esse divini [...] Proprium et principium cognitionis est lux [...] Lux in omni vivente est principium motus et vitae.[7]

The attempt to unify the constitution and activity of the universe around light is such that the *De intelligentiis* is the best (and perhaps the only) example of a "light metaphysics" *sensu stricto*, that is, a unified metaphysical system based upon a conception of light that is not merely metaphorical or analogical. This is an important point, for it must be recognised that many philosophers and theologians traditionally identified with the "metaphysics of light" use natural light as a source of images, metaphors, and analogies; and, in so doing, they frequently appeal to the reader to recognise the non-literal status of their imagery.[8] As a result, it becomes very difficult to accept that these authors are "light metaphysicians" at all.

---

[7]See *De intelligentiis*, props. VI, VII, VIII, IX, X, ed. Clemens Baeumker, in *Beiträge* 3/2 (1908), pp. 7-11: "The first of all substances is light. From which it follows that other substances participate in the nature of light [...] Every substance which exercises influence over another is by essence light or has the nature of light [...] However much something has of light, it has as much of divine being, if light is by its essence a divine entity, as has been shown, to participate in light means to participate in divine being [...] The essential quality and origin of cognition is light [...] In every living thing light is the principle of movement and of life." Several of these propositions are found in condensed form in Vincent of Beauvais, see *Speculum maius*, lib. II, cc. 33-38 (I, cols 100-103).

[8]E.g. Bonaventure, *In II Sent.*, d. 13, a. 2, q. 1, ad. 4 (Quaracchi, II, 318b): "[...] lux spiritualis est communis creatori et creaturae secundum analogiam" ("spiritual light is common to the creator and creatures by analogy"); Plotinus, *Enneads*, V, 1, 6; VI, 4, 7, ed. A.H. Armstrong, 6 vols. (London: Heinemann, 1966-1988), V, p. 31; VI, pp. 294-296; Pseudo-Dionysius, *De divinis nominibus*, c. 4, § 1 (*PG* 3, col. 693B; Chevalier, I, p. 146); idem, *De caelesti hierarchia*, c. 13, § 3 (*PG* 3, col. 301A; Chevalier, II, pp. 948-950). On the metaphorical status of light motifs in

Further difficulties arise due to the fragmentary treatment of light to be found in many authors subsumed within the category of "light metaphysics." It is hard to appreciate how Saint Augustine, for example, can be regarded as a representative of this tradition when he does not elaborate his ideas about light into a coherent philosophy and employs them in at least ten different categories, most of which are not metaphysical in any strict sense of the word.[9] It would perhaps be more appropriate, though hardly more revealing, to refer to writers such as Augustine as exponents of a wide-ranging Christian light symbolism. And yet, even with this kind of qualification, there is still a risk of pressing a superficial sense of uniformity on different thinkers with different interests and different approaches. This is hardly surprising given that "light metaphysics" is often used to designate a vast field of inquiry which extends from ancient Neoplatonists and the early Fathers of the Church to scholastic theologians and Renaissance philosophers. But even within the realm of a single tradition, there are often very marked contrasts, and especially great caution must be exercised with writers who appear to share common intellectual and religious concerns: one must be prepared to make fine distinctions between the principles, terminology, and general schemes of ideas that underpin their doctrines. It is also important to recognise that, as is the case with Augustine, conceptions about the nature of light formed a heterogeneous matrix of ideas and were very rarely worked into a systematic body of doctrines, even in thirteenth-century writers for whom light was of undoubted importance.

Plotinus, see Rein Ferwerda, *La signification des images et des métaphores dans la pensée de Plotin* (Groningen: Wolters, 1965), pp. 57-61, esp. the list of metaphors given in n. 7, p. 59. See also Ch. 6, n. 14.

[9] Joseph-François Thonnard, "La notion de la lumière en philosophie augustinienne," *Recherches Augustiniennes* 2 (1962), 125-175. The ten categories proposed by Thonnard are: external physical light; lower psychological light; intellectual light of divine illumination; the eye ray; substantial light of angels and spirits; supernatural light of faith; light of reason co-operating with faith; mystical light of the Holy Spirit; uncreated light of divine truth; light of the Word in the Trinity. For additional references and bibliography on light in Augustine, see below Ch. 7, nn. 9-11.

156

Above all, one must be aware that ancient and medieval writers (and Dante is no exception) employed light motifs in an extremely wide range of contexts. Luminous phenomena and the properties of natural light provided an extraordinarily rich source of metaphors for medieval philosophers, theologians, mystics, and poets. The theme of light pervaded almost all areas of speculative endeavour and human activity in the Middle Ages, from painting and manuscript illumination to architecture and theories of beauty,[10] from the liturgy and religious literature to political tracts and vernacular verse.[11] The ubiquity of the light image has prompted some critics to resolve the "metaphysics of light" into separate areas according to the context and purposes of the discussion. Thus, David Lindberg has proposed a four-fold division corresponding to the use of light imagery as a metaphor for human knowing (light epistemology), as a hermeneutic aid for the theologian (light theology), as part of a scientific interest in light radiation (the physics or aetiology of light), and as an important element in creation doctrines (the cosmogony or metaphysics of light).[12] The only scholar to consider similar distinctions as being relevant to Dante studies is Cesare Vasoli, who, in the introduction to his authoritative commentary on the *Convivio*, has noted "gli

[10]On light *topoi* in medieval and Renaissance Italian art, see Hills, *The Light of Early Italian Painting*; Barasch, *Light and Colour in the Italian Renaissance Theory of Art.* On architecture and stained glass prodcution, see respectively von Simson, *The Gothic Cathedral* and David Evans, *Mediaeval Optics and Stained Glass*, unpublished Ph.D. dissertation, University of Birmingham, 1979; John Gage, "Gothic Glass: Two Aspects of the Dionysian Aesthetic," *Art History* 5 (1982), 36-58. For "l'esthétique de la lumière," see de Bruyne, *Études d'esthétique médiévale*, esp. III, pp. 1-26.

[11]For light motifs in the liturgy and medieval mystical tradition, see Hedwig, *Sphaera Lucis*, pp. 24, 27, 32-33, 58-60, 84-85, 243-255; B.I. Mullahy, "Liturgical Use of Light," in *New Catholic Encyclopaedia* 8 (1967), 751-754; Margot Schmidt, "Lumière: au moyen-âge," in *DS* 9 (1976), pp. 1162-1173. For further references to the pervasive influence of light in medieval thought, art, and literature, see Hedwig, *Sphaera Lucis*, pp. 15-16.

[12]Lindberg, *Theories of Vision*, pp. 95-99; idem, *Roger Bacon's Philosophy of Nature*, introd., p. 1.

evidenti interessi nutriti da Dante per la teoria della «perspectiva», e per la «fisica», «metafisica», e «teologia» della luce."[13]

This more discriminating contextual approach to medieval light traditions is an important development, and the chapters that follow will use some of these subdivisions to organise the Dantean material. But a classification such as Lindberg's does not solve all the difficulties, and it even raises a few of its own. First, one should be careful about treating the separate categories in strict isolation from one another. Medieval writers integrate light with religious beliefs, scientific concerns, and philosophical doctrines in divergent ways to produce complex results. Second, a revised scheme with a contextual focus still does not give sufficient emphasis to individual thinkers and the divergences between them. To do justice to the rich complexity of medieval writings on light and their possible influence on Dante requires a closer examination of the writings of important figures, such as the Pseudo-Dionysius, Grosseteste, and Bonaventure, authorities to whom Dantists have repeatedly referred.[14]

Having outlined the principal reasons for my general critique of "light metaphysics," it is now essential to apply this to the work of scholars and critics who have established a close relationship between Dante's thought and poetry and a so-called "light-metaphysics" tradition.

In the essays collected in *Saggi di filosofia dantesca* (1930), Bruno Nardi was one of the first to consider several aspects of this tradition in relation to Dante. One of the most important doctrinal areas identified by Nardi was emanation, a concept which Neoplatonic writers expressed through light metaphors and used to explain the derivation of the material world from the intelligible realm. In his essay

---

[13]Vasoli, *Il Convivio*, introd., p. lxxiii.

[14]This type of approach is in accordance with Spitzer's criticisms of the unit-idea, see Leo Spitzer, "Geistesgeschichte vs. History of Ideas as Applied to Hitlerism," *Journal of the History of Ideas* 5 (1944), 191-203, esp. pp. 194, 198-200, 202-203. On the need for detailed attention to individual elaboration, see Quentin Skinner, "Meaning and Understanding in the History of Ideas," *History and Theory* 8 (1969), 3-35.

158

on "Dante e le macchie lunari," Nardi showed how a Neoplatonic conception of "causalità per via d'emanazione" developed in a monotheistic setting and was transmitted in works such as the *Liber de causis*. For Nardi, the hierarchical cosmos presented in the *De causis* provided Dante with the essential premises upon which to build the sequence of ideas expounded by Beatrice in *Paradiso* II.[15] Nardi reaffirmed this view in other essays, most notably in discussing Dante's doctrine of creation, which he summarised in the essay, "Dante e Pietro d'Abano." Scholars have accepted Nardi's view that Dante's ideas about the cosmological system are permeated with elements of Neoplatonic doctrine. Indeed, it is largely because Dante's Neoplatonising orientations have been so thoroughly endorsed that a widespread, and at times rather uncritical, acceptance of his indebtedness to the "metaphysics of light" has resulted.[16] In much contemporary Dante studies, it now seems as if any aspect of his thinking which involves light is envisaged as part of his debt to a Neoplatonic "metafisica della luce." One critic has even allowed himself to make exaggerated claims of Plotinian influences (which are only very

---

[15]*Saggi di filosofia dantesca*, pp. 3-39. For the influence of the *De causis* on *Par.* II, see ibid., pp. 17-18, 22-23, 97-98, 105. Nardi's earliest references to the "metafisica della luce" were in his first published work on Dante, see *Sigieri di Brabante nella "Divina Commedia" di Dante Alighieri e le fonti della filosofia di Dante* (Spianate, 1912), pp. 33-37, 66. At the very end of his career, he continued to affirm the validity of the tradition for Dante studies, see "Sulla comunicazione di P.E. Guidubaldi", in *Atti del congresso internazionale di studi danteschi*, 2 vols. (Florence: Sansoni, 1965-1966), II, p. 257.

[16]See e.g. Boyde, *Dante Philomythes*, pp. 210-211; Corti, *Percorsi dell'invenzione*, pp. 157-160; Fallani, *Dante e la cultura figurativa*, pp. 123-124; Fengler and Stephany, "The Visual Arts: A Basis for Dante's Imagery in the *Purgatory* and the *Paradise*," p. 136; Pompeo Giannantonio, "Struttura e allegoria nel *Paradiso*," *Letture classensi* 11 (1982), p. 64; Romano Guardini, *Studi su Dante*, 2nd edn (Brescia: Morcelliana, 1979), pp. 278-279; Adriana Oliviero, "La composizione dei cieli in Restoro d'Arezzo e in Dante," in *Dante e la scienza*, ed. Boyde and Russo, pp. 360-362; Daniel Radcliff-Umstead, "Dante on Light," *Italian Quarterly* 9 (1965), 30-45; Luigi Santoro, "Dante's *Paradiso* (Canto I) and the Aesthetics of Light," in *Dante Readings*, ed. Eric Haywood (Dublin: Irish Academic Press, 1987), pp. 107-122. A more critical essay which distinguishes between light symbolism and more conceptual uses of light is Vincent Cantarino, "Dante and Islam: Theory of Light in the *Paradiso*," *Kentucky Romance Quarterly* 15 (1968), 3-35. For a dissenting view, see Attilio Mellone, "Luce," in *ED* III, pp. 712-713.

remote and superficial) and derivations (which are impossible) for *Paradiso* II.[17] Given such imprecisions, it should be clear that Dante's attitude to light needs to be defined with greater rigour and his intellectual relationship to medieval sources must be analysed with more discrimination.

In another essay republished in *Saggi di filosofia dantesca*, Nardi argued that Dante's doctrine of the Empyrean bears close similarities to the ideas of Neoplatonic writers associated with the "metaphysics of light."[18] In the first part of this essay, Nardi demonstrated how Aristotle's doctrine of natural place, with its concomitant notion of movement towards place, led to an apparently contradictory view according to which the eighth sphere, though without a place of its own, was ascribed a rotating movement. Nardi then traced speculations about light back to Eastern religious sects and Gnostic writings such as the *Chaldaean Oracles*, and he showed how the Alexandrian Neoplatonist, Proclus, employed light as the *locus* of the universe to provide a solution to the problem of the place of the eighth sphere. In the following section, he connected this Proclean doctrine with the Plotinian scheme of emanation in which the third hypostasis (soul) formed a connective link between the intelligible and sensible by giving off a pure light that encircles the eighth sphere. Later in the essay, this notion of light as the *vinculum universi* is applied to *Paradiso* XIII and XXX to support Nardi's claim that "[n]ella dottrina neoplatonica di Proclo [...] abbiamo già [...] tutti gli elementi caratteristici che ci bastano per chiarire il concetto dantesco dell'Empireo."[19] Unlike his study of the "macchie lunari," this essay is open to a number of criticisms because Nardi gives an extremely brief account of the medieval views

---

[17]Rudof Palgen, "Gli elementi plotiniani nel «Paradiso»," in *Atti del convegno internazionale sul tema: Plotino e il Neoplatonismo in Oriente e in Occidente* (Rome: Accademia nazionale dei Lincei, 1974), p. 524, relates *Par.* II, 121-123 to Plotinian terminology, by arguing that "[il] passo contiene quattro termini tecnici del plotinismo."

[18]Nardi, "La dottrina dell'Empireo nella sua genesi storica e nel pensiero dantesco," in *Saggi di filosofia dantesca*, pp. 167-214.

[19]Ibid., p. 181.

160

about the nature of light and his use of Proclus and Plotinus to interpret *Paradiso* XIII, 76-86 is controversial.[20]

Nardi also suggested that ideas characteristic of Baeumker's "Lichtmetaphysik" had influenced Dante's views on the origin, nature, and workings of the human intellect. In his essay, "La conoscenza umana," Nardi outlined the doctrinal context of Dante's thought on this matter by reviewing thirteenth-century theories of knowledge. He discussed three solutions to the problem of how man forms concepts, each of which employed the light metaphor. First, Aristotle's doctrine of the active intellect (*intellectus agens*) according to which a mental power abstracts intelligible forms from sensory impressions in a manner akin to that in which light renders colour visible in act; second, Augustine's so-called theory of illumination in which man's grasp of true concepts was made dependent on a divine light shining in the mind; and third, Islamic conceptions of the *intellectus agens* which transformed Aristotle's intellect into a separate cosmic Intelligence that was believed to irradiate intelligible forms into human intellects. Nardi then showed how a group of thirteenth-century theologians melded Augustinian ideas with the Arabo-Aristotelian explanation of human knowing, thereby identifying divine light with Aristotle's agent intellect.[21]

Because of the scope of the topic, the problems raised by discussions of human intellection in Dante will not be examined here. The subject was one of the most widely-discussed areas in medieval thought, and although Dante's treatment

---

[20]For recent supporters of this view of the Empyrean in the *Comedy*, see Bortolo Martinelli, "La dottrina dell'Empireo nell'Epistola a Cangrande (capp. 24-27)," *Studi danteschi* 57 (1985), 49-143; Vasoli, "Dante e l'immagine enciclopedica del mondo nel *Convivio*," pp. 64-72. The opposite view is found in Attilio Mellone, "Empireo," in *ED* II, pp. 668-671.

[21]Nardi, "La conoscenza umana," in *Dante e la cultura medievale*, pp. 135-172. Nardi argued (pp. 153-154) that Dante's position on the human intellect was representative of a Neo-Augustinian school (Roger Marston, William of Auvergne, Jean de la Rochelle, Roger Bacon) insofar as he regarded the first principles of human knowing as irradiated into the soul (see also n. 33 below).

161

is fragmentary, it would undoubtedly reward further investigation.[22] For the purposes of this chapter, it is important to recognise that thinkers with very different philosophical orientations used the light metaphor to talk about the way in which the intellect functions. Hence, radically different doctrinal positions may be concealed by similarities in expression.

Joseph Mazzeo's work on Dante and the so-called "light-metaphysics" tradition is one of the most detailed contributions to the subject so far. In *Medieval Cultural Tradition in Dante's "Comedy"* (1960), Mazzeo expanded an earlier article into two separate chapters: one dealing with the tradition itself; and the other with the ways in which Dante elaborates its doctrines.[23] His account of the tradition in this work remains a useful introduction in English to ideas about light in Plato, Augustine, Bonaventure, Grosseteste, Bartholomew of Bologna, and the author of the *De intelligentiis*. "Light metaphysics" is, however, an important element in Mazzeo's earlier volume, *Structure and Thought in Dante's*

---

[22]For important recent contributions that call for modifications to, and qualifications of, Nardi's views, see M.L. Führer, "The Contemplative Intellect in the Psychology of Albert the Great," in *Historia philosophiae Medii Aevi: Studien zur Geschichte der Philosophie des Mittelalters*, ed. Burkhard Mojsisch and Olaf Pluta (Amsterdam and Philadelphia: Grüner, 1991), pp. 305-319; idem, "The Theory of Intellection in Albert the Great and Its Influence on Nicholas of Cusa," in *Nicholas of Cusa in Search of God and Wisdom: Essays in Honor of Morimichi Watanabe*, ed. Gerald Christianson and Thomas M. Izbicki (New York and Leiden: E.J. Brill, 1991), pp. 45-56; Patrick Lee, "Aquinas and Avicenna on the Active Intellect," *The Thomist* 45 (1981), 41-53; Steven P. Marrone, *William of Auvergne and Robert Grosseteste: New Ideas of Truth in the Early Thirteenth Century* (Princeton, NJ: Princeton University Press, 1983); James McEvoy, "La connaissance intellectuelle selon Robert Grosseteste," *Revue de philosophie de Louvain* 75 (1977), 5-48; Jesse Owens, "Faith, Ideas, Illumination, and Experience," in *The Cambridge History of Later Medieval Philosophy*, ed. Kretzmann, pp. 440-459. On the Arabic background, see esp. the important studies in Herbert A. Davidson, *Alfarabi, Avicenna, and Averroes: Their Cosmologies, Theories of Active Intellect and Theories of Human Intellect* (Oxford and New York: Oxford University Press, 1992).

[23]Mazzeo's original essay was published as "Light Metaphysics, Dante's *Convivio* and the Letter to Cangrande della Scala," *Traditio* 14 (1958), 191-229; see also idem, "Light, Love, and Beauty in the *Paradiso*," *Romance Philology* 11 (1957-1958), 1-18. The relevant chapters in *Medieval Cultural Tradition in Dante's "Comedy"* (Ithaca: Cornell University Press, 1960) are: Ch. 2: "The Light-Metaphysics Tradition," pp. 56-90; and Ch. 3: "Light Metaphysics in the Works of Dante," pp. 91-132.

162

"Paradiso" (1958), and this work must be considered first if we are to appreciate the general import of his thesis.

In *Structure and Thought*, Mazzeo offered a general interpretation of the *Paradiso* that placed considerable emphasis on Platonic and Neoplatonic patrimonies of thought, and hence, like Nardi, set him apart from the exclusively Thomist-Aristotelian *forma mentis* imposed on Dante by an earlier generation of Italian scholars. In the first essay, "Dante and the Phaedrus Tradition," Mazzeo advanced his central thesis, namely, that "[i]t is light, spiritual or natural, in its role as beauty, in its role as the correlate of love, which gives us the key to this amorous journey through higher and higher levels of reality."[24] Drawing on the work of Edgar de Bruyne, Mazzeo put forward a reading of the *Paradiso* in which light is viewed as a principle of beauty that leads Dante-*personaggio* to a supernatural contact with the divine source of Beauty. In the following essay on poetic expression, Mazzeo developed his argument still further by suggesting that the images and concepts of the "light-metaphysics" tradition allowed Dante to resolve the fundamental problem of the *Paradiso*: the reduction of objects of thought to objects of vision.[25]

The first essay in *Medieval Cultural Tradition* acts as a foundation for the others and sets out a clear statement of Mazzeo's methodological principles. Without disapproving of the historical contextualisation and corpus of facts provided by Nardi's scholarly studies, Mazzeo argued that a historiographical approach to the *Comedy* has certain limitations, because it does not consider the function of ideas within the poem. Having raised the important question of how scholarship such as Nardi's can serve the literary critic, Mazzeo believed that an answer was to be found in the concept of the model. Medieval culture provided

---

[24]*Structure and Thought in the "Paradiso"* (Ithaca: Cornell University Press, 1958), p. 18; cf. preface, p. vii.

[25]Ibid., pp. 44-47, esp. p. 42: "a translation [sc. the *Paradiso*] into terms of sensible light of a timeless vision of a spiritual or intellectual light."

163

Dante with models for organising complex experience, defining characters, and ordering the cosmos; and the critic can use such models to identify and appreciate the structuring principles of the *Comedy*. The central idea which overarches Mazzeo's set of essays is that of hierarchy, and the model of light (or the "light-metaphysics" tradition) finds its justification within this framework. Put simply, Mazzeo believed that "light metaphysics" provided Dante with a way of organising the universe into a "hierarchy of lights" and of allowing him, as poet, to structure the final phase of his "vision" as a journey from material to divine light.

Although the principle of the model will not be questioned here (one thinks of C.S. Lewis's felicitous *The Discarded Image*), several criticisms can be levelled against the overall structure and specific features that Mazzeo included in his model. If a model is used too loosely and its features are too diverse and disparate, it may simplify a complex set of traditions to the point of distortion. Throughout his essay, Mazzeo assumed that the "light metaphysicians," whom he placed in a chronological order following Baeumker's survey, formed a clearly-defined intellectual tradition and presented a *doctrina communis*. As was argued earlier, there is little value in stating that a philosophical or theological tradition existed, unless one is prepared to define this tradition in its many-faceted complexity and to pay close attention to specific authors. These *lacunae* are particularly apparent in the one brief paragraph Mazzeo devoted to Judaeo-Arabic speculations on light. Without any supporting evidence, he reached the conclusion that as a result of Latin translations of Judaeo-Arabic works, "a strongly Neoplatonic doctrine of light penetrated the whole intellectual atmosphere of thirteenth-century Europe."[26]

The belief that light speculation was exclusively Neoplatonic in character is also potentially misleading, because light imagery and doctrines associated with light were prevalent in almost all ancient philosophical traditions and in a wide

---

[26]*Medieval Cultural Tradition*, p. 61; this assumption reveals one of Mazzeo's sources, Edgar de Bruyne's *Études d'esthéthique médiévale*, III, p. 18: "Au XIIIe siècle se répand donc une «doctrine commune» fortement néoplatonicienne qui imprègne toute l'atmosphère intellectuelle."

164

variety of medieval writings. Many of the doctrines which Mazzeo regarded as characteristic of Neoplatonic "light metaphysics" are thus too general to be used to define a specific tradition.[27] For example, the ideas that intellection and life occur through some form of light have in fact foundation in Aristotle.[28]

Mazzeo's discussion of Aquinas' attitude to the light image also requires considerable qualification and some revision. After detailed examination of the Thomist commentary on *De anima*, Mazzeo concluded that Aquinas displayed a general mistrust of light metaphors. From his analysis of this commentary, he then inferred that "the speculations of St. Bonaventure [...] provide a better key to some elements of great importance in the actual architecture of Dante's universe, especially the *Paradiso*, than does the work of Aquinas."[29] As we shall see in Chapter 7, Bonaventure is a recurrent point of reference for Dantists concerned with the study of light in Dante. However, a single text does not provide sufficient evidence to support Mazzeo's preference for Bonaventure over Aquinas, and he undoubtedly failed to take into account the important role which Aquinas assigned to light throughout his writings.[30] If Mazzeo had used the Thomist commentary on the Gospel of St. John, for example, he might have reached rather different conclusions. In this commentary, Aquinas argues that sensible light is analogous to intellectual light, employs light to discuss the Neoplatonic doctrine of participation,

---

[27]See *Medieval Cultural Tradition*, p. 75. Mazzeo reduces the "light-metaphysics" tradition to the following principles: (i) all the positive principles of the universe are reduced to light; (ii) light is the substantial form of the universe; (iii) light occupies an intermediate place between body and soul; (iv) light is the principle by which the intellect understands; (v) these ideas are set into a hierarchically-ordered universe.

[28]For light in the process of knowing and in the generation of man, see Aristotle, *De anima*, III, 5, 430a 16 and *De gen. animalium*, II, 3, 736b 29-36 respectively.

[29]*Medieval Cultural Tradition*, p. 66.

[30]For this view, see *Structure and Thought*, p. 11; *Medieval Cultural Tradition*, pp. 66, 112. For similar views in Guidubaldi, see "S. Tommaso e la metafisica della luce," *Dante Europeo*, III, pp. 355-363, esp. p. 356, where he speaks of Aquinas' "costante atteggiamento antiluministico." Cf. Nardi, *Saggi di filosofia dantesca*, p. 204.

165

and expresses the concept of divine omnipresence through light imagery.[31] While Aquinas was intensely preoccupied with using language accurately and argued that light was neither a substantial form nor a spiritual substance, he frequently adapted light images for the purposes of analogy. In fact, many of the ideas relating to causation, human knowing, and beauty that are discussed by Mazzeo can be found in Thomist writings; and Aquinas also made special use of light in elaborating ideas related to human knowing such as the doctrines of the *intellectus agens* and the *lumen gloriae*.

In examining the influence of the "light-metaphysics" tradition on the *Comedy*, Mazzeo concentrated, as Nardi had done, on doctrinal passages in the *Paradiso*, especially the discourses in cantos II and XIII (ll. 52-82).[32] Following Nardi, he also believed that Dante's epistemology was Augustinian and rested upon some form of divine light shining into the human mind.[33] One new suggestion he advanced was to argue that Dante conceived of the point of light in *Paradiso* XXVIII, 41 as the substantial form of the universe.[34] In the middle of this chapter, however, Mazzeo reiterated his earlier Platonising reading of the *Paradiso* by describing the phases in Dante-*personaggio*'s erotic flight to the divine source of beauty in terms of the poet's use of imagery related to light, wings, and the eyes.[35] This section of his essay is therefore concerned with

[31]Quotations from this commentary are reproduced in Ch. 7, pp. 231-232. For an indication of the widespread use of light in Thomist writings, see the entries under *lumen* and *lux* in the computerised *Index Thomisticus*, ed. Roberto Busa (Stuttgart: Frommann-Holzboog, 1975), sec. II, concordantia prima, vol. 13, pp. 17-42 and 52-66.

[32]*Medieval Cultural Tradition*, p. 107.

[33]Ibid., p. 114: "As Nardi demonstrated, Dante's epistemology is Augustinian and posits some form of divine illumination as the actualizing principle in the process of human knowing." This assertion requires qualification given that, in the *Comedy*, Dante shows man as subject to impressions from the senses and assigns an important role to sensory perception in forming concepts.

[34]Ibid., pp. 108-109. But on the inadequacy of this view, see Kenelm Foster, Review of *Medieval Cultural Tradition*, in *Modern Language Notes* 76 (1961), p. 943.

[35]*Medieval Cultural Tradition*, pp. 113-132. In these pages Mazzeo's discussion depends on the notion of physical light as an exemplar of spiritual beauty developed earlier on pp. 78-89. For the

166

matters that have very little direct bearing on ideas related to emanationism, causality, and cosmology, and it is highly questionable whether any of this section, which makes up nearly half the chapter, strictly pertains to "light metaphysics" as it has been defined by historians of medieval philosophy.

The most wide-ranging survey of light as image and concept in Dante, Egidio Guidubaldi's three-volume work, *Dante Europeo*, is unfortunately the most open to criticism of all.[36] Guidubaldi identified the most significant "moment" in the history of "light metaphysics" as the transition from using light as metaphor to the intuition that light was the constitutive essence of reality, no longer a mere symbol, but a fundamental metaphysical principle. According to Guidubaldi, the central figure in this radical change of emphasis is Robert Grosseteste, whom he calls "il primo organico sistematizzatore della metafisica della luce." And yet, while Guidubaldi's concern, in a Dantean context, with Grosseteste is welcome, his analysis of his philosophy of light is largely inaccurate.[37] Guidubaldi listed six essential features of Grosseteste's theory of light. The first two points concern light as the first corporeal form and as the princpal cause of motion, and as such they can be corroborated in his *opusucla* and are discussed by all historians of Grosseteste's thought.[38] But the other four features, namely, that light forms the basis for sphericity, colour, sound, and rhythm, are either *loci communes* that are not distinguishing features of Grosseteste's philosophy or they simply lack any foundation in his writings.[39] Light's capacity to diffuse itself spherically is found in

structuring elements in Dante's "universal vision," see: (i) light imagery (pp. 119-125); (ii) wing imagery (pp. 125-126); and (iii) the function of eyes (pp. 126-130).

[36]*Dante Europeo*, II, pp. 171-456; III, pp. 245-351. See also idem, *Dal "De Luce" di R. Grossatesta all'islamico "Libro della Scala": Il problema delle fonti una volta accettata la mediazione oxfordiana* (Florence: Olschki, 1978), esp. pp. 55-82.

[37]*Dante Europeo*, II, pp. 257-283, 288-300; III, pp. 268-287. For the importance of this emphasis, see Nardi, "Sulla comunicazione di P.E. Guidubaldi," p. 257.

[38]On light as the substantial form of the universe and as the primary efficient cause of motion, see *De luce* (Baur, pp. 51-58) and *De motu corporali et luce* (Baur, pp. 70-72).

[39]For these "sei punti-chiave," see *Dante Europeo*, II, pp. 267-280; III, p. 269. Guidubaldi later applies these points to an analysis of light in Dante, see ibid., II, pp. 283-295; III, pp. 271-282.

167

Arabic works and many medieval writings, whereas the definition of colour as "lux incorporata perspicuo" is a commonplace derived from Aristotle.[40] Guidubaldi's suggestion that Grosseteste conceived light as a rhythmic quality seems to be no more than his own attempt to relate Grossetestean passages to other dominant motifs in the *Paradiso* such as music. Grosseteste did use ideas related to light to devise a theory of sound, but his works provide absolutely no evidence that he envisaged light as an "esigenza ritmica portata a risolversi in gesto e danza."[41]

As a result of this idiosyncratic approach to light in Grosseteste, much work remains to be done to provide an adequate assessment of his relationship to Dante. Grosseteste viewed light as the *prima forma corporeitatis*, the single source of energy that brings all corporeal things into existence and gives them material extension. The behaviour of light was also, in Grosseteste's view, directly analogous to the way in which he believed efficient causes to operate; and light-rays were intrinsic to his theories of sensation, movement, tides, heat, and sound. In Part One, we saw that Grosseteste's high regard for the science of optics contrasts sharply with Dante's own understanding of "perspettiva." In the next chapter this assessment will be developed further by using Grosseteste's writings as one of several points of reference and comparison for assessing Dante's own use of concepts of light.

---

[40] On the sphericity of light, see the passage from Averroës quoted in Ch. 2, n. 42. On colour, see Grosseteste, *Hexaëmeron*, p. II, c. 10, § 2, ed. Richard C. Dales and Servius Gieben (London: Oxford University Press, 1982), p. 99: "lux namque incorporata perspicuo humido color est" ("for colour is light incorporated in the wet transparent medium"); *De colore* (Baur, p. 78): "Color est lux incorporata perspicuo" ("colour is light incorporated in the transparent medium"). This is essentially Aristotle's own definition, see *De sensu*, 3, 439a 28-439b 15. Broadly Aristotelian definitions are found widely in medieval sources, see Albert, *De sensu et sensato*, tr. II, c. 1 (Jammy, V, p. 21b); Aquinas, *In De anima*, lib. II, lect. 14, § 425 (Pirotta, p. 106); Vincent of Beauvais, *Speculum maius*, lib. II, c. 40 (I, cols 117-118). See also the passages quoted in Ch. 4, n. 66.

[41] *Dante Europeo*, II, p. 267, cf. pp. 279-281; III, pp. 269, 280-282. On the role of light in Grosseteste's theory of sound, see below Ch. 6, n. 48.

168

A number of points concerning Dante and the "metaphysics of light" should now be clear. Dante scholars and critics regard its principal doctrines to be emanation, the role of light in God's creative act, and the Empyrean; and they consider Dante's principal authorities to be Neoplatonists, the authors of the *De causis* and *De intelligentiis*, Grosseteste, and Bonaventure. The term "light metaphysics" is, however, often used uncritically and, given the diversity of contexts and differing strands of thought subsumed within this heading, it is misleading to regard medieval ideas about light as forming a unified, coherent, or even exclusively Neoplatonic body of doctrine.

This summary of critical approaches to "light metaphysics" in Dante helps us to isolate some areas that require further inquiry in line with the context-oriented approach proposed earlier in this chapter. First of all, emanation and the act of creation as described in *Paradiso* VII and XXIX have drawn differing conclusions which need to be re-examined and clarified. Second, very few scholars have investigated the extent to which light informs Dante's cosmology in general and his use of important concepts related to light such as celestial influence and efficient causality. And finally relatively little coverage has been given to Dante's intellectual relationship to, and poetic treatment of, a wide range of theological writings on light, including those dealing with the Empyrean heaven. These first two doctrinal areas, the place of light in Dante's cosmos and in his conception of God's creative act, will be examined in the Chapter 6. Chapter 7 will present a more wide-ranging analysis of analogies drawn from light in medieval theology and in the *Paradiso*, before re-considering in its final section the importance of light in Dante's doctrine of the Empyrean. The division between Chapters 6 and 7 is of course largely an artificial one: Dante's pronouncements upon the act of creation, for example, do not fit neatly into either category, but this distinction has been retained to organise the material according to context. In order to re-examine critically some of the prevailing views about Dante's indebtedness to certain authors in his use of light as image and concept, both chapters aim to undertake a close study of light theories in a wide range of writings. It is my contention that, rather than specific authors, Dante relies on more general ideas which he rethinks, syncretically blends with other sources, and transforms with consummate skill and artistry in his poetry. As is the case with his optics, we will see that the key to Dante's knowledge and use of medieval writings on light lies in his assimilation and often highly innovative presentation of traditional materials.

CHAPTER 6

Light in the Cosmos and in God's Creative Act

This chapter attempts to provide a more contextualised study of two areas of philosophical and scientific inquiry in which light was an important, if not central, concern for Dante: conceptions of the structure and operations of the physical universe; and explanations of the creative act. Dante's thought on these subjects draws on an extremely rich intellectual patrimony that involves concepts such as efficient causality, celestial influence, and emanation as well as the disciplines of astrology and cosmology.[1] As a first step towards understanding his use of such ideas, it is therefore necessary to preface the Dantean material with a brief historical introduction.

Efficient Causality and Celestial Influence

To appreciate the way in which medieval writers (including Dante) expressed causal action through a variety of nouns and verbs related to light calls for some discussion of theories of physical and celestial causation in antiquity and Islam.

---

[1]While no discipline by the name cosmology existed in the Middle Ages, statements about the universe and its laws as an ordered whole were fundamental to almost all medieval thinkers. For a near comprehensive treatment of medieval cosmology, see Grant, *Planets, Stars and Orbs: The Medieval Cosmos, 1200-1687*. A good introductory survey is Edward Grant, "Cosmology," in *Science in the Middle Ages*, ed. Lindberg, pp. 265-302. For recent studies of Dante's cosmology and astrology, see below nn. 38-39.

172

In the *Physica*, Aristotle discussed the nature and function of efficient causality in his so-called theory of four causes. He argued that an efficient cause initiated motion and/or produced change (both of which involved the passage from potency to act) in bodies.[2] And he maintained that this process could only occur when an efficient cause was itself in act and thus able to actualise forms in the potency of matter. In other words, Aristotle regarded efficient causality as involving the communication of something already possessed by the cause to its effect, an idea which historians of his thought define as a transmission theory of causality.[3] Aristotle applied his view of efficient causality as maker or mover by transmission of likeness to the celestial bodies in order to explain how the heavens brought about changes in the sublunary world. In Book II, Chapter x of *De generatione et corruptione*, he argued that the sun, by moving along the ecliptic, was the efficient cause of coming-to-be on Earth. Aristotle also held that the sun exercised other types of causal activity: through its movement, it regulated the seasons, brought about rains, and had a role in human generation.[4] In his *Meteorologica*, he advanced a similar line of argument when he maintained that the eternally moving bodies were the source of change in the terrestrial world.[5] As

---

[2]For Aristotle's discussions of the four causes, see *Physica*, II, 3, 194b 24-195b 30; II, 7, 198a 14-198b 9. He defines efficient causality as the source of change in *Physica*, II, 3, 194b 29-31; 195a 23. On Aristotle's causal theory, see J. Peterson, "Aristotle's Incomplete Causal Theory," *The Thomist* 36 (1972), 420-432.

[3]See e.g. A.C. Lloyd, "The Principle that the Cause is Greater than the Effect," *Phronesis* 21 (1976), 146-156; Richard Sorabji, *Time, Creation and the Continuum: Theories in Antiquity and the Early Middle Ages* (London: Duckworth, 1983), p. 307.

[4]*De generatione et corruptione*, II, 10, 336b 16-19. Note also the role Aristotle attributes to the sun in human generation in *Physica*, II, 2, 194b 13-14: "homo enim hominem generat ex materia et sol," Graeco-Latin trans. For further references to the types of causality exercised by the sun in Aristotle, see Edward Grant, "Medieval and Renaissance Scholastic Conceptions of the Influence of the Celestial Region on the Terrestrial," *Journal of Medieval and Renaissance Studies* 17 (1987), p. 6.

[5]*Meteorologica*, I, 2, 339a 31-32.

many scholars have argued, these passages were fundamental in establishing astrology in the scholastic tradition.[6]

Aristotle did not, however, put forward a systematic theory of celestial causation; and the historical development of ideas relating to such conceptions exemplifies how later writers opened up the light image to philosophical development. Ptolemy redefined and expanded Aristotle's views on the causality of the heavens into a more detailed system. He accepted the traditional view that heavenly bodies exercised causal power on what was below them by movement, but he enlarged the extent to which the heavens co-operated in this process, and gave specific explanations of the way in which each planet imparted its effects. With regard to the process of transmission, Ptolemy added an important new suggestion by stating that a force "flowed" from the ethereal substance of the stars and descended to the sublunary world:

[...] a certain power emanating from the eternal ethereal substance is dispersed through and permeates the whole region of the earth, which throughout is subject to change, since, of the primary sublunary elements, fire and air are encompassed and changed by the motions of the ether, and in turn encompass and change all else.[7]

---

[6]For the importance of these Aristotelian passages to medieval astrological lore, see Baeumker, *Witelo*, pp. 454-456; Grant, "Medieval and Renaissance Scholastic Conceptions," pp. 1-2; J.D. North, "Medieval Concepts of Celestial Influence: A Survey," in *Astrology, Science, and Society: Historical Essays*, ed. Patrick Curry (Woodbridge, Suffolk: Boydell 1987), pp. 5-17, esp. p. 5; McEvoy, *The Philosophy of Robert Grosseteste*, p. 281. See further J.D. North, "Celestial Influence-the Major Premiss of Astrology," in *"Astrologi hallucinati": Stars and the End of the World in Luther's Time*, ed. Paola Zambelli (Berlin and New York: de Gruyter, 1986), pp. 45-100, with a brief section on Dante, pp. 82-85.

[7]*Tetrabiblos (Opus quadripartitum)*, lib. I, c. 2, ed. and trans. F.E. Robbins (London and Cambridge, MA: Harvard University Press, 1940), pp. 4-6, cited by Grant, "Medieval and Renaissance Scholastic Conceptions," n. 3, p. 3.

174

This is the point at which the idea of celestial influence begins to emerge as a concept bearing close etymological connections with the idea of light flowing forth from a luminous source. The process of illumination, after all, presents quite precise analogies with the way in which Aristotle believed efficient causality to operate: a medium is illuminated (that is, its potential transparency is actualised) when light transmits a likeness of itself to this medium. After Ptolemy, light images were increasingly associated with ideas of celestial causation, and these metaphors produced new ways of thinking about the causality of the heavens. Following Ptolemy's lead, celestial influence with all that this word once implied became a major theme in Arabic philosophy and medieval astrology. For example, Abu Ma'shar (786-866), the most important Arabic authority on astrology, used radiation to link the heavens to the Earth.[8] By the late thirteenth century, almost all the major Western philosophers and theologians expressed celestial and physical causation with words related to light such as *influere*, *fluere*, *diffundere*, and *multiplicare*.[9] Although the verbs of inflowing and multiplying were widely used, the extent to which the light image was conceptualised varied from author to author: unlike the author of the *De intelligentiis* very few scholastics directly identified light with causal action itself.

Arabic thinkers also devised several new species of efficient cause to account for the different stages by which the heavens regulated generation and established diversity on Earth. Efficient causality was divided into four types: immediate or remote, depending on the position of a causal agent in a chain of efficient causes; dispositive or perfective, according to whether the cause prepared matter or induced forms into pre-disposed matter. Perfective causality was

---

[8]See Richard Lemay, *Abu Ma'shar and Latin Aristotelianism in the Twelfth Century: The Recovery of Natural Philosophy through Arabic Astrology* (Beirut: American University of Beirut, 1962), pp. 65-67, 231-234.

[9]For the use of this terminology in theologians such as William of Auvergne, Alexander of Hales, and St. Bonaventure, see Jacques G. Bougerol, "Le rôle de l'*influentia* dans la théologie de la grâce chez S. Bonaventure," *Revue de théologie de Louvain* 5 (1974), esp. pp. 273-277.

especially important, for, like the idea of influence, it ran counter to Aristotle's conception of indwelling causes and his belief that celestial bodies exercised power by movement. Once again, the concept and the image upon which it rested became increasingly difficult to distinguish, with Latin writers expressing the action of perfective causality by terms such as *fluxus* and *influxus*.[10]

The light of the heavenly bodies was a central feature in other Arabic doctrines, especially those concerning the generation of life on Earth. These ideas had some basis in Aristotle, who, in his *De generatione animalium*, had argued that the reproductive vital heat is a spirit which bears some relation to the fifth element or aether.[11] Arabic writers developed theories in which they taught that the heavens induced life through the influence of their light and that heavenly light prepared man to receive the rational soul.[12]

Emanation and Creation Doctrines

The Neoplatonic theory of emanation, which accorded a central place to the light image, attempts to resolve the fundamental problem with which Plotinus (c. 205-270 A.D.), and indeed all subsequent Neoplatonists, struggled. In Plotinus' words:

---

[10]The classical elaboration of the difference between perfective and dispositive causality is found in Avicenna, see *Sufficientia*, lib. I, c. 10 (Venice, 1508; reprint, Frankfurt-am-Main: Minerva, 1961), fol. 19ra: "Principium motus aut est praeparans aut est perficiens. Sed praeparans est id quod praeparat materiam [...] Perficiens est id quod tribuit formam constituentem species naturales" ("The origin of movement is either dispositive or perfective. But the dispositive is that which prepares matter [...] Perfective is that which endows a form which constitutes a natural species"). This idea contrasts with the traditional Aristotelian scheme according to which "causa principalis movet [...] causa instrumentalis movet mota" ("the primary cause moves [...] the instrumental cause, having been itself moved, moves").

[11]Reference above in Ch. 5, n. 28.

[12]Avicenna provides a succinct exposition of the complexion doctrine in his *De medicinis cordialibus* (Van Riet, I, p. 190), cited in McEvoy, *The Philosophy of Robert Grosseteste*, p. 281 and n. 67.

"How do all things come from the One, which is simple and has in it no diverse variety, or any sort of doubleness?"[13]

To explain and justify the passage of multiplicity from the One, Plotinus taught that the universe came into existence because of an outflowing of being from the first cause, the One. On the assumption that this first principle was complete in its own perfection, Plotinus argued that the One overflowed its own goodness spontaneously. He expressed this overflowing action by recourse to a range of metaphors drawn from natural phenomena with the capacity for self-diffusion, such as water, fire, perfume, and light.[14] For Plotinus, the One, by irradiating itself outwards, gave rise to the second principle, Mind, which gained order by turning back to the One, its cause.[15] Soul, the principle below Mind, was then produced by a further emanation from, and reversion to, this second principle. The emanative process proceeded in this way until a chain of interconnected causes, issuing from the plenitude of the One, was produced. Plotinus thus combined a radically non-Aristotelian conception of being as eternally generated with a wholly Aristotelian transmission model of causality.[16] He had in short applied Aristotle's causal theory to those logical categories of thought (the One, Mind, and Soul) which he believed to be the ultimate principles of spiritual reality.

While causation in Aristotle had largely been a question of actualising a potency in matter, Plotinus now made it an activity that was essentially productive. Causes were not indwelling, but descended into their effects from above; and hence

---

[13]Plotinus, *Enneads*, V, 2, 1 (Armstrong, V, p. 59).

[14]For an analysis of the concept, related imagery, and its sources in late Stoicism, see A.H. Armstrong, "'Emanation' in Plotinus," *Mind* n.s. 46 (1937), 61-66; R.T. Wallis, *Neoplatonism* (London: Duckworth, 1972), pp. 61-69. For important examples of Plotinus' use of light metaphors, see *Enneads*, V, 1, 6; V, 3, 12 (Armstrong, VI, pp. 31, 115). As Armstrong demonstrates in the above article, Plotinus was very conscious that these images were metaphorical expressions (see also above Ch. 5, n. 8).

[15]*Enneads*, III, 8, 7-11 (Armstrong, III, pp. 381-401).

[16]For Plotinus' adaptation of the Aristotelian causal model, see *Enneads*, IV, 3, 10 (Armstrong, IV, p. 69).

the emanative process became central to causation, since all effects were produced by emanation from above. Although later Neoplatonists, such as Iamblichus (d. c. 330) and Proclus (412-485) multiplied the number of intermediaries between the One, Mind, and Soul, they nonetheless accepted the essentials of Plotinus's teachings. The universe proceeds outwards from the One and forms an ordered hierarchy which descends from Mind to matter by communication of likeness.[17] Post-Plotinian Neoplatonic writers usually presented this causal scheme within the triad of remaining (*monê*), downward procession (*próodos*), and upward return (*epistrophê*).[18]

From Plotinus' use of light similes to explain emanation, it is relatively easy to appreciate how later writers in Neoplatonic traditions developed a pervasive light symbolism.[19] Plotinus had quite clearly given light an important role in the constitution of the spiritual and physical universes: light was, in a sense, the principle which unified and connected the spiritual and corporeal realms. And it is in Plotinus' *Enneads* that one also finds the first clear exposition of the idea that light is a principle of form in the material world.[20]

---

[17]For Proclus' adaptation of the Aristotelian transmission model and his use of Plotinian principles in the *The Elements of Theology*, see *The Elements of Theology: A Revised Text, with Translation, Introduction, and Commentary*, ed. E.R. Dodds, 2nd edn (Oxford: Clarendon, 1963), prop. 29, p. 35: "All procession is accomplished through the likeness of the secondary to the primary" and prop. 27 (Dodds, p. 32): "Every producing cause is productive of secondary existents because of its completeness and superfluity of potency." For emanation in Proclus, although the light analogies are far less pronounced, see *Elements of Theology*, props. 26-27 (Dodds, pp. 30-32); and see the editor's commentary on pp. 213-214. For theories of causality in other Neoplatonic writers, see Stephen Gersh, *From Iamblichus to Eriugena: An Investigation of the Prehistory and Evolution of the Pseudo-Dionysian Tradition* (Leiden: E.J. Brill, 1978).

[18]The best-known, and most succinct, enunciation of this triad is Proclean, see *Elements of Theology*, prop. 35 (Dodds, p. 39): "Every effect remains in its cause, proceeds from it, and reverts upon it."

[19]On metaphors of emanation in later Neoplatonists, see Gersh, *From Iamblichus to Eriugena*, pp. 17-26.

[20]On the importance of light as a unifying principle in Plotinus, see Armstrong, "'Emanation' in Plotinus," p. 63; Lindberg, "The Genesis of Kepler's Theory of Light," p. 10. The notion of light as *vinculum universi* was most notably developed later by the Renaissance Neoplatonists, Marsilio Ficino (1423-1499) and Francesco Patrizi (1529-1597), see Adele Anna Spedicati, "Sulla teoria della luce in F. Patrizi," *Bollettino di storia di filosofia dell'università di Lecce* 5 (1977), 243-

178

Light images had, of course, a particularly deep resonance for Christian writers, who used light symbols to express many Christian concepts and doctrines such as divine love and omnipresence, the Trinity, creation, and the resurrection of the body (see Chapter 7). In so doing, these authors frequently adapted Neoplatonic ideas and images to Christian contexts. There are many examples of this kind of assimilation. Plotinian emanation was combined with traditional views on creation. Sun-similes, images of emanation, and other Neoplatonic light imagery were widely used to talk about the divine nature and to describe the derivation of things from God. The triad of remaining, procession, and return was employed to explain the operation of divine love.[21]

The Pseudo-Dionysius (fl. 500) and John Scottus Eriugena (810-875) are undeniably the key figures in assimilating Neoplatonic structures of thought, imagery, and terminology to a Christian setting. Eriugena's challenging brand of Christian Platonism was extremely influential amongst twelfth-century thinkers, and even though his works were later condemned, his translations of the Pseudo-Dionysius continued to be read in the thirteenth century.[22] As regards the

---

263. For light as form, see *Enneads*, I, 6, 3 (Armstrong, I, p. 241), cited by Lindberg, "The Genesis of Kepler's Theory of Light," p. 11.

[21] For the use of Plotinian emanation and other emanative schemes by Christian writers, see John M. Gay, "Four Medieval Views of Creation," *Harvard Theological Review* 66 (1963), 243-273; Harry Austryn Wolfson, "The Identification of *Ex Nihilo* with Emanation in Gregory of Nyssa," *Harvard Theological Review* 63 (1970), 53-60; idem, "The Meaning of *Ex Nihilo* in the Church Fathers, Arabic and Jewish Philosophy, and St. Thomas," in *Mediaeval Studies in Honor of J.D.M. Ford* (Cambridge, MA: Harvard University Press, 1948), pp. 355-370. It should be remembered that the Nicene creed's doctrine of Christ as *lumen de lumine* is a phrase which consciously echoes Plotinian light terminology, see *Enneads*, IV, 3, 17 (Armstrong, IV, p. 89). On the importance of the Neoplatonic triad to Christian thinkers, see I.P. Sheldon-Williams, in *The Cambridge History of Later Greek and Early Medieval Philosophy*, ed. A.H. Armstrong, reprint (Cambridge: Cambridge University Press, 1970), p. 431; Sorabji, *Time, Creation and the Continuum*, pp. 307-318.

[22] On Eriugena the seminal study is Maïeul J. Cappuyns, *Jean Scot Érigène: Sa vie, son oeuvre, sa pensée* (Brussels: Culture et civilisation, 1964; first edn 1933). But see also *Eriugena Redivivus: Zur Wirkungsgeschichte seines Denkens im Mittelalter und im Ubergung zur Neuzeit*, ed. Werner Beierwaltes (Heidelberg: Carl Winter, 1987); Dermot Moran, *The Philosophy of John Scottus Eriugena: A Study of Idealism in the Middle Ages* (Cambridge: Cambridge University Press, 1989). On light in Eriugena, see James McEvoy, "Metaphors of Light and

179

authority of the Aeropagite, whom the Latin West esteemed as the disciple of St. Paul, this was not seriously questioned until the fifteenth century.[23] The Pseudo-Dionysius is a key figure in medieval thought, and he exercised a strong and distinctive influence on medieval mysticism in general and light speculation in particular. Having absorbed the complex patrimony of Athenian Neoplatonism with its technical vocabulary and logical categories, he applied it to biblical and liturgical themes. He thus spoke of Trinitarian distinctions as processions, applied the concept of overflow to God, and adapted the Proclean triad to divine love.[24] It is vital, of course, to make the appropriate distinctions between Athenian Neoplatonism, on the one hand, and the Pseudo-Dionysius' and Eriugena's highly personal syntheses of Neoplatonic thought with Christian doctrine and Scripture, on the other. Unlike Plotinus, the Pseudo-Dionysius believed that all things had a direct causal relation to God, and he never described one thing as emanating directly from another.

The emanative language adopted by these Christian authorities is especially prominent in the generation of scholars, cosmologists, and philosophers associated with the so-called Chartrian school. The luminous hierarchies elaborated by poets such as Bernard Silvestris and Alain of Lille owe much to these traditions of thought. In Silvestris' *Cosmographia*, for example, the World Soul is described as deriving from the Nous by emanation.[25] But the writings of the Pseudo-Dionysius

---

Metaphysics of Light in Eriugena," in *Begriff und Metapher: Sprachform des Denkens bei Eriugena*, ed. Werner Beierwaltes (Heidelberg: Carl Winter, 1990), pp. 149-167.

[23]The only dissenting voice was Peter Abelard. For an introduction to Dionysian thought, see Paul Rorem, *Pseudo-Dionysius: A Commentary on the Texts and an Introduction to Their Influence* (Oxford and New York: Oxford University Press, 1993); I.P. Sheldon-Williams, in *The Cambridge History of Later Greek and Early Medieval Philosophy*, ed. Armstrong, ch. 30, pp. 457-472. For more detailed primary and secondary material on the Pseudo-Dionysius and his putative influence on Dante, see below Ch. 7, pp. 239-245.

[24]References in order: (i) "processions," *De divinis nominibus*, c. 2, § 4 (*PG* 3, col. 640D; Chevalier, I, p. 74); (ii) concept of overflow, c. 4, § 9 (*PG* 3, col. 708A; Chevalier, I, p. 200); (iii) Proclean triad applied to divine love, c. 4, § 14 (*PG* 3, col. 705B; Chevalier, I, p. 193).

[25]For the use of emanative metaphors in twelfth-century writers, see Gregory, *Anima Mundi*, p. 92 and n. 3; idem, "The Platonic Inheritance," p. 73. For important poetic elaborations of

180

and Eriugena were not the only means by which medieval authors gained access to doctrines of emanation.[26] Islamic philosophers not only developed astrological ideas based on celestial influence, but they also put forward more metaphysical doctrines based on emanation.[27] These philosophers adopted emanation doctrines primarily because Aristotle had not established a creative god in Book XII (Lambda) of his *Metaphysica*. In this book, Aristotle attempted to demonstrate the existence of forty-seven (or fifty-five) immaterial Intelligences that he held responsible for moving the celestial spheres, and he argued that these Intelligences were presided over by one unmoved mover, the Prime Mover.[28] Yet he did not assign efficient causality (i.e. the production of motion) to his Prime Mover, since he believed that this first mover exercised final causality over all things. More crucially still, he did not treat the Prime Mover as a creative cause – this would have been a logical impossibility in his cosmos where being was not created but existed eternally.[29] For Islamic thinkers, Aristotle's inability to explain the origin of the universe was a serious flaw, one which led Avicenna to revise Aristotle's concept of efficient causality as mover or maker through motion by extending its

---

emanation which often harbour echoes of the Pseudo-Dionysius and Eriugena, see Alain de Lille, *Anticlaudianus*, lib. I, 500, ed. R. Bossuat (Paris: Vrin, 1955), p. 138; Bernard Silvestris, *Cosmographia (De mundi universitate)*, lib. I, c. 2, § 13; lib. II, c. 5, § 3, ed. Peter Dronke (Leiden: E.J. Brill, 1978), pp. 102, 128.

[26]The only version of Plotinian emanation (taken from *Enneads*, I, 1, 8) available to Latin writers before 1500 was found in Macrobius's *Commentarium in Somnium Scipionis* (see Ch. 4, n. 31).

[27]On emanation in Islamic philosophers, see Abd al-Rahman Badawi, *Histoire de la philosophie en Islam*, 2 vols. (Paris: Vrin, 1972), (Alkindi), II, pp. 417-418, (Alfarabi), II, pp. 538-545, (Avicenna), II, pp. 648-653; T.-A. Druart, "Al-Farabi and Emanationism," *Studies in Medieval Philosophy*, ed. John F. Wippel (Washington, DC: Catholic University of America, 1987), pp. 23-44. For emanationism in later Jewish and Muslim writers, see Arthur Hyman, "Maimonides on Creation and Emanation," *Studies in Medieval Philosophy*, ed. Wippel, pp. 45-61; Barry S. Kogan, "Averroes and the Theory of Emanation," *Mediaeval Studies* 43 (1981), 384-404.

[28]Aristotle's account of the movers in the *Metaphysica* seems difficult to reconcile with his view that the aether is the cause of circular motion in *De caelo*, I, 2, 269b 1-15.

[29]*Metaphysica*, XII, 7, 1072a 24-29, 1072b 14.

181

frame of reference to include creative causality.[30] In his own *Metaphysica*, Avicenna developed a doctrine which placed emanation firmly within a cosmological setting and emphasised the creative powers of efficient causes. He expounded the entire cosmological system contained in Book Lambda through an intricate system of successive emanations. God emanates a first Intelligence that reflects upon itself and in turn emanates a second Intelligence, a celestial sphere and a soul:

Igitur ex prima intelligentia, inquantum intelligit primum, sequitur esse alterius intelligentiae inferioris ea, et inquantum intelligit seipsam, sequitur ex ea forma caeli ultimi et eius perfectio et haec est anima.[31]

In this way, emanations proceed triadically until ten spheres, ten souls, and nine Intelligences have been produced.

While Avicenna's model of emanation was particularly noteworthy, it should not be forgotten that Arabic thinkers put forward a variety of other emanation models, many of which combined emanationist conceptions with more traditional ideas about creation.[32] These versions of emanation either formed part

---

[30]Avicenna, *Avicenna latinus: liber de philosophia prima*, tr. VI, c. 1 (Van Riet, II, p. 292): "... intentio efficientis est inducere esse et dare ipsum post non esse" ("the intention of the efficient cause is to induce being and to produce this after non-being"). On the originality and importance of this revision, see Etienne Gilson, "Notes pour l'histoire de la cause efficiente," *AHDLMA* 29 (1962), 7-31; Michael E. Mamura, "The Metaphysics of Efficient Causality in Avicenna (Ibn Sina)," in *Islamic Philosophy and Theology: Studies in Honor of George F. Hournai*, ed. Michael E. Mamura (Albany: State University of New York Press, 1984), pp. 172-187. For Albert's use of this scheme, see W.B. Dunphy, "St. Albert and the Five Causes," *AHDLMA* 33 (1966), 7-21.

[31]Avicenna, *Avicenna latinus: liber de philosophia prima*, tr. X, c. 4 (Van Riet, II, p. 483): "Thus, out of the first intelligence by its reflecting upon the first cause there proceeds the being of another intelligence below it, and by its reflecting upon itself, there proceeds out of it the form of the last heaven and its perfection and this is soul."

[32]See Wolfson's essays cited in n. 21. See also Harry Austryn Wolfson, "The Meaning of *Ex Nihilo* in Isaac Israeli," *The Jewish Quarterly Review* n.s. 50 (1959), 1-12; idem, "The Platonic,

182

of a metaphysical system (the One, Mind, Soul) or were located within a cosmological setting. In Islamic thought, it is possible to distinguish two principal types of emanative scheme: one presents the universe as proceeding directly from the will and power of God; the other posits a series of intermediaries that assist the first cause in creation. The *De causis* is a particularly good example of the second type of doctrine. This pithy treatise on the first causes of the universe and their interrelations was translated by Gerard of Cremona *circa* 1167-1187 and it became immensely important to the metaphysical speculations of the Latin West.[33] Like their Arabic counterparts, thirteenth-century philosophers felt that this work filled the *lacuna* left by Aristotle's inability to explain the origin of being in the *Metaphysica*. In the course of its thirty-one (or thirty-two) propositions the *De causis* gave a structured exposition of the universe in which the first cause, which is said to be a pure light, was shown to exercise its power over all others causes, whilst being assisted in its creative activities by the Intelligences:

[...] causa prima non cessat illuminare causatum suum et ipsa non illuminatur a lumine alio, quoniam ipsa est lumen purum super quod non est lumen.[34]

Aristotelian and Stoic Theories of Creation in Hallevi and Maimonides," in *Essays in Honour of the Very Rev. Dr. J.H. Hertz, Chief Rabbi of Great Britain* (London, 1942), pp. 427-442.

[33] See Cristina d'Acona Costa, *Recherches sur le Liber de Causis* (Paris: Vrin, 1995), pp. 195-258; the essays collected in *Pseudo-Aristotle in the Middle Ages: The Theology and Other Texts*, ed. Jill Kraye, W.F. Ryan, and Charles B. Schmitt (London: Warburg Institute, 1986); H.D. Saffrey, "L'état actuel des recherches sur le *Liber de causis* comme source de la métaphysique au moyen âge," *Miscellanea Mediaevalia* 2 (1963), 267-281.

[34] *De causis*, prop. V (VI), ed. Adriaan Pattin, in "Le *Liber de causis*. Édition établie à l'aide de 90 manuscrits avec introduction et notes," *Tijdschrift voor filosofie* 28 (1966), p. 105: "the first cause does not cease illuminating what it has caused and it is not illuminated by another light, since this is the pure light beyond which there is no light." On the doctrinal elements in the *De causis*, see Theodore Sweeney, "The Doctrine of Creation in *Liber de Causis*," in *An Etienne Gilson Tribute*, ed. C.J. O'Neil (Milwaukee: Marquette University Press, 1959); Cristina d'Acona Costa, "Le fonti e la struttura del *Liber de Causis*," *Medioevo* 15 (1989), 1-38; idem, *Recherches*, pp. 53-194.

183

Emanation and creation doctrines were also developed by Jewish philosophers such as Isaac Israeli (900-955), Avicebron (c. 1021-1058), and Petrus Alfonsi (fl. 1120). Alfonsi and Israeli both presented doctrines in which emanation was reconciled with aspects of Aristotelian cosmology; and all three writers blended emanation with more traditional ideas on creation.[35] Israeli traced the origin of all things to two simple substances (matter and first form) and argued that their combination gave rise to an intellect, three rational souls, and a sphere whose movement produces the four elements. Following a similar scheme, Avicebron believed that God created universal matter and universal form *ex nihilo* and a series of emanations then proceeded to diversify these forms.[36] Given this rich doctrinal heritage, it is hardly surprising that many twelfth-century and thirteenth-century thinkers were attracted by emanationist modes of thought.[37]

[35]For Alfonsi's and Israeli's development of emanation within a cosmological framework, see Israeli, *Liber de definicionibus*, ed. T.J. Muckle, in *AHDLMA* 11-12 (1937-1938), pp. 312-318; Alfonsi, *Dialogus*, titulus primus (*PL* 157, cols 555-561). For the doctrinal background to these emanative schemes which in many respects are similar to Avicebron's, see M.-T. d'Alverny, "Pseudo-Aristotle, *De elementis*," in *Pseudo-Aristotle in the Middle Ages*, ed. Kraye, pp. 63-83, esp. p. 74; Giuseppe Battisti Saccaro, "Il Grossatesta e la luce," pp. 50-52. An earlier study is A. Altmann and S.M. Stern, *Isaac Israeli: A Neoplatonic Philosopher of the Early Tenth Century* (London and Oxford: Oxford University Press, 1958).

[36]For the divine creation of these two primordial principles, see Avicebron, *Fons vitae*, lib. V, c. 12, ed. Clemens Baeumker, in *Beiträge* I/2-4 (1892-1895), p. 279. For important sections dealing with emanation, creation, and related light imagery, see *Fons vitae*, lib. III, cc. 35, 52-53 (Baeumker, pp. 160-161, 195-198); lib. IV, cc. 14, 19 (Baeumker, pp. 242-245, 252-256). The Arabic sources of Avicebron's emanation doctrine are discussed by Jacques Schlanger, *La philosophie de Salomon Ibn Gabirol: Étude d'un néoplatonisme* (Leiden: E.J. Brill, 1968), pp. 190-209.

[37]For emanative doctrines in Gundissalinus, see *De processione mundi*, ed. Georg Bülow, in *Beiträge* 24/3 (1925), pp. 39-41; idem, *De unitate*, ed. Paul Correns, in *Beiträge* I/1 (1891), p. 5. For another late twelfth-century work which used Avicennan ideas on emanation, see the *Liber de causis primis et secundis*, in *Notes et textes sur l'avicennisme latin aux confins des XIIe-XIIIe siècles*, ed. Roland de Vaux (Paris: Vrin, 1934), pp. 97-102. Albert the Great explains the origin of being and the production of effects by first causes with emanationist ideas in his *De causis et processu universitatis*. The whole treatise, especially Book I, is a sustained exposition of Arabic views on *fluxus, processio*, and *emanatio*, see Alain de Libera, "Albert le Grand et Thomas d'Aquin, interprètes du *Liber de Causis*," *Revue des sciences philosophiques et théologiques* 74 (1990), 347-378; for Albert's use of Algazeli and Israeli in this work, see also pp. 364-365, 376-377. On the Neoplatonic elements in Albert's treatise, see further Theodore Sweeney, "'Esse

184

Light in Dante's Cosmos

On the basis of the preceding discussion, this section aims to determine the place of light in Dante's cosmology[38] and astrology,[39] areas of his thought which continue to receive critical attention, though not from this particular point of view. Because of the connection between the behaviour of light and causal action, special emphasis will be given to Dante's views on efficient causality and celestial influence. Such views will also be considered with regard to their bearing upon his more general conception of the cosmos.

Dante makes numerous statements on the transmission of "vertude celestiale" (he is more sparing in his use of the synonym, "influenza"), and a good starting-point is *Convivio* II, a book which is centred upon the idea that the angelic Intelligences have an important role to play in determining human dispositions. Dante devotes several chapters of this book to a digression aimed at assimilating the characteristics of the seven liberal arts to the various types of influence found in, and exercised by, the celestial bodies. An early chapter in his literal exposition of the *canzone*, "Voi che 'ntendendo il terzo cielo movete," reveals Dante's reason for this excursus. In describing how his mind was won over by thoughts of the

---

Primum Creatum' in Albert the Great's *Liber de Causis et Processu Universitatis*," *The Thomist* 44 (1980), 599-646.

[38]On Dante's cosmology (the *Questio* is, after all, a learned disquisition on cosmological matters), see Boyde, *Dante Philomythes*, part 1; Alison Cornish, "Dante's Moral Cosmology," in *Cosmology: Historical, Literary, Philosophical, Religious and Scientific Perspectives*, ed. Norriss S. Hetherington (New York and London: Garland, 1993), pp. 201-215; James Dauphiné, *Le cosmos de Dante* (Paris: Les Belles Lettres, 1984), with non-technical sections on light, pp. 35, 107-124; Alessandro Ghisalberti, "La cosmologia del Duecento e Dante," *Letture classensi* 13 (1984), 33-48. The editors of the *Enciclopedia dantesca* acknowledge Dante's cosmology as an "elemento di rilievo" (*ED* II, p. 235), although they treat his cosmological ideas under separate headings such as "Astronomia" and "Fisica."

[39]Ideale Capasso and Giorgio Tabarroni observe that Dante's references to astrological lore provide "strutture essenziali all'intelaiatura del tutto il poema e in particolare del *Paradiso*," see "Astrologia," in *ED* I, p. 431. For recent studies of Dante's astrology, see Richard Kay, "Astrology and Astronomy," in *The Divine Comedy and the Encyclopaedia of the Arts and Sciences*, ed. di Scipio and Scaglione, pp. 147-162; idem, *Dante's Christian Astrology*; Ernesto Travi, *Dal cerchio al centro: Studi danteschi* (Milan: Vita e pensiero, 1990), pp. 11-33.

185

"Donna Gentile," Dante says that he was overcome by the effect on him of a "spirito" descending from Venus. The details he provides are revealing:

Dico anche che questo spirito viene per li raggi de la stella: per che sapere si vuole che li raggi di ciascuno cielo sono la via per la quale discende la loro vertude in queste cose di qua giù. (II, vi, 9)

In Book III, while drawing an interesting analogy between the illumination of the Earth by the Sun and the irradiation of God's goodness into all things, Dante reiterates the idea that "virtue" is transmitted in light rays:

Ove è da sapere che discender la virtude d'una cosa in altra non è altro che ridurre quella in sua similitudine, sì come ne li agenti naturali vedemo manifestamente; che, discendendo la loro virtù ne le pazienti cose, recano quelle a loro similitudine [...] Onde vedemo lo sole che, discendendo lo raggio suo qua giù, reduce le cose a sua similitudine di lume, quanto esse per loro disposizione possono da la [sua] virtude lume ricevere [...] Ove ancora è da sapere che lo primo agente, cioè Dio, pinge la sua virtù in cose per modo di diritto raggio, e in cose per modo di splendore reverberato; onde ne le Intelligenze raggia la divina luce sanza mezzo, ne l'altre si ripercuote da queste Intelligenze prima illuminate. (III, xiv, 2-4)

In both these passages, then, Dante shows that the planets and stars possess celestial "virtude," which radiates from them in the form of light rays. This is a doctrine which is reaffirmed in his poetry, especially the *Comedy*, where he repeatedly describes "virtue" as shining out of the heavenly bodies in the form of planetary light.[40]

---

[40]See the following passages where "virtue" is often, though not exclusively, connected with the sun's rays: *Rime* LXXXIII, 96-101; *Con.* IV, i, 11; *Purg.* XXXII, 52-57; *Par.* II, 143; VIII, 2-3;

186

Light is not the only means by which Dante explains the passage of celestial influence to the Earth, however. In *Convivio* II, Chapter xiii, he returns to the subject in a discussion of substantial generation and shows that he was well disposed to other explanations of how the heavens impart their influence. He now gives two further solutions: the spirits that move the spheres; and the stars themselves (para. 5). Dante repeatedly affirms that these spirits, the Intelligences, are the principal factors in the process of celestial causation,[41] endorsing fully the more narrowly Aristotelian view that celestial movement, which is derived from the action of the Intelligences on their spheres, is a primary cause of generation on Earth.[42]

Interestingly, in *Paradiso* VII, the poet combines both movement and light rays, when, referring to the Intelligences, he describes "lo raggio e 'l moto de le luci sante" (l. 141) as the instruments by which the stars impress forms into matter. In this section of the canto the idea that sublunary forms are generated by the "virtù informante" of the stars is unusual, but the way in which this is achieved (by radial transmission) is not. Dante's view that light rays are an instrument of celestial causation can in fact be found in a wide range of scholastic writers, including his more traditional authorities.[43]　Albert and Aquinas, for example, discuss celestial influence in a way which bears quite close similarities to Dante's own comments.　Hence, in a treatise known directly to Dante, the *De causis*

XII, 46-51; XIII, 61-63, 69; XXI, 14-15; XXII, 112-113; XXV, 54, 70; XXVII, 144. The idea is implicit in *Par.* I, 37-45.

[41]For Dante's references to the angelic movers as celestial causes, see *Con.* II, vi, 6; *Par.* II, 127-129; VIII, 109-111, 127-128.

[42]References to movement as a source of celestial causation include: *Rime* CVI, 93; *Con.* II, vi, 5; II, viii, 5; III, xv, 14-17; IV, ii, 6-7; *Purg.* XVI, 67-73; XX, 13-15; *Par.* XIII, 66.

[43]On p. 96 of his *Dante's Angelic Intelligences: Their Importance in the Cosmos and in pre-Christian Religion* (Rome: Edizioni di storia e letteratura, 1983), Stephen Bemrose argues that light is "[p]erhaps the least controversial feature of his [sc. Dante's] theory." For pertinent earlier observations of this point, see Boyde, *Dante Philomythes*, p. 253; Nardi, *Saggi di filosofia dantesca*, p. 353.

proprietatum elementorum, Albert makes the following digression in discussing the reasons for tides:

[...] licet omnes planetae communiter habeant in inferioribus effectum, tamen sol et luna praecipui planetae sunt, quorum proprietates et virtutes sequuntur inferiora propter tres causas, quarum prima est quantitas luminis ipsorum; alii enim planetae lucentia quidem corpora sunt et motu et lumine movent inferiora, sed radios et umbras notabiles non emittunt super inferiora; luminaria autem, que sunt sol et luna, et motu et lumine et radiis movent.[44]

Similarly, in discussing the way in which the heavens induce forms in his *De caelo et mundo*, Albert notes that:

[...] forma, quam explicat [sc. the orb] per motum, non determinatur per motum tantum, sed etiam per situm et luminis sui virtutem. Et ideo orbes inferiores in astronomia dicuntur iungi superioribus et non e converso, quia inferiores accipiunt influentiam a superioribus et non e converso; et hoc praecipue est in sole, qui iungitur multis superioribus et magnam habet virtutem ex lumine suo.[45]

---

[44]Albert, *De causis proprietatum elementorum*, lib. I, tr. 2, c. 10 (Hossfeld, p. 80): "although all planets commonly affect lower bodies, the sun and the moon are however principally the planets whose properties and powers influence lower bodies. This is due to three causes, the first of which is the quantity of light in them. The other planets are certain lucent bodies and they move what is below them by their light and movement, but they do not project notable rays and shadows onto lower bodies; the luminaries, however, which are the sun and the moon, move what is below them by their movement, light, and rays"; cf. lib. I, tr. 2, c. 4 (Hossfeld, p. 67): "[...] virtutes caelestes transportantur per aërem in radiis stellarum" ("celestial virtues are transported through the air in the rays of stars"). In the accompanying treatise, *De natura locorum*, tr. I, c. 4 (Hossfeld, p. 7), Albert refuses to identify *virtus* with *lumen*, but he does accept that "virtue" descends in light rays.

[45]Albert, *De caelo*, lib. II, tr. 3, c. 14 (Hossfeld, pp. 175-176): "form, which is produced by movement, is not determined by movement alone but also by the site and virtue of its light. And hence astronomers say that the lower orbs are joined to the higher ones and not the converse, since the lower ones accept influence from the higher ones and not the converse; and this is especially so with the sun, which is joined to many higher bodies and has great power due to its light." In his commentary on *Super Dionysius De divinis nominibus*, c. 2, q. 32 (Simon, p. 65),

188

Many relevant discussions of related matters are also found in Aquinas' writings as is revealed by the following comments which are taken from two of his later works:

[...] lumen agit quasi instrumentaliter in virtute corporum caelestium ad producendas formas substantiales [...] Unde dicimus, quod sicut corpora elementaria habent qualitates activas, per quas agunt, ita lux est qualitas activa corporis caelestis, per quam agit.[46]

It is worthwhile to compare Dante's views on this subject with the work of other thirteenth-century writers, especially Grosseteste and the "perspectivists." A number of Arabic and Jewish philosophers had maintained that efficient causality in the sublunary world was closely based upon the action of light.[47] In the thirteenth century, Grosseteste, Bacon, and Witelo adhered to this tradition by arguing that straight-line patterns of light were the primary efficient cause of a wide range of natural phenomena. Both Dante and Grosseteste express the commonplace view

---

Albert states that light is the instrument of the Primum Mobile. On Albert's knowledge and use of astrology, see Paolo Zambelli, "Albert le Grand et l'astrologie," *RThAM* 49 (1982), 141-158.

[46]Aquinas, *ST*, Ia, q. 67, a. 3, ad 3 (Caramello, p. 329): "light acts as it were instrumentally in the virtue of celestial bodies to produce substantial forms"; idem, *In De anima*, lib. II, lect. 14, § 420 (Pirotta, p. 106): "Hence, our own conclusion is that, just as the corporeal elements have certain active qualities through which they affect things materially, so light is the active quality of the heavenly bodies; by their light these bodies are active" (trans. Foster and Humphries, p. 268) For a study of Aquinas' views on celestial causality with 130 relevant texts, see Thomas Litt, *Les corps célestes dans l'univers de St. Thomas d'Aquin* (Louvain and Paris: Peeters and Vrin, 1963), pp. 110-124. Similar examples are found widely elsewhere, see e.g. Bartholomew of Bologna, *Tractatus de luce*, pars IV, c. 7 (Squadrani, p. 478); Pseudo-Aquinas, *In De generatione et corruptione*, lib. II, lect. 10, § 260 (Spiazzi, p. 575); Vincent of Beauvais, *Speculum maius*, lib. XV, c. 48 (I, col. 1121). For an important poetic treatment of "influance," "rais," and "li cors celeste," see Jean de Meun, *Roman de la Rose*, ll. 17473-17496, 19482-19488 (Lecoy, III, pp. 24-25, 85).

[47]The idea that all causal action follows the pattern of light is found in certain Arabic and Jewish philosophers, see Alkindi's *De radiis* (quoted in Ch. 1, n. 34); Avicebron, *Fons vitae*, lib. III, c. 52 (Baeumker, p. 196): "Essentiae substantiam simplicium non sunt defluxae, sed vires earum et radii, haec sunt quae defluunt et effunduntur" ("The substance of simple essences are not emitted but it is their powers and rays that issue forth and are poured out").

that light rays are causal determinants in the production of foliage, the rainbow, and colour. But in marked contrast to Grosseteste, Dante does not base his ideas about movement, sound, volcanoes, and tides upon light.[48] As noted in Chapter 1, Grosseteste and other specialist optical writers also believed the lines, angles, and figures formed by light rays could be used to analyse the entire causal system.[49] Yet there is no evidence in Dante's writings that he ever accepted this doctrine; he neither assigns causation exclusively to light rays, nor does he place his knowledge of geometrical optics at the service of his astrology.

Apart from the notion of radial transmission, the passage quoted earlier from *Convivio* III, xiv is interesting in many other respects. First, these paragraphs show that Dante had thoroughly assimilated the idea that efficient causality is based on the transmission of likeness: a fundamental principle which he develops for a variety of purposes in many of his writings. Second, it is important to recognise

---

[48]For light as the single efficient cause of motion and sound, see Grosseteste, *De motu corporali et luce* (Baur, p. 90): "[...] motio corporalis est vis multiplicativa lucis" ("the movement of bodies is due to the power of light to multiply itself"); *Comm. in Post. analy.*, lib. II, c. 4 (Rossi, pp. 385-386): "Substantia enim soni est lux incorporata in subtilissimo aere" ("The substance of sound is light which has been incorporated within the most subtle parts of air"). For further discussion and documentation, see James McEvoy, *The Philosophy of Robert Grosseteste*, pp. 169-172, 293-295; idem, "The Sun as *res* and *signum*: Robert Grosseteste's Commentary on *Ecclesiasticus*, ch. 43, vv. 1-5," *RThAM* 41 (1974), 38-92. See also Vescovini, *Studi sulla prospettiva*, p. 20, who argues that Grosseteste developed "un concetto della causalità fondato sulle proprietà e sulle caratteristiche della luce." For similar ideas in the "perspectivists," see Bacon, *Opus maius*, lib. IV, dist. 2, c. 1 (Bridges, I, p. 112); idem, *Opus tertium*, cc. 36-37 (Brewer, pp. 117-119); Pecham, *Perspectiva communis*, lib. I, prop. 27 (Lindberg, p. 108); Witelo, *Perspectiva*, prol. (Risner, pp. 1-2). The author of the *De intelligentiis* gives an extreme version of the doctrine of light as universal efficient cause, see prop. VII (Baeumker, p. 8): "Omnis substantia influens in aliam est lux in essentia vel naturam lucis habens" ("Every substance which exercises influence over another is by essence light or has the nature of light"). Cf. *Ep.* XIII, 70: "[...] cum omnis vis causandi sit radius quidam influens a prima causa que Deus est" ("since all causal power is like a ray of influence which descends from the first cause, which is God"); and see Bortolo Martinelli, "La dottrina dell'Empireo," pp. 121-124.

[49]On the use of geometrical optics to analyse pyramids and rays of light incident on the Earth, see Grosseteste, *De lineis* (Baur, pp. 64-65); idem, *De natura locorum* (Baur, pp. 66-68). For the "perspectivists," see Bacon, *De multiplicatione specierum*, pars II, cc. 9-10 (Lindberg, pp. 162-172); idem, *Opus maius*, lib. IV, dist. 3, c. 2 (Bridges, I, pp. 123-124); Pecham, *Perspectiva communis*, lib. I, props. 6 and 21 (Lindberg, pp. 64 and 98); Witelo, *Perspectiva*, lib. II, prop. 52 (Risner, p. 109).

traces of a number of propositions from the *Liber de causis* in Dante's discussion.[50] Third, and perhaps most significant of all, the entire passage illustrates Dante's keen interest in intuiting and developing analogies between natural and divine light. Although the question of direct influence is problematic, the Pseudo-Dionysius' writings contain many similar analogies, some of which will be explored in the following chapter. Fourthly, the sequence from *Convivio* III, xiv also reveals one of the more philosophically distinctive aspects of Dante's causal theory – his preoccupation with the angelic Intelligences as secondary causes. As is now well known, Dante was deeply concerned with the role of the Intelligences as God's instruments in the transmission of effects.[51] In almost all of his writings, he displays a deep-seated fascination with tracing the path of efficient causes that move the universe, showing causal transmission to take place across the celestial orbs and their movers.[52] In an earlier chapter of *Convivio* III, where Dante puts forward the idea of love as a unitive force and from here affirms the natural instinct to love Lady Philosophy, it is the *De causis* which once again provides the framework of ideas upon which his argument is articulated

[50]For the relevant propositions (I, IV, IX, and XVIII), see Vasoli, *Il Convivio*, pp. 454-455. Dante's borrowing from the *De causis* is indicated by a distinctive terminological echo of prop. IX: Dante renders the Latin "absque medio" as "sanza mezzo."

[51]Kenelm Foster, "Tommaso d'Aquino," in *ED* V, pp. 637-638; Bruno Nardi, *Dal "Convivio" alla "Commedia" (Sei saggi danteschi)* (Rome: Istituto storico italiano per il Medio Evo, 1960), pp. 37-47. More recently, see the study by Bemrose, *Dante's Angelic Intelligences*, esp. pp. 90-113. See also Ideale Capasso and Giorgio Tabarroni, "Cielo," in *ED* I, p. 1001: "I cieli [...] divengono strumenti di attuazione dell'azione providenziale e creatrice di Dio [...] divengono mediatori del processo discendente che va di Dio alle essenze animate e inanimate del mondo."

[52]Albert provides an important discussion of this idea in a treatise known to Dante, see *De intellectu et intelligibili*, lib. I, tr. 1, c. 4 (Jammy, II, p. 241b): "[...] omne ultimum in ordine causatorum non exit a prima causa nisi per causas quae in medio sunt. Ultima autem causata sunt formae generabilium et corruptibilium. Mediae autem causae sunt motores orbium caelestium, quos intelligentias caelestes vocaverunt Philosophi" ("every final effect in a causal order only issues from the first cause insofar as it comes through intermediary causes. The final caused effects are the forms of generated and corruptible things. The intermediary causes are the motor-powers of the celestial orbs which philosophers called the celestial intelligences"), quoted by Nardi, *Saggi di filosofia dantesca*, p. 68.

191

Ciascuna forma sustanziale procede da la sua prima cagione, la quale è Iddio, sì come nel libro Di Cagioni è scritto, e non ricevono diversitade per quella, che è simplicissima, ma per le secondarie cagioni e per la materia in che discende. Onde nel medesimo libro si scrive, trattando de la infusione de la bontà divina: «E fanno[si] diverse le bontadi e li doni per lo concorrimento de la cosa che riceve».[53] Onde, con ciò che sia cosa che ciascuno effetto ritegna de la natura de la sua cagione – sì come dice Alpetragio quando afferma che quello che è causato da corpo circulare ne ha in alcuno modo circulare essere –, ciascuna forma ha essere de la divina natura in alcun modo: non che la divina natura sia divisa e comunicata in quelle, ma da quelle è participata, per lo modo quasi che la natura del sole è participata ne l'altre stelle.[54]

Here, Dante adopts the principle that effect resembles cause to show that the human form is the noblest of all those that proceed from God. To avoid identifying cause with effect, he takes care to explain how diversity arises: this is a result of the imperfection of secondary causes and matter.[55] He then integrates these arguments with a doctrine of participation: all being is said to come from God, and beings can only acquire existence by sharing in this divine nature. Clearly, the language and thought of these paragraphs express a richly Neoplatonic view of the relation between Creator and created, although there are some more traditional

---

[53]See De causis, prop. XIX (XX), 157 (Pattin, p. 177): "Prima enim bonitas influit bonitates super res omnes influxione una; verumtamen unaquaeque rerum recipit ex illa influxione secundum modum suae virtutis et sui esse" ("Goodnesses flow into all things from the first goodness in one influx; however, everything receives that influx of things according to the mode of its virtue and being"). Interestingly, Vasoli has also related this passage to Dionysian texts, see Il Convivio, p. 302, where he cites De divinis nominibus, c. 4, § 4 (PG 3, col. 697C; Chevalier, I, p. 162).

[54]Con. III, ii, 4-5. For the Neoplatonic elements in this passage, Dante's use of the De causis, and the reference to Alpetragius, see Nardi, Saggi di filosofia dantesca, pp. 106-108, 161-166.

[55]For the idea of matter as the cause of imperfection, see Con. III, iv, 7; III, vi, 6. Dante assigns earthly imperfection to the heavens and matter in Par. XIII, 73-78.

192

elements here too.[56] As far as light terms are concerned, there are possible echoes of the language of the *De causis* in the words ("procede," "discende," "infusione," "comunicata") that Dante uses to describe the descent of substantial forms.

Light imagery, verbs related to light, and technical nouns such as "infusione" and "influenza" also play an important part in Dante's digression on human generation in *Convivio*, IV, xx-xxiii.[57] In this section of chapters, Dante affirms the importance of celestial influence on man by giving an extended treatment of the relationship between the "cielo" and "umana nobilitade," a fundamental theme of Arabic philosophy. Concepts built around light imagery are one of the principal ways in which Dante develops this concern, and it is revealing that he begins his exposition by citing a celebrated biblical text from the Book of James (Jam 1:17) which contains a strong suggestion of a downward flow of light.[58]

Nardi has shown that Dante borrows from Albert's *De natura et origine animae* for general and specific points of doctrine in these chapters,[59] and it thus seems reasonable to point out Albertine derivations for Dante's use of nouns and verbs related to light. In this treatise, and also in his *De causis et processu*

---

[56]The important theme of reception *per modum recipientis* is developed throughout the *De causis*, see props IX (X), 99 (Pattin, p. 160); XVIII (XIX), 149 (Pattin, p. 174); XIX (XX), 157-158, 161 (Pattin, pp. 177-178, 80); XXI (XXII), 170 (Pattin, p. 182); XXIII (XXIV), 177-180 (Pattin, pp. 185-187). Although the doctrine of participation has Platonic foundations, the idea of substantial forms is Aristotelian.

[57]Dante uses the following terms in *Con.* IV, xx-xxi: ("discende"), xx, §§ 6, 8, 10; xxi, § 1; ("infusione"), xx, § 7; ("raggio divino"), xx, § 8; ("procedere"), xxi, § 2; ("multiplicare"), xxi, § 8; ("influenza"), xxi, § 9. All these words are applied either to light (the sun or divine rays) or to God's "bontade" or "vertude" which is elsewhere (cf. *Con.* III, vii, 2-5; III, xiv, 2-4) identified symbolically with light. See also *Ep.* XIII, 61: "[...] divinum lumen, id est divinam bonitatem" ("divine light, that is, divine goodness").

[58]Jam. 1:17: "Omne datum optimum, et omne donum perfectum de sursum est, descendens a Patre luminum" ("Every perfect gift and every good gift is from above, and comes from the father of lights"). This passage is used by the Pseudo-Dionysius to enunciate the theme of procession and return in his *De caelesti hierarchia*, c. 1 (*PG* 3, col. 120A; Chevalier, II, p. 727).

[59]Bruno Nardi, "L'origine dell'anima umana secondo Dante," in his *Studi sulla filosofia medievale* (Rome: Edizioni di storia e letteratura, 1960), pp. 9-68; idem, *Dante e la cultura medievale*, pp. 207-224.

193

universitatis, Albert constantly uses verbs such as *fluere*, *influere*, *multiplicare*, *diffundere*, and *procedere* to refer to the way in which first causes produce their effects.[60]

In adopting "influire" and "infondere" elsewhere in his writings, Dante is very often translating Latin verbs from propositions found in the *De causis*. By the late thirteenth century it had become common for scholastic writers to express the *causalitas causae efficientis* with a variety of verbs related to water and, by extension, to light (*diffundere*, *influere*, etc.)[61] Aquinas is no exception, and he repeatedly describes the behaviour of light with verbs such as *diffundere*, *procedere*, and *multiplicare*.[62] Dante's use of a verb such as "multiplicare" with

---

[60]Albert's *Liber de natura et origine animae*, ed. Bernhard Geyer (Münster: Aschendorff, 1955) is a key text for Dante's soul doctrine in general and for his use of concepts such as *influere* in particular. See esp. tr. I, c. 5 (Geyer, p. 14): "complementum ultimum, quod est intellectualis formae, non per instrumentum neque ex materia, sed per suam lucem influit intellectus causae purus et immixtus" ("the final complement, which is the complement of the intellectual form, is not produced instrumentally nor does it come out of matter, but through its light the pure and unmixed intellect flows out causes"), quoted by Vasoli, *Il Convivio*, p. 767. For Albert's treatment of verbs such as *irradiare*, *fluere formas*, *multiplicare*, and the nouns *expansio luminis* and *fluxus*, see e.g. *Liber de natura et origine animae*, tr. II, c. 4 (Geyer, pp. 23-24). In Albert's *De causis et processu universitatis a prima causa*, verbs such as *influere*, *fluere*, and their cognate forms are especially prevalent, see lib. I, tr. 4, ed. Winfried Fauser (Münster: Aschendorff, 1993), pp. 42-58, and most notably, cc. 1-3 (Fauser, pp. 42-46), entitled: "Quid sit fluxus"; "Quid sit influere"; "De modo fluxus et influxus."

[61]For Dante's other uses of "infondere" and "infusione," either as applied to light and "divina bontade," and/or as translated from the *De causis*, see *Con*. III, ii, 4; III, vi, 12; III, xiii, 9; IV, v, 13; IV, xx, 7; *Par*. XIII, 44. For "influenza" and "influire," terms which are translated from the *De causis* and/or used to express cosmological and astrological ideas, see *Rime* LXXXIII, 96-101; *Con*. III, iv, 18; III, xv, 15; IV, xxi, 9; *Par*. IV, 59; *Questio*, 67 and 70; cf. *Ep*. XIII, 56-57.

[62]For Aquinas' use of *influentia* and *influxus*, see *Index Thomisticus*, ed. Busa, sec. II, conc. prima, vol. 11, pp. 757-765. For specific examples, see *ST*, Ia, q. 21, a. 4, resp. (Caramello, p. 124), where Thomas follows *De causis*, prop. I, and refers *influere* to the action of the first cause. See also Aquinas, *In librum de causis expositio*, lect. 1, § 35, ed. Ceslai Pera, 2nd edn (Turin: Marietti, 1972), p. 6, where he states that "causalitas causae efficientis consideratur secundum influentiam." For *influxus*, see *Quaestiones quodlibetales*, qd. 3, a. 2, resp. ed. Raymund M. Spiazzi, 9th edn (Turin: Marietti, 1956), p. 45: "[...] superiores angeli possunt agere in inferiores angelos et in animas nostras, sicut id quod est in actu, agit in id quod est in potentia; et huiusmodi actio dicitur influxus" ("higher angels can act on lower ones and on our souls, as something in act can act on what is in potency; and this kind of action is called influx"). See also the Thomist comparison of God's creative *influxus* to the action of physical light in his *In librum de causis*, lect. 20, § 362 (Pera, p. 111). On the multiplication of light, see Aquinas, *In De caelo*, lib. I, lect. 10, § 393 (Spiazzi, p. 194): "plures species rerum generantur ex virtute radiorum solis

194

reference to light thus provides no evidence of a distinctive debt to Grosseteste or Bonaventure, as is sometimes assumed.

As Dante prepares to discuss how man receives his rational soul, he refers to Guinizzelli's "Al cor gentil rempaira sempre Amore," borrows Guinizzelli's comparison of "virtue" descending into a precious stone, and adapts this for his own purposes in order to indicate the two-phase process by which man acquires a rational soul. This allusive web of citations and light comparisons underscores Dante's familiarity with the scholastic distinction between dispositive and perfective causes:

Dice [sc. a line from "Le dolci rime") adunque che Dio solo porge questa grazia a l'anima di quelli cui vede stare perfettamente ne la sua persona, acconcio e disposto a questo divino atto ricevere [...] onde se l'anima è imperfettamente posta, non è disposta a ricevere questa benedetta e divina infusione: sì come se una pietra margarita è male disposta, o vero imperfetta, la vertù celestiale ricever non può, sì come disse quel nobile Guido Guinizzelli in una sua canzone, che comincia: *Al cor gentil ripara sempre Amore.*[63] Puote adunque l'anima stare non bene ne la persona per manco di complessione, o forse per manco di temporale: e in questa cotale questo raggio divino mai non risplende. (IV, xx, 7-8)

In the *Monarchia*, Dante reworks many of his ideas on celestial causation and human generation, using them as the basis for his own brand of closely

---

et stellarum, quae per reflexionem circa terram multiplicantur" ("many species of things are generated out of the virtue of the rays of the sun and the stars, which are multiplied by reflection on the Earth"); *In II Sent.*, dist. 13, q. 1, a. 3, ad 3, in *Opera omnia*, VI, p. 500b: "non sequitur calor sicut ex confrictione corporum, sed quia ex concurrentibus multis radiis ad unum punctum oportet multiplicari lumen, et ex hoc multiplicatur calor" ("heat is not produced by the friction of bodies, but because light is necessarily multiplied by the concurrence of many light rays to one point, and it is due to this that heat is generated").

[63]For Guinizzelli's well-known *canzone*, see *Poeti del Duecento*, ed. Contini, II, p. 461, esp. ll. 11-14: "Foco d'amore in gentil cor s'aprende / come vertute in petra preziosa, / che da la stella valor no i discende / anti che 'l sol la faccia gentil cosa." See also ll. 16-20, 31-42.

195

reasoned political polemics. In *Monarchia* I, Chapter ix, he starts with the premise that mankind is the offspring of the heavens ("Humanum genus filius est celi"). With propositions from the *De causis* in mind, he then argues that man as the effect produced by a higher causal principle will possess a greater state of perfection if he assimilates himself as fully as possible to his first celestial cause, the mover of the Primum Mobile. It is now possible for Dante to draw a correspondence between the single cause regulating the cosmos and the single ruler who should regulate man in his earthly life:

Humanum genus filius est celi, quod est perfectissimum in omni opere suo: generat enim homo hominem et sol, iuxta secundum *De naturali auditu*. Ergo optime se habet humanum genus cum vestigia celi, in quantum propria natura permictit, ymitatur. Et cum celum totum unico motu, scilicet Primi Mobilis, et ab unico motore, qui Deus est, reguletur in omnibus suis partibus, motibus et motoribus [...] humanum genus tunc optime se habet, quando ab unico principe tanquam ab unico motore, et unica lege tanquam unico motu, in suis motoribus et motibus reguletur.[64]

Despite its cosmological framework, this tightly argued passage contains no suggestion of the downward flow of divine causal power, and its accompanying light imagery, which Dante repeatedly describes in the *Convivio*. In other sections

---

[64]*Mon*. I, ix, 1-2: "Mankind is the son of heaven, which is quite perfect in all its workings: for man and the sun generate man, as we read in the second book of the *Physics*. Therefore mankind is in its ideal state when it follows the footsteps of heaven, insofar as its nature allows. And since the whole sphere of heaven is guided by a single movement [i.e. that of the Primum Mobile], and by a single source of motion [who is God] in all its parts, movements and causes of movement [...] mankind is in its ideal state when it is guided by a single ruler [as by a single source of motion] and in accordance with a single law [as by a single movement] in its own causes of movement and in its own movements," trans. Prue Shaw, in *Dante's "De Monarchia"* (Cambridge: Cambridge University Press, 1995), pp. 19-21. For the Aristotelian source of the idea that *homo enim hominem generat et sol*, see above n. 4. Dante repeatedly draws connections between the advent and/or demise of Universal Monarchy and the disposition of the heavens, see *Con*. IV, v, 13; *Mon*. I, viii, 2; III, xv, 12; *Par*. XXX, 137-138.

196

of the *Monarchia* and in the *Questio*, though, he does express God's causality as a descent of divine power and a diffusion of goodness, and, in so doing, reworks images modelled on the flow of water from a source.[65] The causal scheme discussed at length in *Monarchia* II, Chapter ii is a fine example of how Dante conceived divine power to act through a chain of efficient causes. In this chapter, the principle that effect resembles cause again underpins a dense chain of argumentation by which Dante asserts the divinely-willed origin of the Roman Empire. Given that God is the first cause and that terrestrial effects bear similarity to their celestial first cause, it follows that the Empire resembles God to some degree. The first cause produces its effects by secondary agents, and any imperfect realisation is assigned to matter, not to matter and the heavens together, which is Dante's position in the *Convivio* and later again in the *Paradiso*:[66]

Est enim natura in mente primi motoris, qui Deus est; deinde in celo, tanquam in organo quo mediante similitudo bonitatis ecterne in fluitantem materiam explicatur [...] si contingat peccatum in forma artis, materie tantum imputandum est [...] et

---

[65]See the following passages in which Dante develops quite similar patterns of thought and imagery: *Mon.* I, viii, 2: "De intentione Dei est ut omne causatum divinam similitudinem representet in quantum propria natura recipere potest [...] 'ad similitudinem' tamen de qualibet dici potest, cum totum universum nichil aliud sit quam vestigium quoddam divine bonitatis" ("It is God's intention that every created thing should show forth his likeness insofar as its own image can receive it [...] after our likeness can be said of anything, since the whole universe is a likeness of divine goodness," trans. Shaw, p. 19); III, xv, 15: "patet quod auctoritas temporalis Monarche sine ullo medio in ipsum de Fonte universalis auctoritatis descendit: qui quidem Fons in arce sue simplicitatis unitus, in multiplices alveos influit ex habundantia bonitatis" ("Thus it is evident then that the authority of the temporal monarch flows down into him without any intermediary from the fountainhead of universal authority; this fountainhead, though one in the citadel of its own simplicity of nature, flows into many streams from the abundance of its goodness," trans. Shaw, p. 149); *Questio*, 46: "cum omnes forme, que sunt in potentia materie, ydealiter sint in actu in Motore celi [...] si omnes iste forme non essent semper in actu, Motor celi deficeret ab integritate diffusionis sue bonitatis, quod non est dicendum" ("since all the forms which are in potency in matter are in act as ideas contained in the Mover of heaven [...] if all these forms were not always in act then the Motor of heaven would fail to ensure the full diffusion of its goodness, which is not possible").

[66]See the passages quoted earlier from *Con.* III, ii, 3 and the references given in n. 55.

quod quicquid est in rebus inferioribus bonum, cum ab ipsa materia esse non possit, sola potentia existente, per prius ab artifice Deo sit et secundario a celo, quod organum est artis divine, quam "naturam" comuniter appellant. Ex hiis iam liquet quod ius, cum sit bonum, per prius in mente Dei est [...] sequitur quod ius a Deo, prout in eo est, sit volitum [...] Et iterum ex hoc sequitur quod ius in rebus nichil est aliud quam similitudo divine voluntatis.[67]

Similarly in the *Questio*, divine providence is shown to operate through a chain of celestial causes that are linked by the transmission of likeness. Dante explains that the position of the sphere of water is the result of the diverse collection of stars in the crystalline sphere. Each of these stars, he argues, differs in magnitude and luminosity, and they therefore impart differing "influences" or "virtues" (interestingly he treats these words as synonyms here) to the Earth. The combined effect of these disparate influences is necessarily irregular. But this irregularity has a felicitous outcome: a habitable land-mass rises above the sphere of water.[68]

---

[67]*Mon.* II, ii, 2-5: "For nature exists in the mind of the first mover, who is God; then in the heavens, as in the instrument by means of which the image of eternal goodness is set forth in fluctuating matter [...] if a flaw appears in the work of art it is to be imputed exclusively to the material [...] and whatever good there is in earthly things, since it cannot come from the material (which exists only as a potentiality), comes primarily from God the maker and secondarily from the heavens, which are the instrument of God's handiwork, which is commonly called 'nature.' From what has been said it is now clear that right, being a good, exists firstly in the mind of God [...] it follows that right is willed by God as being something which is in him [...] And again it follows from this that right is simply the image of divine will" (trans. Shaw, pp. 49-51).

[68]*Questio*, 70: "licet celum stellatum habeat unitatem in substantia, habet tamen multiplicitatem in virtute; propter quod oportuit habere diversitatem illam in partibus quam videmus, ut per organa diversa virtutes diversas influeret" ("although the crystalline sphere has unity with regard to its substance, it nonetheless has multiplicity with regard to its virtue; and on this account it must have that diversity in the parts we see, so that by different organs it can pass on diverse influences"). This passage should be read in close relation to *Par.* II, 115-117. For Dante's useful gloss of "virtus" as synonymous with "influentia," see *Questio*, 67: "Non in celum lune; quia cum organum sue virtutis sive influentie sit ipsa luna [...]" ("Not in the heaven of the moon, because the organ of its virtue or influence being this moon [...]").

198

All of the passages discussed so far show how Dante conceived nature to be a system of secondary causes regulated by a transmission theory cf causality, a system in which light and ideas related to celestial influence played an important role. In the *Comedy*, as we shall now see, Dante gives these ideas their most sustained and original formulation.

One of the earliest examples is found in the first substantive doctrinal section in the poem, Virgil's discourse on "Fortuna" in *Inferno* VII. Fortune is presented as a luminous Intelligence, similar to the other "splendori mondani" who move the spheres, and Virgil explains how God assigns her a special role in the execution of His providential designs in a passage which is noteworthy for its poetic intensification and unexpected anticipation of the motif of God's light permeating the universe.[69] Charles Martel's discourse on differentiation in canto VIII of the *Paradiso* makes no reference to light-mediating Intelligences, but it is even more important to the astrological themes pursued in the *Comedy*. God is again shown to act indirectly through a chain of secondary causes, using the "cielo" to realise the diversity of human dispositions.[70] Yet nowhere is Dante's understanding of the causal process expressed with greater force, set more firmly within a cosmological framework, and made to articulate a finer vision of diversity than in *Paradiso* II. His grandiose arrangement of celestial spheres, heavenly bodies, and angelic Intelligences shows how ideas associated with the cosmos, causality, and the transmission of energy provided Dante with the intellectual inspiration for some of his finest doctrinal poetry. What is more, the discourse also shows us how such concepts lent themselves to the cosmological light imagery that Dante develops in the *Paradiso*. Indeed, this canto can be used to demonstrate the leading role that light plays in many aspects of Dante's cosmology.

---

[69]*Inf.* VII, 73-84, esp. 73-76: "Colui lo cui saver tutto trascende, / fece li cieli e diè lor chi conduce / sì, ch'ogne parte ad ogne parte splende, / distribuendo igualmente la luce."

[70]*Par.* VIII, 109-117, and esp. 118-144. For other passages relevant to Dante's astrology, see the studies referred to in n. 39.

In medieval elaborations of the Ptolemaic-Aristotelian cosmos, the celestial spheres, stars, and planets were all believed to be formed from Aristotle's fifth element, the aether.[71] From Dante's isolated statements about this substance, it can be inferred that he followed the general view according to which aether was rarefied in the spheres[72] and more compact in the stars.[73] Line 32 of *Paradiso* II describes the surface of the moon as "lucida, spessa, solida e pulita," and hence it seems that Dante is following traditional teachings by relating the luminosity of the stars to the concentration of aether in them. This compact aetherial structure was thought to provide a dense and polished surface from which the light of the sun could be reflected.[74] It is therefore highly appropriate that Dante refers to the planets as "specchi" in the *Paradiso* and that his ascent takes place against a glittering backcloth of "splendori," "cristalli," and "margarite."[75]

[71]On the fifth element, see Aristotle, *De caelo*, II, 7, 289a 11-19; cf. I, 3, 270b 1-25.

[72]The transparency of the planetary spheres allows Dante-*personaggio* to look back through the orbs to the Earth in *Par.* XXII, 133-135; XXVII, 73-87.

[73]For the transparency of heavens, see Aristotle, *De anima*, II, 7, 418b 7-19; this passage states that air and water possess a certain substance which is also found in the upper eternal body. For aether as constituting the stars, see *De caelo*, II, 7, 289a 13. For Dante's explicit mentions of aether (both of which are generic), see *Par.* XXII, 132; XXVII, 70.

[74]The view that the density of aether is the cause of resplendence is developed most notably in commentaries on Aristotle's *De caelo*, see Albert, *De caelo et mundo*, lib. II, tr. 2; lib. II, tr. 3, c. 4 (Hossfeld, pp. 107, 149): "causa lucis in stellis est spissitudo earum, et lumen quidem in aliquibus recipitur in profundum ipsarum et in quibusdam diffunditur in superficie [...] necesse est caelum esse spissius et minus spissum, ut diversificetur suum instrumentum, quod lumen est [...] Et quod densitas maior est in ipsa stella, hoc est ideo, ut magis in ipsa commassetur nobilitas luminis, quae dat idoneitatem ad suscipiendum actum motoris" ("the cause of light in stars is their thickness, and light is received by some of them in their depths and in others it is diffused on their surface [...] the heavens must be more and less dense so that their instrument, that is, light, is diversified [...] And greater density in this star is so that the nobility of light is more accumulated in it, for it is light that allows the stars to receive the act of the motor intelligence"); Aquinas, *In De caelo*, lib. II, lect. 10, § 384 (Spiazzi, pp. 190-191): "[...] in corporibus caelestibus, quanto est maior congregatio per modum inspissationis, tanto magis multiplicatur luminositas et virtus activa, sicut patet in ipsis corporibus stellarum" ("[...] in celestial bodies the greater the compactness [of the aether] due to thickening, the more luminosity and active virtue is multiplied, as is clear in the bodies of the stars").

[75]For cosmological mirror-metaphors, see *Par.* XXI, 18; cf. *Purg.* IV, 62. For comparisons of planets to shining jewels and crystals, see *Par.* II, 33-34; VII, 127; XXI, 25.

200

In his earlier writings, Dante had argued that the planets do nct derive their light solely from the rays of the sun, but possessed some luminosity of their own, although the relationship between inherent and derived light is not discussed in detail.[76] In the *Paradiso*, a new and far more fascinating description of light in the heavenly bodies is given at the end of canto II, where Dante dwells with loving emphasis on the luminous relationship between the Intelligence of the eighth heaven, the fixed stars, and the many stars located within this heaven. The whole sequence is fundamental to Dante's cosmological system and his ideas on the transmission of causal power, and as such it deserves more detailed explanation.[77]

The being of the universe ("esser": l. 114) is contained in the universal undifferentiated influence ("virtute": l. 113) of the Primum Mobile, the body which rotates in the Empyrean ("ciel de la divina pace": l. 112). This influence is diversified through different essences in the next heaven, the sphere of the fixed stars ("Lo ciel seguente, c'ha tante vedute": l. 115). These diverse essences ("diverse essenze": l. 116) are in effect the stars and each of them is endowed with a distinctive influence or "virtue." The other seven celestial spheres ("Li altri giron": l. 118) take this task of successive differentiation still further in order to realise on Earth the various forms present within them (ll. 119-120):[78]

[76]For the stars as possessing a light of their own, see *Rime* LXXXI, 77-78; *Con.* III, ix, 11; *Mon.* III, iv, 18-19. Cf. *Con.* III, xii, 7: "Lo quale [sc. sole] di sensibile luce sé prima e poi tutte le corpora celestiali e le elementali allumina." For additional references to the stars receiving light from the sun, see *Rime* LXXXIII, 117; XC, 17; *Con.* II, vi, 9; II, xiii, 15; cf. Bruno Nardi, *Saggi e note di critica dantesca* (Milan and Naples: Ricciardi, 1966), pp. 404-407. For other aspects of the sun as planetary body and poetic symbol in Dante, see Antonietta Bufano, "Sole," in *ED* V, pp. 296-304; Jeannine Quillet, "'Soleil' et 'lune' chez Dante," in *Le soleil, la lune et les étoiles au moyen âge* (Aix-en-Provence: Publications du CUERMA, 1983), pp. 329-337.

[77]For the doctrinal context of the canto, see Nardi, "La dottrina delle macchie lunari nel secondo canto del «Paradiso»," in *Saggi di filosofia dantesca*, pp. 3-39; Leonardo Sebastio, "*Paradiso* II," in *Strutture narrative e dinamiche culturali in Dante e nel Fiore* (Florence: Olschki, 1990), pp. 55-95; Giorgio Stabile, "Navigazione celeste e simbolismo lunare in «Paradiso» II," *Studi medievali* ser. 3a/21 (1980), esp. pp. 125-140; Cesare Vasoli, "Il canto II del *Paradiso*," *Lectura Dantis Metelliana* 2 (1992), 27-51.

[78]For "essere" as the "first of effects" and the notion that determination takes place through lower causes, see *De causis*, prop. IV, 37 and prop. IX (X), 98 (Pattin, pp. 142, 16C): "Primum

Dentro dal ciel de la divina pace
si gira un corpo ne la cui virtute
l'esser di tutto suo contento giace.
Lo ciel seguente, c'ha tante vedute,
quell' esser parte per diverse essenze,
da lui distratte e da lui contenute.
Li altri giron per varie differenze
le distinzion che dentro da sé hanno
dispongono a lor fini e lor semenze. (112-120)

Reworking ideas first elaborated in Book II of the *Convivio*,[79] Beatrice then describes how the differentiated "influences" descend from above and are relayed across the orbs, which she refers to as "organi" of the universe:

Questi organi del mondo così vanno,
come tu vedi ormai, di grado in grado,
che di sù prendono e di sotto fanno. (121-123)

It is here that Dante draws upon two leading ideas, both to be found in the *De causis*, on which much of Beatrice's discourse depends: first, the notion that the universe is arranged in a series of rungs which form a hierarchy; and second, the principle that power is transmitted from the higher to the lower.

In the next tercet, Beatrice moves closer to resolving the *vexata quaestio* of the moon spots, which is her nominal theme, by describing how the Intelligences transmit motion and power ("moto" and "virtù": l. 127) to their heavens. She compares the "santi giri" to hammers and the "beati motor" to artisans (l. 128) – a

---

rerum creatarum est esse et non est ante ipsum creatum aliud [...] intelligentiae secundae proiciunt visus suos super universalem formam, quae est in intelligentiis universalibus, et dividunt eam et separant eam" ("The first of all created things is being and there is no other created being before this [...] the secondary intelligences cast their sight upon the universal form, which is in the universal intelligences, and divide and separate this form"), cited by Nardi, *Saggi di filosofia dantesca*, pp. 97, 104-105.

[79]*Con.* II, xiv, 15: "lo detto cielo [sc. crystalline] ordina col suo movimento la cotidiana revoluzione di tutti li altri, per la quale ogni die tutti quelli ricevono [e mandano] qua giù la vertude" (cf. the text from Albert's *De caelo et mundo* quoted in n. 45).

similar comparison is found in the *Convivio* (IV, iv, 12). In the following tercet, which consciously echoes Boethius' *Consolatio philosophiae*, she suggests that the eighth sphere ("'l ciel cui tanti lumi fanno bello": l. 130) takes on an image of the Intelligence which causes its motion ("la mente profonda che lui volve": l. 131), just as wax receives the imprint of a seal (l. 132):[80]

> Lo moto e la virtù d'i santi giri,
> come dal fabbro l'arte del martello,
> da' beati motor convien che spiri;
> e 'l ciel cui tanti lumi fanno bello,
> de la mente profonda che lui volve
> prende l'image e fassene suggello. (127-132)

The final five tercets of the canto continue to concentrate on the relations between Intelligence and sphere, and these lines are the most difficult and interesting of all. First, Beatrice seems to imply that the heavens are animated, since she compares the relationship between the Intelligence and the eighth heaven to that between a human soul and its body.[81] Just as the soul gives life to the body and endows it with different powers, so the goodness of the Intelligence is multiplied throughout the stars without losing its unity:

> E come l'alma dentro a vostra polve
> per differenti membra e conformate

---

[80]Boethius, *Consolatio philosophiae*, lib. III, metrum ix (*PL* 63, cols 760-761).

[81]The idea that the heavens are animated was a view which was condemned by the Bishop of Paris in 1277, see props. 92 and 102 in *Chartularium Universitatis Parisiensis*, ed. Heinrich Denifle and E. Chatelain, 4 vols. (Paris, 1889-1897), I, pp. 548-549: "Quod corpora celestia moventur a principio intrinseco quod est anima [...] Quod anima celi est intelligentia, et orbes celestes non sunt instrumenta intelligentiarum sed organa, sicut auris et oculus sunt organa virtutis sensitive" ("That celestial bodies are moved by an intrinsic principle which is soul [...] That the word soul is an intelligence and that the celestial orbs are not the instruments of the intelligences but organs, as the ears and eyes are organs of the sensitive power"). On the doctrinal context to this problem, see Richard C. Dales, "The De-Animation of the Heavens in the Middle Ages," *Journal of the History of Ideas* 41 (1980), 531-550; Barbara de Mottoni Faes, "La dottrina dell'«anima mundi» nella prima metà del secolo XIII: Guglielmo d'Alvernia, «Summa halensis», Alberto Magno," *Studi medievali* ser. 3a/22 (1981), 253-297.

a diverse potenze si risolve,
così l'intelligenza sua bontate
multiplicata per le stelle spiega,
girando sé sovra sua unitate. (133-138)

This comparison between soul and body, on the one hand, and between
Intelligence and planetary body, on the other, is fundamental to Dante's revised
explanation of the moon spots. In his minor works, Dante repeatedly describes
"bontade" by analogy to light, but these earlier writings offer no indication that the
relationship between the Intelligences and their celestial bodies might be analogous
to that between the human soul and its body.

In the final ten lines of the canto, Beatrice discusses the "alloy" ("lega": l.
139) formed by the union between the planetary body and the mixed and multiform
"virtue" of the Intelligence of the eighth heaven. As Dante describes the process
by which this "virtue" shines out through this heaven, he places the key verbs
("derivare," "lucere": ll. 142-143) in emphatic rhyme position. The mixed "virtue"
flows from a joyous nature and shines out like happiness from the pupil of a living
eye:

Virtù diversa fa diversa lega
col prezïoso corpo ch'ella avviva,
nel qual, sì come vita in voi, si lega.
Per la natura lieta onde deriva,
la virtù mista per lo corpo luce
come letizia per pupilla viva. (139-144)

A revealing parallel is found in *Purgatorio* XXV, a canto which also combines
intellectually challenging ideas with a wealth of simile and metaphor. In Statius'
discussion of the generation of the human soul in this canto, he describes how a
"spirito novo, di vertù repleto" (l. 72) is breathed into the foetus to make a single
soul, a substance which lives and feels and circles on itself: "[...] e fassi un'alma

204

sola, / che vive e sente e sé in sé rigira" (ll. 74-75).[82] If the union between cosmic soul and planetary body is to be understood in similar terms in *Paradiso* II, then presumably the outshining of "virtù mista" (l. 143) does not derive from the Intelligence as separate substance, but from the Intelligence as substantial form of the "ciel cui tanti lumi fanno bello."

In the concluding lines of the canto, Beatrice returns to her theme and explains the "segni bui" on the moon's surface in terms of the differentiation of "virtù" in the eighth heaven, as described in lines 125-127. The darker and lighter areas on the surface of the moon ("lo turbo e 'l chiaro": l. 148) are analogous to the differing brightnesses of the stars in the eighth heaven. One and the same explanation is valid for all heavens in the cosmological system, including that of the moon. From a narrow point of view, the moon may still possess different densities of aether, but Dante's central concern is to show that these variations are not accidental, not due to the inadequate disposition of matter, and therefore not due to chance. Rather, the moon and its "macchie," which proclaim that the "virtue" of this heaven too is "mista" and "diversa," have their place within the system of the heavens as a whole, a system whose ultimate purpose is, as we later learn in canto VIII, to differentiate individual beings on Earth:

> Da essa vien ciò che da luce a luce
> par differente, non da denso e raro;
> essa è formal principio che produce,
> conforme a sua bontà, lo turbo e 'l chiaro». (145-148)

As has been shown, Beatrice's solution to the moon spots relies philosophically on premises taken from the *Liber de causis* and a general Neoplatonic climate of thought. But Dante's cosmology is not simply a patchwork

---

[82]*Purg.* XXV, 70-75. Note also the light comparison Dante uses to describe how the soul, as soon as it is separated from its body, forms a new "aerial body" by irradiating out its "formative virtue" (see above Ch. 4, p. 138).

205

of Neoplatonic sources; the physical universe of *Paradiso* II is characterised by the poet's urge to create links between macrocosm and microcosm, to rework and develop earlier ideas in a new context, and to stress the analogies between man and the universe in a more pronounced and deliberate way than his "scientific" source-material.[83] The comparison between the Intelligence shining through the stars and light shining through the human eye is the best example of all (l. 144). Dante's playful sense of the interplay possible between scientific ideas, literary allusions, and micro-macrocosmic parallels recalls the poetic cosmologies of twelfth-century poets such as Alain of Lille and Bernard Silvestris. Although Dante's cosmos is ultimately his own highly personal creation, having neither a world-soul nor the hypostatised divinities that are found in the *Cosmographia* and *Anticlaudianus*, the next section will show that Dante's sensitivity towards nature as a system with its own finalities provides evidence of strains of twelfth-century naturalism in both his thought and poetry.[84]

---

[83]For the simile of the artist and that of the hammer and iron, see Albert, *De caelo et mundo*, lib. II, tr. 3, c. 14 (Hossfeld, p. 175), cited by Nardi, *Saggi di filosofia dantesca*, pp. 97-98. But see also Albert's *De intellectu et intelligibili*, lib. I, tr. 1, c. 4 (Jammy, V, p. 242a), and cf. *Con.* IV, v, 15. For the luminous action of the Intelligences, see also *De intellectu et intelligibili*, lib. II, tr. un., c. 11 (Jammy, V, p. 261a): "intelligentia per substantiam separata, lumen diffundit et ingerit per totum orbum sibi subiectum, sicut anima in corpus suum, et hoc lumen ubique praesens proportionaliter efficitur in his quae se extendunt ad ipsum" ("the intelligence through its separated substance diffuses and directs light throughout the whole orb which is subject to it, as the soul does in its body, and this light which is present everywhere takes effect in all those things which extend to it").

[84]For pertinent observations on Dante and twelfth-century philosophy, see Dronke, "Tradition and Innovation in Medieval Western Colour-Imagery," p. 78; Vasoli, *Il Convivio*, p. 401. On twelfth-century thinkers, poets, and cosmologists, see M.-D. Chenu, *Nature, Man and Society in the Twelfth Century: Essays on New Theological Perspectives in the Latin West*, trans. Jerome Taylor and Lester K. Little (Chicago and London: University of Chicago Press, 1968); Winthrop Wetherbee, *Platonism and Poetry in the Twelfth Century: The Literary Influence of the School of Chartres* (Princeton, NJ: Princeton University Press, 1974); idem, "Philosophy, Cosmology, and the Twelfth-Century Renaissance," in *A History of Twelfth-Century Western Philosophy*, ed. Dronke, pp. 21-53. See also the other articles collected in *A History of Twelfth-Century Western Philosophy*, ed. Dronke.

Dante's Creation Doctrine

Dante's expositions of the act of creation in the *Paradiso* have become the subject of critical dispute. Does the role Dante assigns to light in the creative act reflect traditional teachings, or is he instead drawing on more heterodox Neoplatonic ideas? Some disagreements even exist about the number of creation discourses in the *Paradiso*, although all critics accept that sections of cantos VII and XXIX describe the act of creation. Let us consider the relevant discourses and the issues they raise in turn.

In canto VII the images of outflowing and unfolding (ll. 65-66), flowing from a source (l. 70), and radiating (l. 71) are especially prominent, and it is the arrangement of verbs in rhyme position that captures the flow of light and, as the metaphorical pattern shifts, that of water:

> La divina bontà, che da sé sperne
> ogne livore, ardendo in sé, sfavilla
> sì che dispiega le bellezze etterne.
> Ciò che da lei sanza mezzo distilla
> non ha poi fine, perché non si move
> la sua imprenta quand' ella sigilla.
> Ciò che da essa sanza mezzo piove
> libero è tutto, perché non soggiace
> a la virtute de le cose nove. (64-72)

In canto XXIX the imagery is equally explicit. The act of creation is a shining forth (ll. 14-15), an opening out (l. 18), and an irradiation (l. 28) from God:[85]

> Non per aver a sé di bene acquisto,
> ch'esser non può, ma perché suo splendore
> potesse, risplendendo, dir "*Subsisto*",
> in sua etternità di tempo fore,

---

[85]See also *Par.* XIX, 89-90: "nullo creato bene a sé la tira / ma essa [sc. the 'prima volontà'], radïando, lui cagiona."

207

> fuor d'ogne altro comprender, come i piacque,
> s'aperse in nuovi amor l'etterno amore. (13-18)

The outflowing of divine light gives rise to three "new loves" (or "'l triforme effetto" as he calls them collectively in line 28): pure form (the angelic Intelligences), sempiternal combinations of form and matter (the heavens), and the prime matter from which the four elements will be formed. Dante compares the emergence of the "triform effect" to the passage of a ray of light through glass, amber, or crystal:

> E come in vetro, in ambra o in cristallo
> raggio resplende sì, che dal venire
> a l'esser tutto non è intervallo,
> così 'l triforme effetto del suo sire
> ne l' esser suo raggiò insieme tutto
> sanza distinzïone in essordire. (25-30)

The idea that light requires no time to traverse space is found in Aristotle who argued that light cannot strictly be said to "move," since it requires no time to render a medium transparent in act. This concept was to become axiomatic in thirteenth-century scholastic thought, but Dante here creatively reworks the *locus communis* that *illuminatio fit in instanti* in order to help define and visualise God's creative act.[86]

The themes and patterns of imagery deployed in an important doctrinal sequence in canto XIII have close correspondences with those developed earlier in

---

[86]For Aristotelian texts which provide the foundation for the scholastic axiom that *illuminatio fit in instanti*, see *De anima*, II, 7, 418b 20-25; *De sensu*, 6, 446b 27. On scholastic elaborations, see also Peter Marshall, "Nicholas Oresme on the Nature, Reflection and Speed of Light," *Isis* 72 (1981), pp. 358-360. This notion was exceptionally commonplace in scholastic thought, see e.g. Bonaventure, *In II Sent.*, d. 13, a. 3, q. 1, fund. 2 (Quaracchi, II, p. 324a): "lux [...] in instanti ab oriente venit in occidens" ("light moves from east to west instantaneously"); Aquinas, *ST*, Ia, q. 67, a. 3, resp. (Caramello, p. 328). But note also a possible biblical precedent in Matt. 24:27: "Sicut fulgor exit ab oriente et paret in occidente, ita erit adventus filii hominis" ("As lightning comes from the east and appears in the west, so it will be with the coming of the Son of man").

208

canto VII, even though Attilio Mellone was undoubtedly right to question whether this passage in *Paradiso* XIII describes creation as such.[87] In both cantos, Dante examines the role of secondary causes in establishing diversity on Earth by positing a fundamental distinction between generated being (which is corruptible) and created being (which is eternal).[88] In canto XIII, Dante's fictional Aquinas discusses how divine light passes through the angels and descends through the varying disposition of matter to produce a diversity of individual beings on Earth:

> Ciò che non more e ciò che può morire
> non è se non splendor di quella idea
> che partorisce, amando, il nostro Sire;
>   ché quella viva luce che sì mea
> dal suo lucente, che non si disuna
> da lui né da l'amor ch'a lor s'intrea,
>   per sua bontate il suo raggiare aduna,
> quasi specchiato, in nove sussistenze,
> etternalmente rimanendosi una. (52-60)

In elaborating these ideas, Dante gives a prominently central role to light. He once again adopts the imagery of flowing and irradiating, and also uses "splendor" (l. 53) in its technical sense as reflected light. Divine light here behaves with the utmost diffusive capacity which Dante expresses through a verb of flow such as "meare" (l. 55); and yet this divine light is shown to retain its essential unity, even though it extends itself to all things, including the most short-lived contingent beings. The poet has here compressed his ideas concerning the causal system with a description of the Trinity in terms of light – this is one of the points at which

---

[87]Attilio Mellone, "Il concorso delle creature nella produzione delle cose secondo Dante," *Divus thomas* 56 (1953), esp. p. 277: "il discorso non tratta della prima creazione del mondo ma del divenire attuale delle creature."

[88]On the fundamental distinction between created and generated in Dante's thought, see Boyde, *Dante Philomythes*, pp. 260-265.

Dante inextricably fuses his more cosmological conception of light's role in the physical universe with Trinitarian theology.

The question that has been repeatedly asked about all these discourses is how far they run counter to a Christian view of creation, and might reflect Neoplatonic emanationism, a hierarchy of being in which the Many is envisaged as a series of successive emanations from the One. Many critics have seen a potential conflict between the metaphors, especially the light metaphors, Dante uses to describe the first moment of creation and more traditional ways of expressing the creative act such as it being God's handiwork. Arthur Lovejoy even wrote that Dante "verged on heresy" in the discourse in *Paradiso* VII and that he risked presenting creation as a necessary process.[89]

A number of points need to be made in this connection. First, despite the light imagery Dante employs, he takes great care to insist upon the orthodox Christian idea that creation involves divine will. Second, there is no question whatsoever that Dante is adhering to emanationist doctrine, describing, that is, one level of being as an emanation from a higher level.[90] Indeed, it is questionable whether the contrast between emanation and creation is as marked in thirteenth-century thought as some Dante scholars have claimed. It was not unusual for

---

[89] Arthur Lovejoy, *The Great Chain of Being* (Cambridge, MA: Harvard University Press, 1933), p. 69. For less extreme views which nonetheless contrast the necessary aspects of emanation with the volitional component of creation, see Boyde, *Dante Philomythes*, pp. 213-214; Marziano Guglielminetti "*Paradiso* 13," in *L'arte dell'interpretare: Studi critici offerti a Giovanni Getto* (Cuneo: L'Arciere, 1984), pp. 76-77; Kay, *Dante's Christian Astrology*, pp. 131, 259; Mazzeo, *Medieval Cultural Tradition*, pp. 97-101, and n. 37, p. 234.

[90] For articles suggesting this erroneous view, see Rudolf Palgen, "Gli elementi plotiniani nel «Paradiso»," in *Atti del convegno internazionale sul tema: Plotino e il Neoplatonismo in Oriente e in Occidente* (Rome: Accademia nazionale dei Lincei, 1974), pp. 505-524; idem, "Il Paradiso Platonico," in *Letteratura e critica: Studi in onore di Natalino Sapegno*, 5 vols. (Rome: Bulzoni, 1974), I, pp. 197-211. On Dante's non-emanationist creation doctrine, see Bemrose, *Dante's Angelic Intelligences*, pp. 56-57, 188-89; Nardi, *Saggi di filosofia dantesca*, p. 44. Mellone has justifiably argued against Nardi's view (*Dante e la cultura medievale*, p. 248) that traces of Neoplatonic emanation can be found in *Convivio*, II, ii, 4 and III, xiv, 4-6, see Attilio Mellone, "Emanatismo neoplatonico di Dante per le citazioni del 'Liber de causis'?," *Divus Thomas* 54 (1951), 205-212; idem, "De Causis," in *ED* II, pp. 327-329.

210

Christian writers to bring together emanation and creation, and even Aquinas applied the term *emanatio* to the creative act. Since it has become traditional (in Dante studies at least) to dissociate Thomist thought from such terminology, some relevant passages from his *Commentum in libros sententiarum* and his *Summa theologiae* are worth citing:[91]

[...] emanatio creaturarum a Deo est sicut exitus artificiatorum ab artifice; unde sicut ab arte artificis effluunt formae artificiales in materia, ita etiam ab ideis in mente divina existentibus fluunt omnes formae et virtutes naturales [...] non solum oportet considerare emanationem alicuius entis particularis ab aliquo particulari agente, sed etiam emanationem totius entis a causa universali, quae est Deus, et hanc quidem emanationem designamus nomine creationis.[92]

As is well known, there are striking differences between Thomas' thought on secondary causes and Dante's conception of them as imprinting sublunary forms into matter, especially in *Paradiso* VII, 130-138.[93] Aquinas clearly makes no concessions to the manner in which the *De causis* presents the Intelligences as assisting the first cause in creation, an idea which influenced Dante deeply.[94]

---

[91]See Wolson's fundamental essays (nn. 21 and 32).

[92]Quotations from *In II Sent.*, d. 18, q. 1, a. 2, sol. in *Opera omnia*, VI, p. 543b and *ST*, Ia, q. 45, a. 1, resp. (Caramello, p. 227): "the emanation of created beings from God is similar to the way in which the maker or artisan produces the things he makes; hence, as forms of artefacts flow into matter from the art of the maker, so also all forms and powers in nature flow from ideas existing in the divine mind [...] not only must we consider the emanation of some particular entity from some particular causal agent, but also that of all being from the universal cause, which is God, and we refer to this emanation with the term 'creation'." Quaestio 45 of the *Summa theologiae* is entitled "De modo emanationis rerum a primo principio."

[93]See Bemrose, *Dante's Angelic Intelligences*, pp. 97-101, 106-109; Foster, "Tommaso d'Aquino," in *ED* V, pp. 638-639; Nardi, *Dante e la cultura medievale*, pp. 253-255; idem, *Saggi di filosofia dantesca*, pp. 354-355.

[94]A good text for comparison is Aquinas' *Quaestiones disputatae de potentia*, q. 3, a. 4, resp. ed. P. Bazzi, 10th edn (Turin and Rome: Marietti, 1965), p. 46. Here, Aquinas condemns the view that "Deus creavit creaturas inferiores mediantibus superioribus" ("God created lower creatures

211

These Thomist passages do, however, help to show that it was very common for medieval writers to adapt Neoplatonic terms and the language of outflow to describe God's creative act.[95] All in all, it is clear that we must proceed with caution (with "piombo a' piedi" one might say) in attempting to establish sources for Dante's creation doctrine, and this is particularly necessary with writers associated with the "light-metaphysics" tradition.

Scholars and critics have nonetheless made considerable efforts to identify doctrinal correspondences. In *Saggi di filosofia dantesca*, Nardi indicated Avicenna as the "source" for Dante's views on corruptibility and non-corruptibility in canto VII,[96] while in later essays, he somewhat nuanced his position by noting that similar ideas are present in Plato's *Timaeus* and in Chalcidius' commentary.[97] More recently, Dante scholars have shown that these ideas are especially characteristic of writers associated with twelfth-century Platonism.[98] In the *Timaeus*, the task of completing the creation of living beings was entrusted to lesser divinities; and the relevant passage (41A) encouraged twelfth-century philosophers and poets to conceive nature as an autonomous force, which, in

---

by means of higher ones"), that is, the very position which Dante appears to defend in *Par.* VII, 130-138; XXIX, 52-54.

[95]Elsewhere, Aquinas frequently adopts the Neoplatonic language of diffusion and emanation to describe divine causation, see e.g. *ST*, Ia, q. 8, a. 1, resp. (Caramello, p. 36); Ia, q. 104, a. 1, ad. 1 (Caramello, p. 493); idem, *In I Sent.*, d. 43, q. 2, a. 1, ad. 1 in *Opera omnia*, VI, p. 350b.

[96]Nardi, *Saggi di filosofia dantesca*, pp. 44, 350-352. Dante's doctrine is related to the Avicennan principle that "a Deo invariabili nihil variabile immediate progredi poterat."

[97]Nardi's emphasis on Avicenna was criticised by Dino Bigongiari in his important review of *Saggi di filosofia dantesca*, see *Speculum* 7 (1932), p. 151. Bigongiari traces the principle to Proclus, and points out its foundation in the *Timaeus* and Chalcidius.

[98]See Nardi, *Dante e la cultura medievale*, pp. 242, 253-262, esp. pp. 256-257, 259-260. See also Bemrose, *Dante's Angelic Intelligences*, p. 108, who cites Plato and Chalcidius, but also adds William of Conches' *Glosae super Platonem*, c. 37 (Jeauneau, p. 105). For the influence of twelfth-century thinkers on Dante's views on corruptibility and incorruptibility, see Boyde, *Dante Philomythes*, pp. 247 and nn. 25-26, pp. 359-360; Mazzeo, "The Analogy of Creation in Dante," in *Medieval Cultural Tradition*, pp. 133-173, esp. pp. 161-162.

212

assisting the Creator, limited His creative power to a single act.[99] The influence of twelfth-century ideas on this aspect of Dante's thought now seems to be firmly accepted, although the extent of his indebtedness to specific authors remains to be determined more precisely.[100] Given Dante's familiarity with the *De causis*, though, it is well worth noting that this widely-disseminated, short treatise also contains at least one proposition related to the principle employed by Dante in canto VII.[101]

Attempts to unearth sources for the creation discourse in canto XXIX have produced far less satisfactory results. In particular, Rudolf Palgen's repeated assertions that Eriugena's *Periphyseon* (or *De divisione naturae*) was Dante's direct source should be ruled out.[102] Following Nardi's suggestion of a link with ideas found in Proclus and Plotinus,[103] several other critics have regarded the light imagery in canto XIII as characteristic of the "light-metaphysics" tradition, and Mazzeo has even argued that Dante presented light as the "principle of being" in this canto.[104] For reasons that should now be clear, these views seem to overstate

---

[99]See J.M. Parent, *La doctrine de la création dans l'École de Chartres* (Paris and Ottawa: Vrin, 1938), passim, but esp. p. 93. See also Gregory, *Anima Mundi*, p. 215 and the useful n. 29, p. 64 of his article, "The Platonic Inheritance," in which he cites examples of mediated creation from William of Conches, Bernard Silvestris, and Alain of Lille.

[100]For a recent study of William of Conches' putative influence on *Con.* IV, i, 4, see Margherita de Bonfils Tempier, "«La prima materia de li elementi»," *Studi danteschi* 58 (1986), 275-291.

[101]*De causis*, prop. X (XI), 101 (Pattin, pp. 160-161): "si intelligentia est semper quae non movetur, tunc ipsa est causa rebus sempiternis quae non destruuntur [nec permutantur] neque cadunt sub generatione et corruptione" ("if intelligence is that which never is moved, then it is the cause of sempiternal things which are neither destroyed [nor changed] nor fall into the realm of generated and corruptible things").

[102]Rudolf Palgen, "Scoto Eriugena, Bonaventura e Dante," *Convivium* 25 (1957), esp. pp. 4-5. See the criticisms levelled at this and related articles by Marta Cristiani, "Scotto Eriugena," in *ED* V, p. 92.

[103]Nardi, *Saggi di filosofia dantesca*, pp. 212-213, and esp. p. 342: "il modo di rappresentare la creazione [in *Par.* XIII, 52-60] come splendore della Luce divina è d'origine neoplatonica."

[104]Mazzeo, *Medieval Cultural Tradition*, p. 107, 108-114; idem, *Structure and Thought*, pp. 15-16. For critics who also believe that there are distinctive echoes of "light-metaphysics" doctrines in these lines, see e.g. Corti, *Percorsi dell'invenzione*, p. 163; Guglielminetti "*Paradiso* 13," pp. 76-77; Kay, *Dante's Christian Astrology*, p. 131.

213

the case: elements of the Plotinian system were fitted to a Christian view of creation, history, and redemption from a very early date; and references to more traditional sources can easily be produced.[105] There are after all several biblical passages that assign an important role to light in God's creation of the world.[106]

Dante and Grosseteste on Light and the Origin of the Cosmos

Dante's creation doctrines in canto VII and especially XXIX are the most appropriate passages to consider in attempting to compare Dante's views on the role of light in creation with the light theory expounded in Grosseteste's *De luce* (c. 1225-1230). This short treatise is widely regarded as the foremost exposition of the thirteenth-century "metaphysics of light" and it has been claimed that its ideas provided Dante with "il presupposto essenziale che sottende l'intero luminismo paradisiaco."[107]

In his *De luce*, Grossesteste made a highly original attempt to explain the production of the physical universe on the basis of the diffusion of light. God creates a dimensionless *punctum lucis*, which multiplies itself spherically and instantaneously in all directions. In so doing, light extends prime matter according to the geometrical laws of light: the capacity of things to exist in three dimensions

---

[105]Bigongiari, Review of *Saggi*, p. 148: "This passage interpreted according to the traditional view corresponds word for word to the ordinary doctrine of the Church"; Foster, "Summa Contra Gentiles," in *ED* V, pp. 479-480, who refers to *Summa contra Gentiles*, lib. IV, §§ 3461-3479 (Pera, III, pp. 264-270) as a probable source. But see also *Summa contra Gentiles*, lib. IV, c. 12, § 3483 (Pera, III, p. 270): "... convenienter et verbum divinae sapientiae splendor lucis nominatur" ("the word of divine wisdom is fittingly called the the splendour of light").

[106]See esp. Gen. 1:3; cf. Jn. 1:3-4. For the idea of Christ as the exemplar of creation, see I Cor. 8:6; Col. 1:15-16; Hebr. 1:2.

[107]Guidubaldi, *Dante Europeo*, II, p. 291. He quotes the complete text of the *De luce*, in vol. II, pp. 295-300. Guidubaldi repeatedly expresses his conviction that Dante's "luminismo" is "oxfordianamente strutturato," see ibid., II, pp. 132, 271-272; III, pp. 270-276, 465-466, 489, 718: "quella dimensione oxfordiana che non è *un aspetto* del luminismo dantesco, ma *la sua stessa essenza*' (italics his). On Grosseteste and Dante, see also John Leyerle, "The Rose-Wheel Design and Dante's *Paradiso*," *University of Toronto Quarterly* 46 (1977), p. 301.

214

is thus dependent on light. On reaching its maximum degree of rarefaction, this light forms a luminous sphere (the first heavenly sphere), from which *lumen* is then diffused inwards and condenses to form the next sphere. The process of rarefaction and condensation continues until nine celestial and four sublunary spheres are brought into existence. In the *De luce*, then, light is the agent responsible for producing the whole physical system, and, because all bodies derive their corporeity from light, Grosseteste calls light the first corporeal form (i.e. that from which every body derives its capacity to occupy space):

Formam primam corporalem, quam quidam corporeitatem vocant, lucem esse arbitror [...] Corporeitas vero est, quam de necessitate consequitur extensio materiae secundum tres dimensiones, cum tamen utraque, corporeitas scilicet et materia, sit substantia in se ipsa simplex, omni carens dimensione. Formam vero in se ipsa simplicem et dimensione carentem in materiam similiter simplicem et dimensione carentem dimensione in omnem partem inducere fuit impossibile, nisi seipsam multiplicando et in omnem partem subito se diffundendo et in sui diffusione materiam extendendo, cum non possit ipsa forma materiam derelinquere, quia non est separabilis, nec potest ipsa materia a forma evacuari [...] Non est ergo lux forma consequens ipsam corporeitatem, sed est ipsa corporeitas [...] Lux ergo praedicto modo materiam primam in formam sphaericam extendens et extremas partes ad summum rarefaciens, in extima [sic] sphaera complevit possibilitatem materiae, nec reliquit eam susceptibilem ulterioris impressionis. Et sic perfectum est corpus primum in extremitate sphaerae, quod dicitur firmamentum, nihil habens in sui compositione nisi materiam primam et formam primam [...] Hoc itaque modo completo corpore primo, quod est firmamentum, ipsum expandit lumen suum ab omni parte sua in centrum totius [...] lumen quidem gignitur ex prima sphaera, et lux, quae in prima sphaera est simplex, in secunda est duplicata.[108]

On close comparison, Dante's account of the role of light in creation in canto XXIX shows itself to be fundamentally different from that expressed in this work. Unlike Grosseteste, Dante does not start with prime matter and form. Instead, he describes a temporal beginning in which God, in a single act of irradiation, creates matter, pure forms (angelic-Intelligences), and permanently conjoined matter and form (the heavens). Dante does *not* describe:

1. a process of successive rarefaction and condensation;

2. the geometrical behaviour of light in constituting and shaping the universe;

3. the formation of the celestial and sublunary spheres from *lumen*.

---

[108]Grosseteste, *De luce* (Baur, pp. 51, 54-55): "I hold that the first corporeal form, which some call corporeity, is light. Corporeity is that which by necessity is the result of the extension of matter in three dimensions, since both of these, that is, corporeity and matter are simple substance lacking in all dimensions. It was therefore impossible for simple and dimensionless form to bring dimensions into similarly simple and dimensionless matter in every part other than by multiplying and diffusing itself in every part instantaneously and by its diffusion extending matter. This is so because form cannot leave matter behind, and because form is not separable and matter cannot be emptied of form [...] Therefore light is not a form which comes after this corporeity, but it is this corporeity itself [...] Therefore light in the way mentioned has in the furthest sphere perfected the potency of matter and left it no longer susceptible to further impressions, by extending matter into the form of a sphere and rarefying the outermost parts as far as possible. And so the first body is perfected in the extreme of the sphere which is called the firmament, having nothing in its own composition except first matter and first form [...] And so when the first body, the firmament, has been produced in this way, it extends its *lumen* from every part of itself to the centre [...] and *lumen* is begotten from the first sphere, and *lux* which is simple in the first body is duplicated in the second." For good discussions of the light theory propounded in the *De luce*, see McEvoy, *The Philosophy of Robert Grosseteste*, pp. 182, 185, 414, 449-451; Saccaro, "Il Grossatesta e la luce," pp. 45-72. See also Grosseteste, *Hexaemeron*, p. II, c. 10, § 1 (Dales and Geiben, pp. 97-98); idem, *Commentarius in VIII libros Physicorum Aristotelis*, lib. III, ed. Richard C. Dales (Boulder, CO: Colorado University Press, 1963), p. 55: "Forma, ut lux, replicat se et multiplicat infinicies ut se extendat in dimensiones et simul secum rapiat materiam" ("Form, as light, replicates itself and multiplies itself infinitely so that it extends itself into three dimensions and simultaneously draws matter with it"). Witelo advances similar ideas in the prologue to his *Perspectiva* (Risner, p. 1): "Quia itaque lumen corporalis formae actum habet: corporalibus dimensionibus corporum (quibus influit) se coaequat, et extensione capacium corporum se extendit" ("And thus since light possesses the actuality of corporeal form, it makes itself equal to the corporeal dimensions of bodies that it flows into, and extends itself to the limits of capacious bodies").

216

While Dante uses light to express the nature of the creative act, he does not show all other forms to derive from light as the first corporeal form. All in all, it is clear that the chief tenet of Grosseteste's "light metaphysics" – the idea that light is the primary component present in all bodies – has no place in his thought. Dante associated light and form and recognised that light had an informing power, but he did so in a significantly less technical manner than Grosseteste.[109]

---

[109]Dante closely associates light and form in *Purg.* XXV, 89 and *Par.* II, 109-110: "voglio informar di luce sì vivace, / che ti tremolerà nel suo aspetto." For the interesting suggestion that a related idea influenced Dante's "Amor, tu vedi ben," see Robert M. Durling and Roland L. Martinez, *Time and the Crystal: Studies in Dante's "Rime Petrose"* (Berkeley, Los Angeles, and Oxford: University of California Press, 1990), pp. 138-139. However, none of the examples in Dante offer evidence of his absorption of ideas directly associated with either Grosseteste and/or Witelo. The notion of light as an informing power was commonplace in the thirteenth century, see e.g. Thomas of Cantimpré, *Liber de natura rerum*, lib. XVII, c. 7 (Boese, p. 389): "Ipse enim sol lux est universalis, que forma est omnium formarum, quia sine luce nulla res habet formam" ("This sun is a universal light, which is the form of all forms, for, without light nothing has form"). On light as "forma formarum," see also Albert, *De meteoris*, lib. II, tr. 1, c 5 (Jammy, II, p. 48).

CHAPTER 7

Adaptations drawn from Light
in the Imagery and Doctrine of the *Paradiso*

The theme of light is self-evidently central to Dante's *Paradiso*. In the final *cantica*, Dante not only represents souls, the angels, Christ, and God as light, but, as protagonist, he passes through light-filled heavenly spheres, and enters into the luminous bodies of planets. In the heaven of the moon, the blessed souls appear as faint ethereal forms, and their translucency deceives Dante-*personaggio* into thinking they are reflected images. But in the six other planetary spheres, Dante meets groups of souls who appear to him hidden by their own radiance and in a variety of coruscating configurations. His cosmic flight takes him to the eighth sphere in which he witnesses the luminous triumphs of Christ and Mary. And in the ninth sphere, he views God and the angelic hierarchies as a point of light surrounded by nine fiery circles. In canto XXX, his ascent takes him beyond the boundary of the physical universe to a heaven of "pure" light, where his sight penetrates the divine ray and he sees into the true and exalted light of God, "l'alta luce che da sé è vera."[1]

---

[1]References in order: (i) individual souls hidden in radiance, *Par*. V. 124-126; VIII, 52-54; XIII, 48; XIV, 38-39; XV, 52-53; XVII, 36; XVIII, 76; XXVI, 82-83; (ii) souls grouped as lights to form circles, a cross, an eagle, and a stairway, X, 64-66; XIV, 97-102, 109-111; XVIII, 73-78, 97-105; XXI, 28-42; (iii) light of Christ and Mary, XXIII, 28-33, 91-93; (iv) God as point of light, XXVIII, 16-39; (v) divine ray and light of God, XXXIII, 52-54, 82-83, 115-117.

218

There are, of course, several well-known studies that examine Dante's treatment of light in the *Comedy*, and especially the *Paradiso*: Guido di Pino's *La figurazione della luce*; Charles Singleton's essay on the three lights; Irma Brandeis' *A Ladder of Vision*; and more recent essays by Sharon Harwood.[2] These critics are, however, primarily concerned with the ways in which light imagery is employed within the poem, either as part of a structural design or in relation to certain themes, and they provide very little discussion of medieval ideas about the nature and behaviour of light. Such discussion is to be found in critical literature on Dante and "light metaphysics," but, as we have already seen, to use this category is to risk simplifying the complex and demanding range of medieval views about light and to give a distorted view of Dante's intellectual debts. One aim of the present chapter is to address these omissions by examining the intellectual relationship between Dante and theological writings that have come to be regarded as part of a "theology of light." Since the term "theology of light" may lend itself to the same kind of oversimplification found in secondary literature on Dante's so-called "metafisica della luce," it will not be used here to define an homogeneous tradition. Instead, this chapter attempts to consider some of the ways in which the light image was used before Dante by theologians in formulating concepts related to God, the angelic hierarchies, the Empyrean, and beatific vision. As in previous chapters, it will not be suggested that Dante's treatment of light in the *Paradiso* is linked to any of the authors and works that will be discussed in a

---

[2]The best-known study of Dante's light imagery is Guido di Pino, *La figurazione della luce nella "Divina Commedia"*; for light in the *Paradiso*, see pp. 147-181. For studies of Dante's presentation of light imagery and its relationship to thematic and structural aspects of the poem, see also Irma Brandeis, *A Ladder of Vision: A Study of Dante's "Comedy"* (New York: Anchor, 1962), pp. 185-227; Sharon Harwood, "Moral Blindness and Freedom of Will: A Study of Light Images in the *Divina Commedia*," *Romance Notes* 16 (1974), 205-221; idem, *A Study of the Theology and Imagery of Dante's Divine Comedy* (Lewiston, Queenston, and Lampeter: Edwin Mellen Press, 1990), pp. 91-119; Charles S. Singleton, "The Three Lights," in *Dante Studies 2: Journey to Beatrice* (Cambridge, MA: Harvard University Press, 1958), pp. 15-38. For the view that Dante's handling of light imagery conforms to the Augustinian doctrine of *visiones*, see also Newman, "St. Augustine's Three Visions."

219

way which might detract from the richly personal quality of his thought and poetry. My working premise will again be that Dante frequently adapts ideas involving light which had come into a general synthesis in works written before 1300. To this end, a series of general parallels will be advanced between analogies based upon light in medieval theological writings and Dante's creative reworking of light imagery and related doctrines in the *Paradiso*. The chapter will nonetheless close by considering the relationship between a specific topic treated by some thirteenth-century theologians and Dante's most innovative adaptation of all, his doctrine of the Empyrean.

Light in the Bible, the Church Fathers, and the Pseudo-Dionysius

Many of the similarities that can be identified between the work of thirteenth-century theologians and Dante's light imagery in the *Paradiso* rest upon a tendency shared by almost all medieval writers: a marked propensity to devise analogies between natural light and divine light. Since many of these comparisons have some basis in the Bible, it is important to outline some of the major scriptural passages.[3] There are, of course, many sections in the Old Testament where divine revelations are accompanied by luminous phenomena, and, in the first Epistle to St. John and Paul's Epistle to Timothy, God is said to be light.[4] God is also said to create the

---

[3]For more detailed discussion of the theme of light in the Bible, see Sebastiano Agrelo, "El tema biblico de la luz," *Antonianum* 50 (1975), 353-417; Edwyn Bevan, *Symbolism and Belief* (London: Allen & Unwin, 1938), pp. 111-133; Dominique Mathieu, "Lumière: Étude biblique," *DS* 9 (1976), pp. 1142-1149.

[4]I Jn. 1:5-6: "[...] quoniam Deus lux est et tenebrae in eo non sunt ullae" ("since God is light and in Him there is no darkness"); I Tim. 6:16: "Qui solus habet immortalitatem et lucem habitans inaccessibilem" ("Who alone is immortal, residing in inaccessible light"). While God is not directly identified with light in the Old Testament (cf. the "Deus absconditus" of Is. 45:15), there are many passages which associate divine interventions in human history with light, see e.g. Ex. 19:16-19; 20:18; Is. 30:27; 66:15; Bar. 5:9; Ezek. 10:4. See also Ps. 103:1-2: "[...] nimis gloria et decore indutus es [sc. God], amictus luce quasi vestimento" ("you are clothed in very great glory and beauty, covered in light as with a garment"); cf. *Par.* VIII, 52-54: "La mia letizia mi ti tien celato / che mi raggia dintorno e mi nasconde / quasi animal di sua seta fasciato."

220

light referred to in the first chapter of Genesis, and in John's Gospel the divine logos is conceived as the Light that shines in the darkness and that enlightens every man coming into this world. In the New Testament, moreover, the coming of the Messiah is repeatedly couched in light images. Biblical writers frequently expressed key themes such as sight and blindness, charity, and the resurrection of the body through light images.[5] Light in the Bible connotes life, salvation, judgement, truth, delight, and joy; and Dante often makes use of these couplings in his poetry, especially the fundamental associations between light, life, and truth that are found in John's Gospel.[6]

Because the present chapter is concerned with how writers came to use light in theological contexts by the 1280s and 1290s, this is not the place for a detailed consideration of biblical light imagery in the *Paradiso*.[7] To appreciate the

---

[5]See I Cor. 15: 41-42: "Alia claritas solis, alia claritas lunae, et alia claritas stellarum. Stella enim a stella differt in claritate. Sic et resurrectio mortuorum" ("The brightness of the sun and the brightness of the moon and the brightness of the stars are all different from one another. For one star differs in brightness from another. So it is with the resurrection of the dead"). See also Dan. 12:3: "qui autem docti fuerint, fulgebunt quasi splendor firmamenti, et qui ad iustitiam erudiunt multos, quasi stellae in perpetuas aeternitates" ("they who were wise will shine like the splendour of the firmament, and they who make people just will be like stars in perpetuity"); cf. Mt. 13:43.

[6]For the association between light and happiness, see Ecc. 11:7: "dulce est lumen et desiderabile oculis videre lumen" ("light is sweet and the eyes delight in seeing light"); cf. *Inf.* X, 69: "non fiere li occhi suoi lo dolce lume?»." For light and life/truth, see Jn. 1:3-5, 9: "Quod factum est in ipso vita erat, et vita erat lux hominum; et lux in tenebris lucet et tenebrae eam non comprehenderunt [...] Erat lux vera, quae illuminat omnem hominem venientem in hunc mundum" ("What was made in him was life and the life was the light of men; and light shines in the darkness and the darkness did not encompass it [...] It was the true light which illuminates every man coming into this world"); cf. *Con.* II, v, 3; II, viii, 14. For other important associations between light and salvation, and between darkness and evil, see Ps. 4:7: "signatum est super nos lumen vultus tui Domine" ("Lord, the light of your face is marked upon us"); Prov. 13:9: "lux iustorum laetificat, lucerna autem impiorum extinguetur" ("the light of the just gives joy but the lamp of the impious is extinguished"); Jn. 3:20-21: "omnis enim qui mala agit odit lucem, et non venit ad lucem ut non arguantur opera eius. Qui autem facit veritatem venit ad lucem ut manifestentur eius opera quia in Deo sunt facta" ("all who commit evil acts hate the light and do not come into the light in order that their deeds are not made known. They who act righteously come into the light so that their deeds are manifested, for they were done in God").

[7]Some of the biblical light motifs upon which Dante draws are discussed in earlier chapters: his direct allusion to Paul's *circumfulsit* in *Par.* XXX (Ch. 3, n. 18); his similes based on star

221

wide range of metaphorical applications that these authors gave to the light image, it is essential to consider earlier discussions of light in the Church Fathers. Commenting on passages such as Genesis 1:3 and John 1:8, the Latin and Greek Fathers often amplified the biblical narrative with additional material, including contemporary ideas about the nature and properties of light.[8] In his *Hexaemeron* (c. 389), for example, St. Ambrose compared the emergence of the first light to the illumination of the world by natural light:

Sicut enim cito lux coelum, terras, maria illuminat, et momento temporis sine ulla comprehensione retectis surgentis diei splendore regionibus, nostro se circumfundit aspectui; ita ortus eius cito debuit explicari.[9]

Augustine (354-430) used similar ideas in his exposition of creation in Genesis, and he also adopted the light image to express the concept of divine omnipresence.[10]

---

scintillation and the rainbow (Ch. 4, nn. 71, 76 and 78); biblical passages on dazzling light (Ch. 3, nn. 17 and 41).

[8]On the image of light in early Christian thought, with particular reference to Athanasius (c. 296-373), see Jaroslav Pelikan, *The Light of the World: A Basic Image in Early Christian Thought* (New York: Harper, 1962). For patristic and scholastic discussions of Gen. 1:3, see respectively Nicholas H. Steneck, *Science and Creation in the Middle Ages: Henry of Langenstein (d. 1397) on Genesis* (London and Notre Dame: Notre Dame University Press, 1976), pp. 52-55; Edward Randal McCarthy, *Medieval Light Theory and Optics and Dun Scotus' Treatment of Light in Distinction 13 of Book II of his Commentary on the Sentences*, unpublished Ph.D. dissertation, City University of New York, 1976, pp. 128-133.

[9]Ambrose, *Hexaemeron*, lib. I, c. 9, § 33 (*PL* 14, col. 142B): "For, just as light quickly illuminated the heavens, the earth and the seas, and in a moment, without our being aware of it, when the land is unveiled at the splendor of dawn, this light is perceived as it encompasses us, in such a manner should its birth be explained," trans. in John J. Savage, *Saint Ambrose: Hexaemeron, Paradise, and Cain and Abel*, reprint (Washington, DC: Catholic University of America Press, 1977), p. 39. The diffusive properties of light drew the admiration of other Latin and Greek Fathers, see e.g. Augustine, *De Gen. ad litt.*, c. 4, §§ 2-3 (*PL* 34, cols 218-219); Basil, *Homiliae in Hexaemeron*, hom. II, § 7 (*PG* 29, col. 45A-B).

[10]For the concept of divine omnipresence expressed through light imagery, see Augustine, *De Gen. ad litt.*, c. 4, § 16 (*PL* 34, col. 226); idem, *Confessionum*, lib. VII, c. 1, § 2, ed. Verheijen, in *CCSL* 26 (1981), p. 93: "Sicut autem luci solis non obsisterat aeris corpus, aeris huius, qui supra terram est, qominus per eum traiceretur penetrans eum non dirrumpendo aut concidendo, sed implendo eum totum, sic tibi putabam non solum caeli et aeris et maris sed etiam terrae

222

He was particularly interested in establishing analogies between different grades or levels of light, developing these correspondences widely to exemplify his ideas on Trinitarian theology and to elaborate a theory of human knowing.[11]

Another important figure in helping to establish the medieval predilection for light analogies was Dionysius the Aeropagite, the Christian Platonist, who will be given more detailed consideration later in this chapter. The Pseudo-Dionysius repeatedly used extended analogies between natural light, especially the rays of the sun, and divine light in discussing aspects of the deity. In the opening section of chapter four of *De divinis nominibus*, for example, he discussed the universal diffusion of divine goodness by analogy to the sun:

Etenim sicut noster sol, non ratiocinans aut praeeligens, sed per ipsum esse, illuminat omnia participare lumen ipsius secundum propriam rationem valentia, ita quidem et bonum, super solem sicut super obscuram imaginem segregate

---

corpus pervium" ("Just as the body of the air does not impede the passage of the light of the sun and its permeating it, not by breaking it or cutting it, but by filling it up completely, so I considered not only the body of the heavens and the air and the sea, but also that of the Earth to be permeated by You"). For the idea that light by its nature reveals, see *De Gen. ad litt.*, c. 5, § 24 (*PL* 34, col. 229): "recte dicitur lucem qua res quaeque manifesta est" ("that by which something is made manifest is fittingly called light"); cf. Eph. 5:14.

[11]The key Augustinian passages are: (i) Trinitarian analogies, *Sermones*, § 17 (*PL* 38, cols 667-669); *De Trinitate*, lib. IV, c. 20, § 27, ed. W.J. Mountain, in *CCSL* 50A, p. 196: "Ibi autem quod manat et de quo manat unius eiusdemque substantiae est. Neque enim sicut aqua de foramine terrae aut lapidis manat sed sicut lux de luce" ("That which flows out and that out of which it flows are one and the same substance. Nor is it like water which flows out of a hole in the ground or stone, but it is like light which comes from light"); (ii) analogies between intelligible and sensible light to discuss human knowing, *De Trinitate*, lib. XII, c. 15, § 24, ed. Mountain, p. 378; *In Johannis evangelium tractatus*, lib. XV, c. 4, § 19 (*PL* 35, col. 1517): "Iam superior lux, qua mens humana illuminatur Deus est" ("Now the higher light by which the human mind is illuminated is God"). On light in Augustine, see also Norman K. Klassen, *Chaucer on Love, Knowledge and Sight* (Cambridge: Brewer, 1995), pp. 4-12; Athanase Sage, "La dialectique de l'illumination," *Recherches augustiniennes* 2 (1962), 111-123; Thonnard, "La notion de la lumière en philosophie augustinienne."

archetypum, per ipsam essentiam omnibus exisistentibus proportionaliter immittit totius bonitatis radios.[12]

Later in this chapter, he discussed the propriety of using light as a name for the good and drew a series of analogies between the material sun and divine goodness:

Quid dicat quidem aliquis de ipso secundum se solari radio? ex bono enim est illud lumen et imago bonitatis [...] magnus iste et totus splendens et semper lucens sol, secundum multam resonantiam boni, et omnia quaecumque participare ipso possunt illuminat, et super extentum habet lumen, ad omnem extendens visibilem mundum et sursum et deorsum propriorum radiorum splendores.[13]

The Pseudo-Dionysius employed other examples drawn from light to illustrate Christian doctrines such as the circumincession of the Divine Persons, a concept he expressed by describing how many lights in a room could interpenetrate, and yet remain unmixed.[14] In his treatise on angels, *De caelesti hierarchia,* he spoke

[12]Pseudo-Dionysius, *De divinis nominibus,* c. 4, § 1 (*PG* 3, col. 694C; Chevalier, I, pp. 147-148, Latin trans. Sarrazin): "As our sun, without reasoning or choosing but through its being, illuminates all things which are able to participate in its light in their own ways, so it is with the good, which, transcending the sun like an obscure image its distant archetype, sends the rays of its goodness proportionately into all things through its essence."

[13]*De divinis nominibus,* c. 4, § 4 (*PG* 3, col. 697B; Chevalier, I, pp. 161-162, Latin trans. Sarrazin): "What can one say of the sun's rays? That light comes from the good and is an image of goodness [...] This great, ever-shining, ever-luminous sun is the image of divine goodness and it illuminates everything that is able to receive its light without diminishing and extends the brightness of its own rays all around to all the visible world." For a similar comparison, see ibid., c. 5, § 8 (*PG* 3, col. 824B; Chevalier, I, pp. 357-359). The sun is, of course, a biblical image, see Mal. 4:2: "et orietur vobis timentibus nomen meum sol iustitiae" ("the sun of justice is risen for you who fear my name"). The image in the Pseudo-Dionysius also echoes Plato's famous sun simile in the *Republic,* VI, 508B (Cornford, pp. 218-220).

[14]*De divinis nominibus,* c. 2, § 4 (*PG* 3, col. 641B; Chevalier, I, pp. 78-79, Latin trans. Sarrazin): "Quemadmodum lumina luminarium, ut sensibilibus et propriis utar exemplis, existentia in domo una, et tota in se invicem totis sunt, et diligentem habent ad se invicem discretionem proprie subsistentem, unita discretione et unitione discreta" ("I will take an example from sensible and familiar objects, it is in a sense like a house in which the light from each of its lamps is mixing together yet each is distinct, they are distinct in unity and united in distinction").

224

about the way in which God imparts His light to different orders of angels and compared their reception of divine light to a ray of sunlight entering matter of differing densities.[15]

Light Analogies in Medieval Theological Works: An Illustrative Anthology

Almost all medieval theologians used light analogies in commentaries on Scripture, on the *Corpus areopagiticum,* and in their independent writings. Although these writers borrowed extensively from the Fathers, they also fashioned their own comparisons from close scrutiny of the properties of natural light: its splendour and radiative properties, its seeming immateriality, its capacity to diffuse itself, and its power to reveal. It was these properties which helped to stimulate ways of thinking about the deity, divine action, providence, the Trinity, grace, and beatific vision. The idea that a light source diffused itself instantaneously came to be very widely used to illustrate the concept of divine omnipresence and/or the action of grace. Light's capacity to generate a luminous replica of itself in the surrounding medium was used to indicate the relations between the First and Second Persons of the Trinity. The relationship between a light source and its receptors was employed to describe the relationship between God, or some aspect of His power, and His creatures. As Augustine's writings show, light comparisons often formed part of a further analogy between the process of seeing and that of understanding. If the eye could only see its object when it was illuminated, then presumably the

---

This example was widely used in the thirteenth century, see e.g. Bartholomew the Englishman, *De rerum proprietatibus,* lib. VIII, c. 40 (Richter, p. 427); Bonaventure, *In I Sent.,* d. 19, art. unic., q. 4, ad. 2 (Quaracchi, I, 349b).

[15]*De caelesti hierarchia,* c. 13, § 3 (*PG* 3, col. 301A-B; Chevalier, II, pp. 949-950, Latin trans. Sarrazin): "[...] solaris radii distributio ad primam materiam benedistributive vadit, omnibus clariorem, et per ipsam manifestius resplendere facit proprios splendores; accedens autem ad grossiores materias, obscuriorem habet distributivam manifestationem" ("the rays of the sun pass freely into primary matter, which is brighter than all other things and this makes its own splendours shine more clearly. But as it passes into more opaque matter, its ability to illuminate by the spreading out of rays becomes more obscure").

eye of the mind could only understand its object in the presence of a higher form of light. This relationship between the outer and inner eye, between physical and intellectual sight, was a highly important one for medieval writers, and in a variety of ways it helped to guide many of the light analogies that they developed. By the later thirteenth century, moreover, the assimilation of Arabic sources led many theologians to adopt a more distinctive approach, enriching their comparisons with detailed information about the properties of mirrors, geometrical optics, and the relationship between light and colour.

It would be almost impossible to give a comprehensive list of references, or to indicate the ancient and patristic sources upon which many of these comparisons and analogies depend. An anthology of passages will be provided from eight authors whose writings allow us to appreciate the pervasive character of analogies drawn from light in medieval theology. These extracts have been chosen precisely because they exemplify aspects of light to which thirteenth-century theologians repeatedly return. Several parallels have been suggested with passages from the *Paradiso*, but these parallels are intended to indicate Dante's adoption of concepts and analogies widely available in contemporary works of theology, rather than to suggest specific sources.

The first passage is taken from a ninth-century medieval hymn that praises the Virgin Mary and compares her miraculous infecundation to the way in which a ray of light penetrates a glass without detriment to its purity:

Sicut vitrum radio
solis penetratur,
inde tamen laesio
nulla vitro datur,
Sic, immo subtilius,
matre non corrupta,
Deus, Dei filius,

226

sua prodit nupta.[16]

This comparison was widely used by twelfth- and thirteenth-century theologians and is also found in early religious and lyric Italian vernacular poetry.[17] The parallels with Dante's description of his own passage into the luminous body of the Moon are quite clear:

Per entro sé l'etterna margarita
ne ricevette, com' acqua recepe
raggio di luce permanendo unita. (II, 34-36)

The second passage from Eriugena's commentary (c. 866) on the Pseudo-Dionysius' *De caelesti hierarchia* refers to the Trinity in terms of light, develops an important biblical passage from the Epistle of St. James, and places emphasis on the idea that a light source retains its unity, despite distributing itself to all things:

Et hec est trina lux, et trina bonitas, tres in essentia, Pater et Filius et Spiritus Sanctus, unus Deus, una bonitas, unum lumen diffusum in omnia que sunt [...] Et in ipso omnia unum sunt, a Patre itaque luminum omnia lumina descendunt.[18]

---

[16]*Lateinische Hymnen des Mittelalters*, ed. Franz Joseph Mone, 3 vols. (Freiburg: Sumptus Herder, 1853-1854), II, p. 63: "As a glass is penetrated by a ray of light which does not harm it in the slightest, so it is and yet in a more subtle manner still that God produced the Son of God, his offspring, in the inviolate mother." See also ibid., I, p. 63; II, p. 411.

[17]The image can be found in many twelfth- and thirteenth-century theologians, see e.g. Bonaventure (quoted below in n. 67). For its use in religious and lyric poetry, see Ch. 1, n. 64. For an interesting example in a twelfth-century Christmas play, see David Bevington, *Medieval Drama* (Boston: Houghton Mifflin, 1975), p. 187. The phenomenon was also discussed in *Sentence* commentaries as an example of how two bodies could be located in one place, see e.g. Aquinas, *In IV Sent.* d. 44, q. 2, a. 2, qla 2, ob. 2, in *Opera omnia*, VII, p. 1088b: "Sed quaedam corpora nunc ratione suae nobilitatis possunt simul esse cum aliis corporibus, scilicet radii solares" ("Yet by reason of their nobility certain bodies, namely the rays of the sun, are able to occupy the same place together with other bodies").

[18]Eriugena, *Expositiones in Ierarchiam coelestem*, lib. I, §§ 67-75, ed. Jeanne Barbet (Turnhout: Brepols, 1975), in *CCCM* 31, pp. 2-3: "And this triune light and triune goodness which is three in essence, Father, Son and Holy Spirit, one God, one goodness, one light diffused into all things that exist [...] and in itself all things are unified, and so from the father of lights all lights

227

In a similar manner, Dante apostrophes God as "trina luce" in *Paradiso* XXXI, supplicating Him to restore order on Earth:

> Oh trina luce che 'n unica stella
> scintillando a lor vista, sì li appaga!
> guarda qua giuso a la nostra procella! (28-30)

A related example from Richard of St. Victor's *De Trinitate* (c. 1148) puts forward a detailed set of correspondences between the First and Second Person of the Trinity, on the one hand, and a ray of light and the sun, on the other:

Certe radius solis de sole procedit et de illo originem trahit, et tamen soli coaevus exsistit. Ex quo enim fuit, de se radium producit et sine radio nullo tempore fuit. Si igitur lux ista corporalis habet radium sibi coaevum, cur non habeat lux illa spiritualis et inaccessibilis radium sibi coaeternum?[19]

---

descend." See also Grosseteste, *Hexaemeron*, p. VIII, c. 3, § 1 (Dales and Gieben, p. 220): "Quod autem Deus sit in personis trinus, inde sequitur quod Deus est lux [...] Lux autem gignens et splendor genitus necessario sese amplectuntur mutuo, et spirant sese mutuum fervorem" ("From the fact that God is present in three Persons it follows that God is light [...] The light which begets and the splendour which is produced are necessarily linked in each other's light and they breath out reciprocal fervour"). Medieval writers were more interested in this aspect of light symbolism than the Pseudo-Dionysius whose main concern was with relations between the One and the many. For other examples of Dante's light Trinitarianism, see *Par.* XIII, 52-57; XXXIII, 115-117, 124-126. Cf. Tasso, *Gerusalemme Liberata*, IX, 56, 5-6: "e de l'Eternità nel trono augusto / risplendea con tre lumi in una luce."

[19]Richard of St. Victor, *De Trinitate*, lib. I, c. 9, in *Richard de St. Victor: La Trinité*, ed. Gaston Salet, in *SC* 63 (Paris: Éditions du Cerf, 1959), p. 80: "It is certain that a ray of light proceeds and draws its origin from the sun and so it is coeval with the sun. Since as soon as the sun existed it produced from itself its ray, it never existed without any ray. If therefore this corporeal light has a ray which is of the same time to it, why does that spiritual and inaccessible light not have a ray coeval to it?" Light imagery was, of course, exceptionally common in this context (see n. 11). For important medieval elaborations, see Bonaventure, *Itinerarium mentis ad Deum*, c. II, § 7 (Quaracchi, V, p. 301b): "illa lux aeterna generat ex se similitudinem seu splendorem coaequalem, consubstantialem, et coaeternalem" ("that eternal light generates out of itself a likeness or splendour which is equal to it, and shares in both one substance and eternity"); Grosseteste, *De ordine emanandi causatorum a Deo* (Baur, pp. 147-150). Dante represents the first two Persons of the Trinity as "luce" and "splendore" in *Par.* XIII, 53-55.

228

My fourth passage is from Hugh of St. Victor's commentary (c. 1140) on the Pseudo-Dionysius' *De caelesti hierarchia* and contains a rather different example: that which is light by its nature (God) is distinguished from that which becomes light when illuminated (God's creatures):

Sicut enim duo sunt: lumen et quod suscipit lumen corpus: et ex his duobus unum efficitur lucens, et ipsum lucens imago quodammodo est, et similitudo luminis, in eo quod lucet sicut ipsum lumen: ita et Deus noster lumen est, et verum lumen est, et ipsum lumen rationales animi mundi et puri concipiunt: et ex eo lucentes fiunt, et non sunt ipsi imago luminis in eo quod sunt: sed in eo quod lucent ex lumine, sicut ipsum lumen lucet [...][20]

Similarly, Dante uses the idea that there is a direct relationship between a light source and its illuminated objects when speaking about God and lesser goods in canto V:

> Io veggio ben sì come già resplende
> ne l'intelletto tuo l'etterna luce,
> che, vista, sola e sempre amore accende;
> e s'altra cosa vostro amor seduce,
> non è se non di quella alcun vestigio,
> mal conosciuto, che quivi traluce. (7-12)

In canto VII, he reworks this idea to suggest how mankind darkens itself through sin. Next to Christ and the angels, man is the most light-infused of God's

---

[20]Hugh of St. Victor, *Commentariorum in hierarchiam coelestem*, lib. II (*PL* 175, col. 955AB): "As there are then two: light and a body which receives light, and out of these two one light is produced, and this lucent body is a kind of image and a likeness of light insofar as it shines like the former light. So it is that God is our light, and is the true light, and pure and clean rational souls look into this light and become luminous because of it. And they are not in themselves an image of this light, but they are so insofar as they shine due to another light, as this light shines."

229

creations, but sin renders human beings dissimilar to the supreme good and they thus receive less of its light:

> Solo il peccato è quel che la disfranca
> e falla dissimile al sommo bene,
> per che del lume suo poco s'imbianca. (79-81)

In the Eagle's discourse on predestination in canto XIX, Dante again uses a related light image within a more epistemological frame of reference. The poet defines human understanding in relation to divine prescience by implying that the divine mind is a source of light while man's own intellectual vision is a derivative or secondary form of this light:

> Dunque vostra veduta, che convene
> essere alcun de' raggi de la mente
> di che tutte le cose son ripiene. (52-54)

Similar imagery is also used in canto XXVI to illustrate how goodness resides primarily in God and secondarily in other creatures. The distinction between a ray and its derivative light provides Dante with a fitting comparison for the essential link which connects God as supreme good with His lesser goods:

> Dunque a l'essenza ov' è tanto avvantaggio,
> che ciascun ben che fuor di lei si trova
> altro non è ch'un lume di suo raggio. (31-33)

The next four examples in the anthology are all taken from thirteenth-century works and illustrate some of the refinements and technical innovations which had been synthesised by this later period. The fifth passage comes from Albert's commentary (c. 1250) on John's Gospel, a work which gave rise to a rich exegetical tradition and stimulated both ancient and medieval writers to devise

230

elaborate analogies between natural and divine light in their glosses upon it.[21] Albert is no exception:

Unde sicut tripliciter corpora se habent ad lucem solis: sic tripliciter habent se corda ad lucem istam. Quaedam enim corpora tenebris coniuncta non recipiunt lucem, nisi ad manifestationem nigredinis et turpitudinis suae, sicut nigra, terra, et opaca. Quaedam autem recipiunt eam ad exteriorem sui illustrationem et pulchritudinem. Quaedam recipiunt eam in sui profundum, sicut perspicua ut lapides pretiosi: et efficiuntur quasi quaedam vasa lucis quae luminaria vocantur. Ita homines [...][22]

Thomas Aquinas' commentary on John (c. 1272) contains three noteworthy passages in which the properties of natural light are applied to higher forms of light

---

[21]Among the more influential earlier commentaries on this text are: Augustine, *In Joahnnis evangelium tractatus* (*PL* 35, cols 1579-1976); Eriugena, *Homélie sur le prologue de Jean*, ed. Edouard Jeauneau, in *SC* 151 (Paris: Éditions du Cerf, 1969); idem, *Commentaire sur l'Évangile de Jean*, ed. Edouard Jeauneau, in *SC* 180 (Paris: Éditions du Cerf, 1972).

[22]Albert, *In Evangelium secundum Ioannem expositio*, c. 1 (Jammy, XI, p. 9a-b): "Hence as bodies are related to the light of the sun in three ways, so there are three ways in which hearts are related to this light. Some bodies in darkness do not receive light other than to reveal their black colour and their ugliness, such as black objects, the earth, and opaque bodies. Other bodies receive light in such a way as to illuminate their exterior and appear beautiful. Others, such as transparent bodies and precious stones, receive light within themselves, and become like certain vessels of light, which are called luminaries. And so it is with men [...]." Albert's writings are especially rich in this type of light-based analogy; see esp. his own extended comparisons in *Super Dionysium De caelesti hierarchia*, c. 3, ed. Wilhelm Kübel and Paul Simon (Münster: Aschendorff, 1993), pp. 45-46; *Super Dionysium De divinis nominibus*, c. 4, comms. 22 and 66 (Simon, pp. 130, 175). See also Eriugena, *Expositiones in Ierarchiam coelestem*, lib. IX, § 408-410 (Barbet, p. 145): "non tamen [sc. the sun symbolising divine goodness] ab omnibus quibus equaliter diffunditur equaliter participatur: vitrum penetrat, silice repercutitur, purum aera pervolat, aquosa nube resultat" ("the sun is diffused equally to all things, but these things do not participate in its light in the same way: the sun penetrates glass, is rebounded from rock, passes through air, shines back from watery clouds"). Of course, this type of comparison was not restricted to theological contexts, see e.g. Albert, *De intellectu et intelligibili*, lib. I, tr. 1, cc. 3-5 (Jammy, V, pp. 241a, 242a, 243a); lib. II, tr. un., c. 10 (Jammy, V, 260a-b); Aquinas, *In librum de causis*, prop. XX, lect. 20, § 363 (Pera, p. 111).

231

and/or the deity. First, he discusses the Manichaean heresy and describes the proper relationship between sensible and intelligible light:

Lux autem ista sensibilis, imago quaedam est illius lucis intelligibilis [...] sicut autem lux particularis habet effectum in re visa, inquantum colores facit actu visibiles, et etiam in vidente, quia per eam oculus confortatur ad videndum, sic lux illa intelligibilis intellectum facit cognoscentem. Quia quidquid luminis est in rationali creatura, totum derivatur ab ipsa suprema luce.

The idea that human intelligence is derived from the supreme intelligible light of God is one which, as we have already seen, Dante gives fullest expression to in the Heaven of Jupiter, where human "veduta" is said to be "alcun de' raggi de la mente / di che tutte le cose son ripiene" (XIX, 52-54).

In the same commentary, Aquinas later makes a distinction between a light source and illuminated objects to express the idea of participation:

Esse enim lucem est proprium Dei, alia vero sunt lucentia, idest participantia lucem; sed Deus lux est per essentiam.

And finally, using more Dionysian language, he compares God's presence in the universe to the way in which the sun diffuses its light to all created things:

Deus hoc modo se habet ad nos ut lux ad homines. Lux autem ubique diffunditur sole existente super terram. Et licet lux sit cum hominibus, non tamen omnes sunt in luce solis, sed tantum eam videntes. Sic ergo cum Deus sit ubique, est cum

232

omnibus qui sunt ubicumque; sed tamen non omnes sunt cum Deo, nisi qui ei coniunguntur per fidem et dilectionem.[23]

The opening lines of the *Paradiso*, which describe the divine light as it penetrates and shines back from the created universe, pay homage to this theme, one which is then re-developed in canto XXXI as Dante-*personaggio* gazes upon the celestial rose and the poet is led to comment:

> ché la divina luce è penetrante
> per l'universo secondo ch'è degno,
> sì che nulla le puote essere ostante. (22-24)

The next example in my anthology of passages is taken from St. Bonaventure's *Sentence* commentary (1250-1252), and although it still rests upon the principle of analogy and again uses a biblical light motif, it puts forward a more technical doctrine of grace:

[...] comparatur ipsa *gratia creata* influentiae luminis et principium eius comparatur soli. Unde et Scriptura vocat Deum sive Christum *solem iustitiae*, quia, sicut ab isto sole materiali influit lumen corporale in aëra, per quod aër formaliter illuminatur; sic a sole spirituali, qui Deus est, influit lumen spirituale in animam, a quo anima formaliter illuminatur et reformatur et gratificatur et

---

[23]References in order: *Super evangelium S. Ioannis lectura*, ed. R. Raphaelis Cai, 6th edn (Rome and Turin: Marietti, 1972), c. 8, lect. 8, § 1142; c. 12, lect. 8, § 1713; c. 17, lect. 6, § 2258, pp. 215, 320, 425: "This sensible light is a certain image of that intelligible light. As particular light has an effect upon the visible object in that it makes colours actually visible, and it also affects the spectator, for by it the eye is strengthened to see, so it is that intelligible light makes the intellect able to know. For, whatever is of light in rational creatures, is entirely derived from this supreme light [...] Being is a light proper to God. While other things shine, that is, they participate in light, God is light by His essence [...] The relationship between God and us is similar to that between light and men. Light is diffused everywhere that the sun is present over the Earth. And although light is present with men, not all of them are in the light of the sun, but they only see it. In a similar way, then, since God is everywhere, He is with everyone wherever they are, but not everyone is with God, unless they are joined to him by faith and love."

vivificatur. Unde inter omnia corporalia maxime assimilatur gratiae Dei luminis influentia. Sicut enim haec est quaedam influentia, quae assimilat corpora ipsum suscipientia ipsi fonti luminis [...] sic gratia est spiritualis influentia quae mentes rationales fonti lucis assimilat et conformat.[24]

The final extract, taken from Roger Bacon's *De utilitate mathematica* (c. 1266), is perhaps less representative. Bacon also discusses the action of grace, but he does so by drawing on his technical knowledge of optical science which he develops into an interesting optical allegory:

gratiae infusio maxime manifestatur per lucis multiplicationem [...] in bonis perfectis infusio gratiae comparatur luci directe incidenti et perpendiculari, quoniam non reflectunt a se gratiam, nec frangunt per declinationem ab incessu recto [...] Sed infusio gratiae in imperfectos, licet bonos, comparatur luci fractae [...] Peccatores autem, qui sunt in peccato mortali, reflectunt et repellunt a se gratiam Dei, et ideo gratia apud eos comparatur luci repulsae seu reflexae.[25]

---

[24]Bonaventure, *In II Sent.*, d. 26, art. unic., q. 2, concl. (Quaracchi, II, p. 636a): "this created grace is compared to the influence of light and its origin is compared to the sun. Hence, Scripture calls God or Christ, the 'sun of justice,' since as corporeal light flows into the air, and air is formally illuminated by it, so it is that from the spiritual sun, who is God, spiritual light flows into the soul and from this the soul is formally illuminated and reformed and filled with grace and life. Hence, from amongst all corporeal things, God's grace is especially assimilated to the influence of light. Just as therefore this certain influence, which assimilates bodies that receive it to this source of light [...] so grace is a spiritual influence, which assimilates and conforms rational minds to the source of light"; cf. q. 6, c.: "Gratia enim est sicut quaedam influentia procedens a luce superna, quae semper habet coniunctionem cum sua origine, sicut lumen cum sole" ("Grace is like a certain influence, which, proceeding from the supernal light, is always connected to its origin, just as light in the medium is connected to the sun").

[25]Bacon, *Mathematica in divinis utilitas* (Bridges, I, pp. 216-217): "the infusion of grace is most especially shown through the multiplication of light [...] in the good who have been perfected the infusion of grace is compared to light which is directly and perpendicularly incident, for the good do not reflect grace from themselves, nor refract it from a straight path [...] But the infusion of grace in those who are not perfected, though still good, is compared to refracted light [...] Sinners who are in mortal sin, however, reflect and send back from themselves the grace of God, and hence in them grace is compared to light which is repercussed or reflected"; see also *Opus tertium*, c. 58 (Brewer, p. 227): "gratia tenet incessum rectum in hominibus perfectis, et incessum

234

As a philosopher, Dante did not possess Bacon's knowledge of refraction, but in the *Paradiso* he nonetheless draws on more general ideas about the behaviour of natural light to describe grace as a ray of light that multiplies itself and shines in human souls. In *Paradiso* X, for example, Thomas Aquinas addresses a *captatio benevolentiae* to Dante-*personaggio* which includes the following statement:

> lo raggio de la grazia, onde s'accende
> verace amore e che poi cresce amando,
> multiplicato in te tanto resplende,
> che ti conduce su per quella scala
> u' sanza risalir nessun discende. (83-87)

Bartholomew of Bologna and Dante

Bartholomew of Bologna's *Tractatus de luce* (c. 1280) provides an excellent example of the extent to which doctrines and analogies based on light were popular amongst theologians in Italy in the 1280s and 1290s. Since this treatise contains perhaps more information about light than any other single medieval work of its kind, it is worthwhile to give a brief description of its contents.[26]

The *Tractatus de luce* consists of six sections and is devoted to elucidating a single verse in John's Gospel: "Ego sum lux mundi, qui sequitur me non ambulabit in tenebris, sed habebit lucem vitae" (Jn. 8:12). In the first section,

---

fractum in imperfectis, et reflectitur a malis" ("grace follows a straight line in men who have been perfected and a bent path in the non-perfected; in the evil it is reflected"). Elsewhere, Bacon argues that geometrical optics is required for a proper understanding of Scripture, see *Opus maius*, lib. V, p. iii, d. 3, c. 1 (Lindberg, pp. 321-324). Peter of Limoges also uses optical knowledge of refraction to define the action of grace, see the passage from his *De oculo morali*, c. 5, a. 1, quoted by Clark, in "Optics for Preachers," nn. 50-51, p. 340; cf. Grosseteste, *Hexaëmeron*, p. II, c. 10, § 4 (Dales and Gieben, p. 100). For less optical applications of technical light comparisons in thirteenth-century theologians, see Jacques G. Bougerol, "Le rôle de l'*influentia*." Cf. *Con.* III, xiv, 5.

[26]On this treatise as one of the foremost examples of medieval "light theology," see Lindberg, *A Catalogue of Medieval and Renaissance Optical Manuscripts*, pp. 43-44; Vescovini, *Studi sulla prospettiva*, pp. 24-28.

235

Bartholomew draws upon technical sources to advance five reasons for preferring *lux*, and not the other terms for light (*lumen*, *radius*, and *splendor*), as a designation for Christ. In the second section, he explains how spiritual and intellectual rays emanate from divine light, providing an example of this emanation by analogy to the reception of natural light in different substances. Section three contains a lengthy discussion of the conditions by which human minds can reach higher truths when illuminated by a light known as the *lumen fidei*. Section four gives an extended analysis of the operations of divine light in the human mind, emphasising its power to illuminate, conciliate, diffuse, reveal, and beautify.[27] The final two sections provide some additional reasons for using other light terms to denote Christ.[28]

The *Tractatus'* suggestive title and the fact that Bartholomew was head of the Franciscan *studium* in Bologna until his death in 1292 have excited interest in the possible connections between this work and Dante.[29] And yet, it would be

[27]*Tractatus de luce*, pars 3, c. 1 (Squadrani, pp. 357-358): "per lumen fidei existens in pupilla mentalis oculi ipse intellectus disponitur ut ab aeternae lucis irradiatione abundanter illustretur" ("intellect is disposed so that it is abundantly illustrated by the irradiation of eternal light due to the light of faith which exists in the pupil of the mental eye"). This light was believed to be a disposition which was added to the human intellect to allow access to *altiora intelligibilia*, see Aquinas, *ST*, Ia-Iae, q. 109, a. 1, resp. (Caramello, II, p. 529). The doctrine of illumination was widely used in discussions of faith. For Dante's use of the term *lumen gratie*, see *Mon*. III, iv, 20: "regnum temporale non recipit esse a spirituali, nec virtutem que est eius auctoritas, nec etiam operationem simpliciter; sed bene ab eo recipit ut virtuosius operetur per lucem gratie quam in celo et in terra benedictio summi Pontifici infundit in illi" ("the temporal realm does not owe its existence to the spiritual realm, nor its power [which is its authority], and not even its function in an absolute sense; but it does receive from it the capacity to operate more efficaciously through the light of grace which in heaven and on Earth the blessing of the supreme Pontiff infuses into it," trans. Shaw, p. 112).

[28]References in order from Bartholomew's *Tractatus de luce*: (i) on the propriety of *lux* as a name for Christ, pars II, cc. 1-5 (Squadrani, pp. 229-238, 337-345); (ii) four grades of illumination, pars II, c. 1 (Squadrani, pp. 347-348); (iii) discussion of *lumen fidei*, pars III, c. 1 (Squadrani, pp. 351-358); (iv) properties of light and optical effects, esp. pars IV, cc. 1-5 (Squadrani, pp. 362-385, 468-472).

[29]For critics who support the view that Dante made direct use of the *Tractatus de luce*, see Gage, *Colour and Culture*, n. 116, p. 280; Guidubaldi, *Dante Europeo*, II, pp. 410-430; III, pp. 346-350; idem, "Bartolomeo da Bologna," in *ED* I, pp. 526-527; Leonardo Olschki, "Sacra dottrina e theologia mystica: Il Canto XXX del *Paradiso*," *Giornale dantesco* n.s. 6 (1933), pp. 17-18;

236

almost impossible to demonstrate conclusively that Dante knew the *Tractatus*, since Bartholomew discusses light images and doctrines that are popular in other medieval writings.[30] It is, however, precisely because many of Bartholomew's analogies are commonplace that it is useful to compare Dante's knowledge and use of light imagery in the *Paradiso* with sections of Bartholomew's treatise. The *Tractatus de luce* provides many passages that present parallels with clusters of light imagery in the final *cantica*, and two lengthy examples deserve to be quoted more extensively. First of all, a passage from section two in which Bartholomew discusses different modes of illumination by distinguishing between shining (*lucere*), illuminating (*illuminare*) and being a source of light (*principium lucis*):[31]

[...] in genere rerum spiritualiter lucentium et illuminantium sunt quattuor gradus [...] Sunt ergo quaedam quae lucent, sed non illuminant: lucent enim quia habent aliquid luminis sibi complantatum, sed tam debile, ut non sufficiant illud aliis communicare, et huiusmodi est vitrum, quod etsi in se luceat, positum tamen in tenebris alia non illuminant. Sunt enim quaedam in secundo gradu, quae lucent in se et illuminant alia circa se posita, ut aliqui lapides pretiosi, qui in tenebris positi, et se et alia faciunt videri. Sunt et alia in tertio gradu, quae lucent in se et illuminant alia, et ita efficaciter illuminant ea, ut illa sufficiant etiam illuminare, tertio loco, alia. Quod patet de igne positio in alabastro: nam ibi primo in se lucet,

---

Simonelli, "Allegoria e simbolo dal 'Convivio'," p. 209, nn. 5-6. More hesitant, but still favourable, judgements will be found in Corti, *Percorsi dell'invenzione*, pp. 161-163; Vasoli, *Il Convivio*, introd. pp. xxvi and lxxiii, also pp. 371-372, 455. See also C. Piana, "Le Questioni inedite 'De glorificatione Beatae Mariae Virginis' di Bartolomeo di Bologna O.F.M e le concezioni del *Paradiso* dantesco," *L'Archiginnasio* 33 (1938), 247-262 (but see Mellone's criticism of this article in "L'Empireo," in *ED* II, p. 669).

[30]Note that, unlike Dante, Bartholomew demonstrates a detailed knowledge of the "perspectivist" tradition; for his references to the twenty-two visible intentions and to Alhazen, see *Tractatus de luce*, pars I, c. 4 and pars II, c. 2 (Squadrani, pp. 340-341, 349).

[31]For similar distinctions between *lux*, *lucere*, and *lumen* in order to differentiate between divine existence, essence, and being, see Anselm, *Monologion*, c. 6 (*PL* 158, col. 151). For their use to discuss the relations between the intellect and the intelligible, see Albert, *De intellectu et intelligibili*, lib. I, tr. 3, c. 1 (Jammy, V, pp. 250a-b).

secundo alabastrum illuminat, et ita efficaciter, ut ipsum etiam alabastrum, tertio loco, illuminet alia extra se posita. In quarto autem gradu est quiddam quod lucet, illuminat, illuminata illuminare facit, et ultra haec omnia est principium luminis, aliis tribuens et ab aliis non recipiens, et huiusmodi est sol.[32]

It is well worth noting that all of these modes of illumination, and indeed some of the specific examples Bartholomew gives, can be found in the *Paradiso*. As was seen in the previous chapter, for example, in order to describe the act of creation in canto XXIX, Dante employs a simile which describes the passage of light "in vetro, in ambra o in cristallo" (l. 25). What is more, the poet compares the luminosity of the blessed souls both to the effect of light on jewels and to the effulgence given off by illuminated alabaster; and he also repeatedly draws on the idea that the sun is a source of light.[33]

[32]*Tractatus de luce*, pars II, c. 1 (Squadrani, pp. 346-347): "In the realm of things which shine and illuminate spiritually there are four grades [...] There are therefore certain bodies which shine but do not illuminate; they shine since they have some light within them, but it is so weak that they are not able to communicate that [light] to others. Glass is one of these kind of things. Although it shines in itself, when it is placed in darkness it does not illuminate other things. There are also certain bodies in the second grade which shine in themselves and illuminate what is placed around them, for example, precious stones, which, when placed in darkness make themselves and other things visible. In the third grade are other bodies which shine in themselves, and illuminate other things, and illuminate so efficaciously that the illuminated objects also illuminate other objects in another place. This is seen in the placing of fire in alabaster, for there it at first shines in itself, then illuminates the alabaster, and it does this to such an effect that this alabaster also illuminates things placed beyond it in another place. In the fourth grade is something which shines, illuminates, makes what it has illuminated illuminate other things, and in addition to all this is a source of light, giving light to others and not receiving from others. And the sun is an example of this kind of thing." The purpose of the analogy is to illustrate how Christian teachings are transmitted in different ways: Christ being the fourth grade.

[33]On the radiant, jewel-like appearance of the blessed souls, see *Par.* IX, 69: "qual fin balasso in che lo sol percuota"; XIX, 4-5: "parea ciascuna rubinetto in cui / raggio di sole ardesse sì acceso." On alabaster, see *Par.* XV, 24: "che parve foco dietro ad alabastro." On the sun as source of light, see *Par.* IX, 8-9: "[...] Sol che la rïempie, / come quel ben ch'a ogne cosa è tanto"; XXIII, 28-30: "vid' i' sopra migliaia di lucerne / un sol [sc. Christ] che tutte quante l'accendea, / come fa 'l nostro le viste superne."

238

There is another passage from part four of the *Tractatus* which provides further parallels with Dante's descriptions of the behaviour of light. In this section of his treatise, Bartholomew describes how light creates beautifying effects in natural objects and is altered when purified and concentrated:

per frequentem subiecti tersionem: huius exemplum patet in quolibet metallo polito et terso [...] per vehementem subiecti inflammationem, ut probatur experimento artis vitrariae [...] per obnubilativi prohibitionem, quod patet in aere, qui eo amplius lumine solis decoratur, quo magis a nubilum opacitate depuratur; per subiecti versus caelum sublimationem; quanto enim lucis subiectum exaltatur versus caelum, tanto, ceteris paribus, lucis natura decorat ipsum. Per maiorem in eo lucis aggregrationem, ut patet in sole respectu aliarum stellarum, et in aliis stellis respectu aliarum partium orbis.[34]

It is again interesting to note that nearly all of these analogies and examples are found in the imagery of the *Paradiso*, where Dante uses analogies to the luminous effects produced by polished metal, molten glass, clouds and vapours, upward movement, and the concentration of light.[35]

---

[34]*Tractatus de luce*, pars IV, c. 5 (Squadrani, p. 472): "[...] through frequent polishing of an object; an example of which is evident in clean, polished metal [...] through exposure of an object to intense heat, as will be proved by the experience of glass-making [...] through the removal of clouds, as is seen in the air, which is made more beautiful by the light of sun the more it is purified from the opacity of clouds; through the raising of an object aloft: all things being equal, the more the luminous object is raised towards heaven, the more it is made decorous by the nature of light. Through increase of the concentration of light in it, as is seen in the sun with respect to the other stars, and in the stars with respect to other parts of the orb."

[35]The parallels with comparisons to the effects of light in Dante include: (i) polished metal, *Par.* XVII, 121-123: "La luce in che rideva il mio tesoro / ch'io trovai lì, si fé prima corusca, / quale a raggio di sole specchio d'oro"; (ii) molten glass, I, 58-63: "Io nol soffersi molto, né sì poco, / ch'io nol vedessi sfavillar dintorno, / com' ferro che bogliente esce del foco; / e di sùbito parve giorno a giorno / essere aggiunto, come quei che puote / avesse il ciel d'un altro sole addorno"; cf. *Purg.* XXVII, 49-51; (iii) cloud/vapour effects, esp. *Par.* V, 133-135; XXVIII, 79-84; (iv) movement upwards resulting in increased luminosity, I, 79-81: "parvemi tanto allor del cielo acceso / de la fiamma del sol, che pioggia o fiume / lago non fece alcun tanto disteso"; (v) concentration of light, I, 61-63; VII, 6; XIV, 68; XXII, 22-24; XXIII, 28-33.

239

The Pseudo-Dionysius and Dante

As has already been mentioned, the Pseudo-Dionysius' use of light analogies and examples greatly influenced the way in which medieval writers responded to, and thought about, light. His writings helped to give intellectual coherence to scholastic definitions of beauty as *claritas*,[36] played some part in the design of Gothic cathedrals, and stimulated the production of stained glass.[37] Several scholars have argued that Dante owed a direct debt to the Pseudo-Dionysius' *De divinis nominibus* and his *De caelesti hierarchia*, especially for his ideas on light.[38] According to Mazzeo and Guidubaldi, Dante had direct knowledge of both these texts, using them as the basis for his own philosophy of light.[39] A Dionysian scholar, Piero Scazzoso, has made even stronger claims of direct Dionysian influences on what he calls Dante's "light mysticism":

[36]The key text in the development of these ideas is *De divinis nominibus*, c. 4, § 7 (*PG* 3, col. 701C; Chevalier, I, pp. 179-180). For scholastic discussions of beauty, see Henri Pouillon, "La beauté, propriété transcendentale chez les scholastiques (1220-1270)," *AHDLMA* 15 (1946), 263-328. One of the most eloquent testimonies to the influence of Dionysian ideas on the aesthetics of light is Abbot Suger's composition on the portal of St. Denis cathedral, see Erwin Panofsky, *Abbot Suger. On the Abbey Church of Saint Denis and its Art Treasures*, 2nd edn (Princeton, NJ: Princeton University Press, 1973), pp. 19-24, 46-48.

[37]Gage, *Colour and Culture*, pp. 69-78; von Simson, *The Gothic Cathedral*, esp. pp. 50-58. On the possible parallels between medieval optics and developments in stained glass, see Evans, *Mediaeval Optics and Stained Glass*.

[38]One of the first modern scholars to give detailed attention to this question was Edmund G. Gardner, *Dante and the Mystics: A Study of the Mystical Aspects of the "Divine Comedy" and its Relation with Some of its Medieval Sources* (London: Dent, 1913), pp. 77-110. The *corpus Aeropagiticum* comprises *De caelesti hierarchia*, *De ecclesia hierarchia*, *De divinis nominibus*, *De theologia mystica* and ten epistles.

[39]Mazzeo believed that Dante followed the Pseudo-Dionysius in using light images to render an intellectual vision, see *Structure and Thought*, pp. 38, 42-45. In *Medieval Cultural Tradition*, Mazzeo recognised his importance in medieval light speculation (see pp. 21, 87-89), but he did not draw any direct connections between his writings and Dante's works. For other supporters of direct influence, see Guidubaldi, *Dante Europeo*, II, pp. 197-215, 258-266; Agostino Petrusi, "Cultura greco-bizantina nel tardo Medioevo nelle Venezie e suoi echi in Dante," in *Dante e la cultura veneta* (Florence: Olschki, 1966), esp. pp. 182-197.

240

la mistica della luce in Dante è di natura pseudo-dionisiana per il modo analogo con cui è espresso, per l'identità frequentissima delle immagini e dei paragoni, per la correspondenza di linguaggio e per la costanza con cui il tema della luce si fonde con altri temi.[40]

More recently, scholars such as Boitani, Gamba, Took, and Vasoli have provided important references to Dante's use of Dionysian light motifs in the *Convivio* and the *Paradiso*.[41]

The evidence internal to Dante's writings does not, however, seem to endorse fully these conclusions. While Dante pays intellectual homage to the Pseudo-Dionysius as a "spirito sapiente" in *Paradiso* X and accepts his ordering of the angelic hierarchies in canto XXVIII,[42] the Pseudo-Dionysius' fictional presence in the poem does not necessarily mean that Dante was acquainted with his work at firsthand.[43] The only explicit direct citation is found in the letter to Cangrande, and, following continued controversy about the paternity of the *Epistola*, this reference must be treated with some caution.[44] The solar analogies

[40]Piero Scazzoso, "Contemplazione naturale e contemplazione soprannaturale confrontate attraverso Plotino e lo Pseudo-Dionigi," in *Lectura Dantis Mystica: Il poema sacro alla luce delle conquiste psicologiche odierne* (Florence: Olschki, 1969), pp. 81-84.

[41]For Vasoli's view that Dante had direct knowledge of *De divinis nominibus*, see *Il Convivio*, introd. p. lxviii. Vasoli supplies over thirty possible references, see: (i) for *De divinis nominibus*, pp. 160, 298, 377, 452-458, 476-478, 526; (ii) for *De caelesti hierarchia*, pp. 160, 305, 470. On the Pseudo-Dionysius' presence in cosmological passages in the *Paradiso*, see J.F. Took, *"L'etterno piacer": Aesthetic Ideas in Dante* (Oxford: Clarendon Press, 1984), pp. 14-16. For Dionysian resonances in the final canto, see Boitani, "The Sibyl's Leaves," pp. 101-102, 113, and nn. 30-47, 64, on pp. 123-125. See also Ulderico Gamba, "'Il lume di quel cero ...': Dionigi Areopagita fu l'ispiratore di Dante?" *Studia Patavina* 32 (1985), 101-114.

[42]*Par.* X, 115-117. Following the Pseudo-Dionysius in *Par.* XXVIII, 129-132, Dante revises the order of the angelic hierarchies that he put forward in *Con.* II, v, 5-11. He also includes the Pseudo-Dionysius as one of several authorities on Christian teaching in lamenting the present-day clergy's disregard for the great doctors in *Ep.* XI, 16.

[43]For the Dionysian order of the angelic hierarchies, see Aquinas, *ST*, Ia, q. 108, a. 5 and a. 6 (Caramello, I, pp. 511-515); idem, *Summa contra Gentiles*, lib. III, c. 80, §§ 2546-2564 (Pera, III, pp. 111-115); Vincent of Beauvais, *Speculum maius*, lib. I, c. 47 (I, cols 50-51).

[44]*Ep.* XIII, 60-61.

241

used in the *Convivio* correspond quite closely to passages in *De divinis nominibus*, but it should be remembered that similar comparisons are found widely in many other medieval works.[45]

Probable echoes of the Pseudo-Dionysius in the light imagery and terminology of the *Paradiso* can be explained by the fact that, by 1250, elements from the Pseudo-Dionysius' thought had become an established part of a common intellectual patrimony. The clearest resonance of a Dionysian light term in the *cantica* comes in Justinian's Latin hymn to God at the beginning of canto VII of the *Paradiso*:

> «*Osanna, sanctus Deus sabaòth,*
> *superillustrans claritate tua*
> *felices ignes horum malacòth!*». (1-3)

Justinian's *superillustrans* is a Dionysian word *par excellence* – nouns and adjectives formed on the prefix *super-* are exceptionally prevalent in the Latin translations of his work. As Heinrich Ostlender first pointed out, Grosseteste is the only medieval translator to use this precise term; and this finding has been enthusiastically accepted to support the thesis that Grosseteste exercised a distinctive influence on Dante.[46] But it does not follow from Dante's use of an eminently Dionysian term that he had read his works extensively, since vocabulary based on *super* is found widely in the Latin and Greek Fathers.[47] Moreover, Dante

[45]For the Dionysian image of sun and its rays used to illustrate the concept of participation, see e.g. Aquinas, *ST*, Ia, 104, a. 1, resp. (Caramello, I, p. 493): "Sicut enim sol est lucens per suam naturam, aer autem fit luminosus participando lumen a sole, non tamen participando naturam solis; ita solus Deus est ens per essentiam suam [...] omnis autem creatura est ens participative" ("As the sun is luminous by its nature, but air becomes luminous by participating in the light of the sun, though it does not participate in the sun's nature; so it is that only God is being by his essence [...] every created thing has its being by participation").

[46]Heinrich Ostlender, "Dantes Mystik," *Deutsches Dante Jahrbuch* 28 (1949), pp. 69-71; followed by Egidio Guidubaldi, "Roberto Grossatesta," in *ED* IV, p. 1006.

[47]For vocabulary formed on the prefix "super" in the Fathers, see Edmond Boissard, "St. Bernard et le Pseudo-Aréopagite," *RThAM* 26 (1959), pp. 222-223.

242

never uses the far more distinctive Dionysian adjectives such as *supersubstantialis* or *superessentialis*.

As far as Dionysian influences in the *Paradiso* are concerned, Dante's presentation of the angelic hierarchies as mirrors of a unified divine light in cantos XIII and XXIX also deserves detailed consideration:

> per sua bontate il suo raggiare aduna,
> quasi specchiato, in nove sussistenze,
> etternalmente rimanendosi una. (XIII, 58-60)
>
> La prima luce, che tutta la raia,
> per tanti modi in essa si recepe,
> quanti son li splendori a chi s'appaia.
> [...]
> Vedi l'eccelso omai e la larghezza
> de l'etterno valor, poscia che tanti
> speculi fatti s'ha in che si spezza,
> uno manendo in sé come davanti».
> (XXIX, 136-138, 142-145)

These sequences and two further passages in the *Paradiso* are particularly interesting,[48] for it was the Pseudo-Dionysius who did perhaps more than any other thinker to organise the cosmos into a cascade of illuminations by presenting the angelic hierarchies as mirrors which received and transmitted divine light. On several occasions in the *De caelesti hierarchia* and *De divinis nominibus*, there are detailed discussions of the idea that angels mediate light in the manner of mirrors:

specula clarissima et munda, receptiva principalis luminis et divini radii; inditae quidem claritatis sacre repleta eamque iterum copiose in ea quae sequuntur declarantia, secundum divinas leges [...] imago dei est angelus, manifestatio occulti

---

[48]*Par.* IX, 61-62: "Sù sono specchi, voi dicete Troni, / onde refulge a noi Dio giudicante."

243

luminis, speculum purum, clarissimum, incontaminatum, incoinquinatum, immaculatum.[49]

Dante's fascination with portraying the angels as "specchi" and his concern to emphasise the unity of the divine light are clearly Dionysian themes, although once again it is important to realise that these ideas had also become commonplace by the late thirteenth century.[50] It is also significant that, unlike Dante, the Pseudo-Dionysius did not assign an active role to the angelic Intelligences in generating change on Earth.[51]

There is one possible indirect echo of the Pseudo-Dionysius' writings in the *Paradiso* that has not, as far as I am aware, been pointed out by Dante scholars. In the opening lines of the *cantica*, Dante meditates upon the way in which God's light penetrates the universe and shines back with different intensities:

[49]*De caelesti hierarchia*, c. 3, § 2 (*PG* 3, col. 165A; Chevalier, II, pp. 787-789, Latin trans. Eriugena): "most bright, clean mirrors receiving the primordial light and divine rays; when they have received the holy clarity fully they pass it on abundantly, in accordance with divine laws, to whatever is below them"; *De divinis nominibus*, c. 4, § 22 (*PG* 3, col. 724B; Chevalier, I, p. 269, Latin trans. Sarrazin): "the angel is an image of God, a manifestation of the hidden light, a pure, bright, unsullied and untarnished and immaculate mirror." For the principle of illumination from above, see also *De caelesti hierarchia*, c. 13, § 4 (*PG* 3, col. 305B; Chevalier, II, p. 970, Latin trans. Sarrazin): "Illucet autem singulis secundis per primas" ("[he] illuminates the second order by way of the first").

[50]For examples of Dionysian angel-*specula* in scholastic literature, see Aquinas, *ST*, Ia, q. 12, a. 4, obj. 1 (Caramello, I, p. 54); Ia, q. 58, a. 4, resp. (Caramello, I, p. 284); Vincent of Beauvais, *Speculum maius*, lib. I, c. 53 (I, col. 54). The theme is especially prevalent in Bonaventure, see e.g. *In II Sent.*, d. 7, p. 2, a. 2, q. 2 arg. (Quaracchi, II, p. 193); idem, *Breviloquium*, lib. II, c. 3 (Quaracchi, V, pp. 220b-221a); idem, *Collationes in Hexaëmeron*, coll. V, c. 9, § 26 (Quaracchi, V, p. 358b); idem, *De sanctis angelis*, ser. I (Quaracchi, IX, p. 613a). Commenting on *De vulgari eloquentia*, I, ii, 3, Mengaldo notes that "il tema della comunicazione angelica *per speculum* è presente in tutta la scolastica della prima metà del Duecento," see *Opere minori*, II, p. 34.

[51]On this important point, which seems to suggest that Dante was both discriminating and independent in his use of Dionysian ideas, see Bemrose, *Dante's Angelic Intelligences*, p. 36 and n. 36; Sofia Vanni Rovighi, "Dionigi," in *ED* II, p. 461. Note also that Dante refutes the Dionysian explanation of the reason for the eclipse at the moment of Christ's death, see *Par.* XXIX, 97-102; cf. Aquinas, *In II Sent.*, d. 13, q. 1, a. 4, ad. 3, in *Opera omnia*, VI, p. 502b.

244

> La gloria di colui che tutto move
> per l'universo penetra, e risplende
> in una parte più e meno altrove. (1-3)

This description of light bears some similarity to the themes enunciated at the very beginning of *De caelesti hierarchia*. In his opening lines, the Pseudo-Dionysius describes the descent of divine light into multiplicity, its varying reception, and its return to the unity of the Godhead:

Omnem divinam lucem, tametsi summa benignitate varie ad creaturas prodeuntem, simplicem tamen manere, et quae illustrat unum efficere.[52]

The notion that the divine ray mediates between the divinity and man is central to the Dionysian corpus. The Pseudo-Dionysius provided his clearest statement of this principle in *De caelesti hierarchia*, where he explains how creatures transmit light to those who are below them (the law of taxiarchy), and, in conferring this light on other beings, raise them to God, the fontal source (the principle of telearchy). In other words, the Pseudo-Dionysius showed how all creatures participate in the divine ray and how light is carried down to lower levels of reality by the mediation of angelic hierarchies. Not all creatures receive the unified divine ray in the same manner: reception of light varies, in true Neoplatonic fashion, according to the capacity of the recipient.[53] Dante makes several references to

---

[52]*De caelesti hierarchia*, c. 1 (*PG* 3, cols 120B-121A; Chevalier, II, p. 727, Latin trans. Eriugena): "Each divine light, though spreading itself variously and with the greatest liberality to creatures, retains its simplicity, and, whatever it illuminates, it makes one." The biblical passage which most commentators cite for Dante's opening verse is Ps. 18:2: "Caeli enarrant gloriam Dei, et opera manuum eius adnuntiat firmamentum" ("The heavens declare the glory of God and the firmament shows the work of His hands").

[53]See Jonathon Scott Lee, "The Doctrine of Reception According to the Capacity of the Recipient in *Ennead* VI. 4-5," *Dionysius* 3 (1979), 79-99.

divine rays in the *Paradiso*, and these passages may provide an additional indication of indirect Dionysian influence.[54]

In summary, then, there are several passages in the *Paradiso* that contain echoes of Dionysian light terms and examples, but there seems to be insufficient evidence to support the view that Dante was *directly* influenced by his writings. Scazzoso's claims are undoubtedly exaggerated, and some Dante scholars may have recently overstated the possible direct borrowings. Even ideas and images that are apparently distinctive to the Pseudo-Dionysius are frequently replicated in other more general medieval sources.[55] New translations and commentaries of his works were produced throughout the thirteenth century; and a sustained interest in his writings led to the compilation of a *Corpus dionysianum* (c. 1230) at the University of Paris.[56] Almost all the major thirteenth-century philosophers and theologians drew on his writings, and his presence is particularly marked in Grosseteste,[57] Aquinas,[58] Albert,[59] and Bonaventure.[60]

---

[54]For references to divine rays or rays from a divine source, see *Par.* III, 37-38; XI, 19-20; XIV, 51; XXIII, 83-84; XXXI, 99; esp. XXXIII, 53 and 77. The question of direct Dionysian derivation is, however, rendered problematic by the absence of any Dantean reference to "supersubstantial" rays.

[55]For Dionysian influence on eleventh- and twelfth-century writers, see Jean Leclerq, "Influence and Non-Influence of Dionysius in the Western Middle Ages," in *Pseudo-Dionysius: The Complete Works* (London and New York: Paulist Press, 1987), pp. 25-32; Paul Rorem, *Biblical and Liturgical Symbols within the Pseudo-Dionysian Synthesis* (Toronto: Pontifical Institute of Mediaeval Studies, 1984), esp. pp. 3-4, 143-147; idem, *Pseudo-Dionysius: A Commentary on the Texts*, passim, but esp. pp. 237-240; Wetherbee, *Platonism and Poetry in the Twelfth Century*, pp. 75-80.

[56]H.-F. Dondaine, *Le corpus dionysien de l'Université de Paris au XIIIe siècle* (Rome: Edizioni di storia e letteratura, 1953). A bibliography of secondary literature until 1972 will be found in Barbara de Mottoni Faes, *Il "Corpus Dionysianum" nel Medioevo: Rassegna di studi: 1900-1972* (Rome: Il Mulino, 1977).

[57]See McEvoy, *The Philosophy of Robert Grosseteste*, esp. pp. 69-142, 243-256, 354-368.

[58]For Aquinas' use of the Pseudo-Dionysius (one scholar has counted 1702 direct citations in Thomist writings), see J.S. Durantel, *St. Thomas et le Pseudo-Denys* (Paris: Alcan, 1919), pp. 60-207; K.F. Doherty, "St. Thomas and the Pseudo-Dionysian Symbol of Light," *New Scholasticism* 34 (1960), 170-189. Vasoli (*Il Convivio*, p. 453) notes that Aquinas' *Summa contra Gentiles* is "un testo fortemente influenzato dalle dottrine dionisiane"; see also p. 526. Given Dante's undoubted familiarity with the *Summa contra Gentiles* (see Foster, "Summa Contra Gentiles," in *ED* V, pp. 479-480), this work may be one of the most likely sources of his

246

Bonaventure's Theory of Light and Dante

Many critics have argued that, in the *Paradiso*, Dante relied on Bonaventurian ideas about light, especially in cantos XIV and XXX which have an especially high frequency of light imagery.[61] However, as is the case with both Bartholomew of Bologna and the Pseudo-Dionysius, it cannot simply be assumed that Dante knew any of Bonaventure's writings without a clear presentation of the possible borrowings. What is more, after amendments to Baeumker's pioneering work, the traditional view of Bonaventure as a principal exponent of medieval "light metaphysics" requires radical revision. In an important essay, V.Ch. Bigi has shown that Bonaventure's doctrine of light does not lend itself to metaphysical speculation.[62] Although light is undeniably an important component of

---

acquaintance with the Pseudo-Dionysius. Aquinas wrote one Dionysian commentary, see *In librum Beati Dionysii De divinis nominibus expositio*, ed. Ceslai Pera (Rome and Turin: Marietti, 1950).

[59]For Albertine works on the Pseudo-Dionysius, see Francis Ruello, "Le commentaire inédit de saint Albert le Grand sur les Noms Divins. Présentation et aperçus de théologie trinitaire," *Traditio* 12 (1956), 231-314; idem, "La *Divinorum nominum reseratio* selon Robert Grosseteste et Albert le Grand," *AHDLMA* 26 (1959), 99-197. Albert wrote commentaries, recently edited in critical editions, on all the Dionysian works, see *Super Dionysium De divinis nominibus*, ed. Paul Simon (Münster: Aschendorff, 1972); *Super Dionysii mysticam theologiam et epistulas*, ed. Paul Simon (Münster: Aschendorff, 1978); *Super Dionysium De caelesti hierarchia*, ed. Wilhelm Kübel and Paul Simon (Münster: Aschendorff, 1994).

[60]For the centrality of the Pseudo-Dionysius in Bonaventure's thought (there are 248 explicit citations), see Jacques G. Bougerol, "St. Bonaventure et la hiérarchie dionysienne," *AHDLMA* 36 (1969), 131-167; idem, "St. Bonaventure et le Pseudo-Denys l'Aréopagite," *Études Franciscaines* 28 suppl. (1968), 33-123; on p. 34, Bougerol notes that Bonaventure was "un des plus dionysiens parmi les grands maîtres du XIIIe siècle." Dionysian influence is especially marked in the negative theology that Bonaventure develops in his mystical treatise *Itinerarium mentis ad Deum* (1259), see e.g. c. 6, § 2 (Quaracchi, V, 310b); c. 7, § 5 (Quaracchi, V, p. 312a).

[61]G. Berretta, "Il Canto XIV del *Paradiso*," *Filologia e letteratura* 11 (1965), 254-269; Giudubaldi, *Dal "De Luce" di R. Grossatesta*, pp. 13, 19-20; Mazzeo, *Medieval Cultural Tradition*, p. 69; Luigi Negri, "La luce nella filosofia del '300 e nella «Commedia»," *GSLI* 82 (1923), pp. 335-336; F. Orestano, "Discontinuità dottrinali nella *Divina Commedia*," *Sophia* 1 (1933), p. 16. More recently, see Riccardo Scrivano, "Poesia e dottrina nel XXX canto del «Paradiso»," *Critica letteraria* 17 (1989), pp. 10-12, who, without fully explaining the putative connections, relates Dante's doctrine in this canto to Bonaventure's discussion of four kinds of light in *De reductione artium ad sanctam theologiam* (Quaracchi, V, p. 319a).

[62]V.Ch. Bigi, "La dottrina della luce in S. Bonaventura," *Divus thomas* 64 (1961), 395-423.

247

Bonaventure's thought, his views are not closely related to either Augustine, or Grosseteste, or other Neoplatonic writers, as has often been assumed.

Bonaventure's principal discussion of light theory covers seven questions in the second book of his *Sentence* commentary.[63] The *quaestio* style of argumentation in this work should make one extremely careful about isolating any of the views he expounds from his general discussion. In providing only selected quotations, Nardi and Mazzeo have presented a partial account of Bonaventure's ideas. Mazzeo, for example, believed that Bonaventure held light to be the intermediary between the body and the soul and that he presented a general theory which was "identical" to that found in the *De intelligentiis*.[64] These claims are not accurate and should be modified in line with the revised findings of Bigi and other scholars.[65] In many respects, Bonaventure took a position that attempted to reconcile opposing scholastic views on the nature and function of light. Thus, he attempted to argue that light was both a substantial and accidental form; and he also paid considerable attention to its nature and effects in the medium.[66]

---

[63]For Bonaventure's principal discussions of light, see *In II Sent.*, d. 13, qq. 1-6 (Quaracchi, II, pp. 310-333); d. 17, a. 2, q. 1 (Quaracchi, II, pp. 419-423). For a commentary on his theory of light in *distinctio* XIII, see McCarthy, *Medieval Light Theory and Optics*, pp. 167-182. For the views expressed on related questions by Aquinas and Albert, see further ibid., pp. 146-153 and 155-164.

[64]Mazzeo, *Medieval Cultural Tradition*, pp. 73-74.

[65]For the idea that sensation is regulated by light, see Augustine, *De Gen. ad litt.*, lib. XIII, c. 16, § 32 (*PL* 34, col. 466); Grosseteste, *Hexaemeron*, lib. II, c. 10, § 1 (Dales and Gieben, p. 98): "Lux igitur est per quam anima in omnibus sensibus agit que instrumentaliter in eisdem agit" ("Light is therefore that by means of which the soul acts in all the senses, and it is that which acts instrumentally in them"). Unlike these writers, Bonaventure did not accord a central role to light in sensation, see James McEvoy, "Microcosm and Macrocosm in the Writings of St. Bonaventure," in *S. Bonaventura 1274-1974*, 5 vols. (Rome: Collegium S. Bonaventura, 1973), II, pp. 309-343; Bigi, "La dottrina della luce," pp. 398-399. Dante clearly does not endorse this doctrine, cf. *Con.* II, xiii, 24: "li spiriti umani, che sono principalmente vapori del cuore." Unlike Grosseteste, Bonaventure did not make light the common form of corporeity, see Bigi, "La dottrina della luce," pp. 403-404; John Francis Quinn, *The Historical Constitution of St. Bonaventure's Philosophy* (Toronto: Pontifical Institute of Mediaeval Studies, 1973), pp. 103-104, and n. 113.

[66]For Bonaventure's position on light as both substantial and accidental form, see *In II Sent.*, d. 13, a. 2, q. 2 (Quaracchi, II, pp. 320a-321a). His discussion of the effects of irradiation in the

248

Like the Pseudo-Dionysius and his own contemporaries, Bonaventure shared the widespread tendency to use light analogies with reference to theological concepts. Bonaventure's keen interest in developing optical imagery in his sermons was noted in Chapter 4, and many other relevant passages on light can be adduced from his works to supplement the passages quoted in the earlier anthology of passages. In a sermon on a verse from Matthew (2:11), for example, Bonaventure provides another variation upon the comparison between the passage of light through transparent bodies and the infecundation of the Virgin Mary:

Unde sicut radius solis per ista pervia munda pertransit sine aliqua ruptura et unit se eis, ita splendor gloriae aeternae, qui est Filius Dei, intravit Mariam puram et mundam sine aliqua ipsius Mariae ruptura et eam fecundavit.[67]

In a later sermon, Bonaventure remarks upon the analogies possible between the material sun and its spiritual counterpart in his exegesis of another verse from Matthew (22:6):

Merito igitur soli Christo et non alii attribuenda est auctoritas officii, ut singulariter *unus Magister* dicatur, eo quod ipse est fontale principium et origo cuiuslibet

---

medium is found at *In II Sent.*, d. 13, a. 3 (Quaracchi, II, pp. 323c-329b), see esp. a. 3, q. 2 (Quaracchi, II, p. 328a): "lumen enim in medio dicit virtutem activam, egredientem a corpore luminoso, per quam corpus luminosum agit et imprimit in haec inferiora; et haec est virtus substantialis ipsi corpori" ("light in the medium connotes active virtue, coming from a luminous body and by which this body acts and impresses itself into what is below it, and this is the substantial virtue of this body").

[67]*Epiphania*, ser. III (Quaracchi, IX, p. 157b): "Hence as a ray of sunlight passes through these clean, transparent bodies without any rupture and joins itself to them, so the splendour of eternal glory, who is the Son of God, entered into pure and clear Mary without any rupture of her and infecundated her." See also *De nativitate B. Virginis Mariae*, ser. I (Quaracchi, IX, pp. 706a-708a).

249

scientiae. Unde sicut unus est sol, tamen multos radios emittit; sic ab uno Magistro, Christo, sole spirituali, multiformes et diversae scientiae procedunt.[68]

And in his *Collationes in Hexaëmeron*, he uses an example taken from the observation of natural light as it passes through coloured glass in order to help to explain one of the means for contemplating the divine:

In quaelibet creatura est refulgentia divini exemplaris, sed cum tenebra permixta; unde est sicut quaedam opacitas admixta luminis [...] Sicut tu vides, quod radius intrans per fenestram diversimode coloratur secundum colores diversos diversarum partium; sic radius divinus in singulis creaturis diversimode et in diversis proprietatibus refulget.[69]

The evidence provided so far in this chapter, then, suggests that Dante's use of analogies based on light was largely a question of reworking ideas and imagery that were widely available in theological works. The final section of this chapter will now show that Dante also made use of a highly specific light analogy employed by certain theologians in analysing beatific vision, an analogy which as poet he radically rethought in order to present his own highly original conception of light's role in this context.

---

[68]*Dominica XXII post Pentecosta*, ser. I (Quaracchi, IX, p. 442a): "The authority of office is therefore fittingly to be attributed to the sun Christ and not to other things due to the fact that he is the fontal principle and origin of all knowledge. Hence as the sun is one, but it sends out many rays; so from one Master, Christ, the spiritual sun, many and diverse teachings proceed." On light as a name for Christ, see also *Collationes in evangelium S. Ioannis*, c. 1, coll. 1 (Quaracchi, VI, pp. 535a-537a).

[69]*Collationes in Hexaëmeron*, lib. IV, coll. 12, § 14 (Quaracchi, V, p. 386b): "In every creature there is a shining back of the divine exemplar, but with the admixture of darkness; hence it is like a certain opaque matter joined to natural light [...] As you see that a ray on entering through a window is variously coloured according to its different parts; so it is with the divine ray which shines back variously in individual creatures and in different properties."

The *Lumen gloriae* and Dante's Conception of the Empyrean

In canto XXX, Dante and Beatrice reach the end of their journey through the physical universe and they enter the Empyrean, a heaven of "pure" light whose nature Beatrice describes upon their arrival:

> ricominciò: «Noi siamo usciti fore
> del maggior corpo al ciel ch'è pura luce:
> luce intellettüal, piena d'amore;
> amor di vero ben, pien di letizia;
> letizia che trascende ogne dolzore. (38-42)

These exquisitely interlaced verses with their step by step repetition of key words in rhyme and at the beginning of each verse seem to suggest that this heaven is not a material body and hence forms no part of the physical universe. With one important exception, this view has been accepted by Dante scholars,[70] and it has become traditional to distinguish Dante's representation of the Empyrean as immaterial in the *Paradiso* from his earlier conception of a "cielo di fiamma" composed of matter in the *Convivio*.[71] In a recent scholarly article, Bortolo

---

[70]Foster, "Tommaso d'Aquino," in *ED* V, p. 641: "Nel poema, infatti, l'Empireo non è più concepito come un corpo, ma come un'entità puramente spirituale"; Etienne Gilson, "À la recherche de l'Empyrée," *Revue des études italiennes* 10 (1965), esp. pp. 160-161: "L'empyrée a perdu son nom [sc. in the *Comedy*] parce qu'il a changé de nature. Il est devenu anonyme en même temps que, comme donnée astronomique distincte, il perdait sa réalité"; Mellone, "Empireo," in *ED* II, pp. 668-671, esp. p. 670: "per Dante teologo nel poema l'Empireo è immateriale, del tutto soprannaturale"; Nardi, *Saggi di filosofia dantesca*, pp. 207-208 and n. 152; idem, *Il punto sull'Epistola a Cangrande* (Florence: Le Monnier, 1960), pp. 31-32. Recent contributions that support the established view include: Chiavacci-Leonardi, "«Le bianche stole»: Il tema della resurrezione nel *Paradiso*," n. 2, p. 250; Stephen Bemrose, "'Una favilla sol della tua gloria': Dante Expresses the Inexpressible," *Forum for Modern Language Notes* 27 (1991), pp. 134-135. See also E.J. Stormon, "The Problems of the Empyrean Heaven in Dante," *Spunti e ricerche: rivista d'italianistica* 3 (1987), 23-33.

[71]The key comparison is between *Par.* XXX, 38-40 and *Con.* II, iii, 8: "Veramente, fuori di tutti questi, li cattolici pongono lo cielo Empireo, che è a dire cielo di fiamma o vero luminoso; e pongono esso essere immobile per avere in sé, secondo ciascuna parte, *ciò che la sua materia vuole*" (italics mine).

251

Martinelli has, however, surveyed Dante's doctrine against contemporary discussions in encyclopaedic works, *Sentence* commentaries, and astronomical writings. There is, he concludes, no historical foundation for the concept of an incorporeal Empyrean. Since it seems implausible that Dante would have countered the whole of scholastic theology in this way, Martinelli rules out the possibility that the "heaven of light" in *Paradiso* XXX is incorporeal.[72]

Despite Martinelli's detailed contribution to this question, I do not believe that he is justified in rejecting an incorporeal Empyrean in *Paradiso* XXX. After all, there is a very clear sense of demarcation between the "maggior corpo" (i.e. the Primum Mobile) in line 39 and the "ciel ch'è pura luce: / luce intellettüal" in lines 39-40. Moreover, in scholastic literature on the Empyrean, this heaven is always discussed under sections dealing with corporeal light, not the intellectual light to which Beatrice refers.[73] More crucially still, Martinelli does not mention the fact that Dante defines the "luce intellettüal" of his Empyrean with the very attributes ("amore," "letizia," and "dolzore") which scholastic theologians had come to use to describe beatific vision.[74] The doctrine of beatific vision is central to Dante's

---

[72]Martinelli, "La dottrina dell'Empireo," esp. p. 112: "il cielo empireo, a norma di quasi tutta la tradizione del XIII e XIV secolo, si deve considerare come un cielo fisico e tale esso è infatti per l'autore dell'epistola e per Dante." Martinelli observes that in *Paradiso* XXX the idea of an incorporeal Empyrean would be "storicamente inattendibile" (p. 112).

[73]For the differences between Dante's "luce intellettüal" and medieval views, see e.g. Albert, *De quattuor coaequaevis*, tr. 3, q. 11, a. 1, sol. (Jammy, XIX, p. 61a): "caelum empyreum aut caelum nobilissimus est inter omnia corporalia simplicia. Et propter hoc dicunt sancti, quod sit igneum, non propter ardorem, sed propter lumen corporale quod habet" ("the Empyrean heaven or the most noble heaven is the simplest of all bodies. And on this account the saints say that it is fiery, not due to ardour but because of the corporeal light it has"); Vincent of Beauvais, *Speculum maius*, lib. III, c. 87 (I, col. 219): "est [sc. the Empyrean] corpus inter omnia corpora simplicia nobilissimum [...] corporale lumen in eo diffusum" ("the Empyrean is the most noble of all simple bodies ... corporeal light is diffused into it"). Alexander of Hales' earlier discussion comes under the section "De luce corporali" in his *Summa theologica*, inq. 3, tr. 3, q. 2, c. 1, a. 1, 4 vols. (Florence: Collegium S. Bonaventura, 1924-1948), II, pp. 327a-328b. Similarly, Bonaventure's discussions of the Empyrean (c. 1250) are in sections dealing with either corporeal light or the created universe, see *In II Sent.*, d. 2, q. 1 (Quaracchi, II, pp. 70-72); *Breviloquium*, lib. II, c. 3 (Quaracchi, V, p. 221a).

[74]See Giovanni Fallani, "Visio beatifica," in *ED* V, pp. 1070-1071: "I tre elementi della visione beatifica definiscono così l'Empireo."

252

poetic treatment of the Empyrean in *Paradiso* XXX, even though it has received relatively little critical attention. The remainder of this chapter will show how Dante draws on this doctrine, a doctrine once again based on a light analogy, in order to present a radically new conception of this heaven in the final cantos of the poem.

The ambiguities presented by conflicting biblical passages on the vision of God helped to give rise to two traditions that put forward opposing views on the question of whether human souls in the afterlife can see God directly.[75] In the Western tradition, Latin Fathers such as Augustine and Gregory the Great categorically affirmed that God was seen directly in his essence,[76] whereas writers in the Eastern tradition, principally the Pseudo-Dionysius and Eriugena, denied direct vision and favoured instead the concept of theophany, the idea that God shows himself to man indirectly through mediated visions.[77] As Dondaine has shown,[78] these two traditions co-existed until a turning point was reached in 1241, when the Bishop of Paris condemned the view that God is *not* seen in his essence by the angels and the blessed.[79] To resolve the problem of how a finite being could see the infinite, Albert the Great transformed the Dionysian concept of theophany into a doctrine known as the *lumen gloriae*. Albert argued that to attain beatific vision, man's created intellect required a special kind of light, a light which

[75]For the biblical tensions between the impossibility of seeing God and the promise of personal vision, contrast, on the one hand, Ex. 33:20; Ezek. 2:1; I Tim. 6:16 with I Jn. 3:2; I Cor. 13:12, on the other.

[76]Augustine, *Ad Paulinum de videndo deo*, c. 13, § 31 (*PL* 33, col. 610).

[77]Pseudo-Dionysius, *De caelesti hierarchia*, c. 4, § 3 (*PG* 3, col. 180C); Eriugena, *Periphyseon*, I, ed. I.P. Sheldon-Williams, 3 vols. (Dublin: Dublin Institute for Advanced Studies, 1968-1981), I, p. 50. On Eriugena's doctrine, see Tullio Gregory, "Note sulla dottrina delle «teofanie» in Giovanni Scotto Eriugena," *Studi medievali* ser. 3/4 (1963), 75-91.

[78]H.-F. Dondaine, "L'objet et le 'medium' de la vision béatifique chez les théologiens du XIIIe siècle," *RThAM* 19 (1952), 60-130. See also P.-M. Contenson, "Avicennisme latin et vision de Dieu au début du XIII siècle," *AHDLMA* 26 (1959), 29-97.

[79]*Chartularium Universitatis Parisiensis*, ed. Denfile and Chatelain, I, p. 170.

acted as a power of the mind, carried the intellect beyond its natural limits, and brought man into supernatural contact with the divine substance:

Ad hunc autem tactum [sc. human intellect with divine substance] proportionatus est intellectus non adhuc per suam naturam, sed per lumen gloriae descendens in ipsum, confortans eum et elevans eum supra suam naturam, et hoc dicitur theophania [...][80]

Aquinas no longer used the word theophany in this sense, but he accepted the essentials of the doctrine of the *lumen gloriae*, and question 12 of the first part of his *Summa theologiae* gives a clear exposition of its features:

Dicendum ergo quod ad videndum Dei essentiam requiritur aliqua similitudo ex parte visivae potentiae, scilicet lumen divinae gloriae confortans intellectum ad videndum Deum; de quo dicitur Psal., In lumine tuo videbimus lumen [...] Facultas autem videndi Deum non competit intellectui creato secundum suam naturam, sed per lumen gloriae, quod intellectum in quadam deiformitate constituit [...] Unde intellectus plus participans de lumine gloriae perfectius Deum videbit. Plus autem participabit de lumine gloriae qui plus habet de charitate; quia ubi est major

---

[80]Albert, *Super Dionysium De divinis nominibus*, c. 13, § 27, sol. (Simon, p. 448): "The intellect is proportioned to this touch not by its own nature but by the light of glory descending into it, strengthening it and raising it beyond its nature, and this light is called theophany." Albert repeatedly deals with *theophania* in this sense in his theological works, in commentaries on the Bible, and in his writings on the Pseudo-Dionysius, see *Summa theologiae, sive De mirabili scientia Dei* (Dionysius, Kübel, and Vogels, p. 99); *Super Dionysium De caelesti hierarchia*, c. 4 (Kübel and Simon, pp. 70-71); *Super Dionysii mysticam theologiam*, c. 2 (Simon, p. 466); *In II Sent.*, d. 4, a. 1 (Jammy, XV, pp. 58a-60b); *Super Isaiam*, c. 8, § 81, ed. Ferdinand Siepmann (Münster: Aschendorff, 1952), pp. 121-122; *Super Matthaeum*, c. 5, § 8 (Schmidt, I, pp. 112-113). My understanding of Albert's development of the doctrine is indebted to J.M. Alonso, "Teofanía y visón beata en Escoto Erigena," *Revista Española de Teología*, 10 (1950), 361-389; Dondaine, *Le corpus dionysien*, pp. 126-128.

254

charitas ibi est maius desiderium, et desiderium quodammodo facit desiderantem aptum et paratum ad susceptionem desiderati.[81]

There can be little doubt that Beatrice refers to this very light in lines 38-42 of canto XXX.[82] This point is confirmed later in the canto when Dante describes how the light that Dante-*personaggio* has seen reveals the Creator to his creatures:

> Lume è là sù che visibile face
> lo creatore a quella creatura
> che solo in lui vedere ha la sua pace.
> E' si distende in circular figura,
> in tanto che la sua circunferenza
> sarebbe al sol troppo larga cintura. (100-105)

These passages refine, extend, and develop ideas already present in earlier discussions of beatific vision in the *Paradiso*.[83] In canto XIV, Dante's Solomon

---

[81]*ST*, Ia, q. 12, a. 2, resp. and a. 6, resp. (Caramello, I, pp. 53 and 56): "It is thus to be said that some likeness is required on the part of the visual power to see the divine essence, namely, the light of divine glory which comforts the intellect, allowing it to see God and of which the Psalm speaks 'In your light shall we see light.' The ability to see God does not belong to the created intellect of its own nature but to the light of glory which makes the intellect in some sense deiform. Hence, the more intellect participates in the light of glory the more perfectly it will see God. They who have more charity will participate more fully in the light of glory, because a greater charity is found where there is a greater desire, and desire in some way predisposes and prepares man to receive what he desires." For further discussion, see also Aquinas, *In Ioah. comm.*, c. 1, lect. 11, § 212 (Cai, p. 42); idem, *Questiones quodlibetales*, quod. 7, q. 1, a. 1, resp. (Spiazzi, p. 134); idem, *Summa contra Gentiles*, lib. III, c. 53, § 2302 (Pera, III, p. 72). On the doctrinal context of the *lumen gloriae*, see Etienne Gilson, "Sur la problématique thomiste de la vision béatifique," *AHDLMA* 31 (1964), 67-88; Dominic J. O'Meara, "Eriugena and Aquinas on Beatific Vision," in *Eriugena Redivivus: Zur Wirkungsgeschichte seines Denkens im Mittelalter und im Ubergung zur Neuzeit*, ed. Werner Beierwaltes (Heidelberg: Carl Winter, 1987), pp. 224-236.

[82]See Nardi, *Saggi di filosofia dantesca*, pp. 267-268. See also Foster, *The Two Dantes*, pp. 71-72; Singleton, *Journey to Beatrice*, pp. 30-38.

[83]Note also that the philosophical foundation of Dante's Universal Monarchy rests on a distinction between two orders that are governed by two lights, one of which allows man to look upon God, see *Mon.* III, xv, 17: "beatitudinem vite ecterne, que consistit in fruitione divini aspectus ad quam propria virtus ascendere non potest, nisi lumine divino adiuta" ("happiness in

255

had explained the relationship between the light of grace and the capacity of souls to see God. When the bodies of these souls are resurrected, the light which God gives freely to these souls will lead to greater powers of vision, to greater love, and to still more radiance:

> per che s'accrescerà ciò che ne dona
> di gratüito lume il sommo bene,
> lume ch'a lui veder ne condiziona;
>   onde la visïon crescer convene,
> crescer l'ardor che di quella s'accende,
> crescer lo raggio che da esso vene. (46-51)

And in canto XXI, Peter Damian announces that he is able to see the divine essence because a light, which derives directly from God, penetrates his soul and raises him beyond his own powers. His opening words adhere very closely to the doctrine as it was elaborated by Albert and Aquinas, although Dante gives their ideas verbal energy and visual power by using unusual verb metaphors ("inventrare" and "mungere": ll. 84 and 87) in rhyme position:

> «Luce divina sopra me s'appunta,
> penetrando per questa in ch'io m'inventro,
>   la cui virtù, col mio veder congiunta,
> mi leva sopra me tanto, ch'i' veggio
> la somma essenza de la quale è munta. (83-87)

In lines 100-105 of canto XXX, however, Dante strikes a crucial note of originality with a synthesis which is as unexpected as it is startling: he equates the *lumen gloriae* with the Empyrean itself.[84] In other words, Dante has taken two light concepts and seamlessly blended them together. Beatific vision still takes place by

---

the eternal life, which consists in the enjoyment of the vision of God to which our powers cannot raise us except with the help of God's light," trans. Shaw, p. 145); cf. *Mon.* I, i, 6; II, i, 7.

[84]Nardi, *Saggi di filosofia dantesca*, pp. 207-208, and n. 152.

256

light in a heaven of light; but, for Dante, this heaven and this light are one and the same. As Martinelli's thorough investigations have shown, medieval literature on the "cielo Empireo" offers no precedents for this conflation of the *lumen gloriae* with the tenth heaven. Yet if Dante's Empyrean may lack historical foundation, this does not mean that we should reinterpret his thought in order to make it comply with tradition. It seems far more appropriate to accept that Dante is instead radically reworking traditional ideas into a new synthesis, that his conception of the Empyrean is one of his most remarkable and original poetic fabrications.

CONCLUSION

Dante's writings provide evidence that he was influenced by a wide variety of contemporary works on optics and light but not to the extent that has been previously thought. As was shown in Part One, he never draws on the more specialised optical doctrines and terminology employed by Alhazen and the authors of the thirteenth-century *perspectivae*. He does not share the "perspectivist" conception of optics as an all-embracing science and shows a notable disregard for some of its leading concepts (e.g. the twenty-two *intentiones visibiles* and the analysis of a visible object into point-forms). Whereas the *perspectivae* synthesised optics and visual theory into a coherent science, Dante does not connect these two lines of inquiry to the same degree. It would even appear that he had relatively little understanding of refraction, one of the principal subject areas in the optical treatises.

And yet, while Dante was not fully aware of the latest developments in thirteenth-century optics, he certainly drew on elements belonging to a common patrimony of ideas on light, vision, and optics. In the *Convivio*, Dante's optical and visual thought may be illustrated adequately without going beyond the commentaries and independent writings of Albert the Great and Thomas Aquinas, thirteenth-century encyclopaedias, and perhaps more popular works such as medieval sermons. The optical content of such works has received relatively little attention from historians of science; but, as this study has shown, such writings were important in transmitting information on optical and visual matters gathered

258

from diverse sources, and they may hold the key to the widespread literary assimilation of optical ideas and concepts related to light in other areas of late medieval and early Renaissance literature. Dante's use of concepts and technical terms relating to vision and optical science can almost invariably be traced to these kind of works, with the commentaries by Albert and Aquinas often providing the most probable sources. The prevailing assumption that the "perspectivist" tradition exercised a direct influence on Dante thus highlights the dangers of mistaking affinities or indirect correspondences for specific sources.

Dante's optics, in the *Convivio* at least, was essentially a geometrical theory of sight concerned with the relation of objects to the observer's eye. He seems to have derived this conception of vision from an attentive reading of the many contemporary discussions of the visual pyramid and rectilinear pathway in vision. He also had a passionate interest in various other aspects of visual theory (from the anatomy of the eye to the psychology of vision), and he held a philosophically sophisticated conception of the visual act which he situated in the context of the internal senses. It is, however, most especially in the *Comedy* that Dante exploits the optical heritage studied in this book, a heritage which he assimilates and develops in his own distinctive ways both as sources of imagery and narrative devices. As poet, he shows a particularly deep concern, both philosophical and narrative, with the means by which the eye could be dazzled, blinded, and deceived, but he was also receptive to other ideas about vision such as the doctrine of beatific vision. And throughout the *Comedy*, but especially in the *Paradiso*, Dante's fascination with light finds its expression in striking luminous effects, optical illusions, and meteorological phenomena. Dante was deeply interested in the ways in which light is modified by interposing media and altered by contact with objects of different shape, colour, and density. His technical similes based on the rainbow have their starting-point in Aristotle's *Meteorologica*, and many of the other details upon which he draws can be found in the sections on optics contained in medieval Aristotelian commentaries. But not all. Dante's optical imagery in the *Comedy* also contains resonances of light and mirror comparisons from the Bible, Neoplatonic writers, and theological works.

The evidence presented in Part Two confirms and extends these findings. To use the label "light metaphysics" is to risk identifying Dante too closely with an important group of thirteenth-century philosophers who also wrote on optics (Bacon, Grosseteste, and Witelo), but whose ideas did not have a marked influence on him. Unlike these writers, Dante did not closely identify the causal action of the heavens with light, nor did he use geometrical optics to analyse the effects of radiation. It is not clear how far he accepts the idea that the mind requires an influx of intellectual light from a supernatural source to form its concepts, but he certainly did not follow Grosseteste in making light the primary component of all bodies and assigning it a central role in theories of sensation, sound, and motion. I have attempted to show that critics who refer to Dante as re-elaborating medieval "light metaphysics" simplify his relationship to a complex and intellectual demanding set of traditions, often without due respect for context and individual elaboration. Although I have not solved all the problems nor dealt with all the issues (e.g. the role of light in theories of intellection), I hope that my general critique and the approach adopted in Chapters 6-7 will at the very least make critics and scholars hesitate before invoking generic critical categories like "light metaphysics."

In the *Paradiso*, a rich heritage of ideas about the nature and function of light nonetheless plays an important part in Dante's poetic disquisitions on creation, cosmology, the angelic Intelligences, and God. Dante conceived the cosmos as created by an irradiation of divine light and believed that the universe was sustained and co-ordinated by a continual flow of luminous energy from God. While Neoplatonic emanationist metaphysics has no place in his thought, Dante repeatedly employs the concept of celestial influence to organise the relations

260

between God, the Intelligences, and man. Dante's use of light imagery in the *Paradiso* may reveal indirect traces of Dionysian influence but what is most apparent is not any specific set of sources but a general debt to the widespread use of light analogies, the kind of comparison which is found in many different categories of medieval writing but is especially common in theological works. There are many interesting parallels between Dante's light imagery in the final *cantica* and discussions of light effects found in commentaries on the Bible and on the *Sentences*, in theological treatises, and in homiletic literature. There is, however, no clear evidence to suggest that the poet owed any specific intellectual debts, as is sometimes asserted, either to Bonaventure's theory of light in his *Sentence* commentary or to Bartholomew of Bologna's *Tractatus de luce*. The recurrent emphasis I have placed upon Dante's own highly innovative adaptations of traditional materials finds its best illustration in his presentation of the Empyrean in *Paradiso* XXX. As I have shown, Dante here draws in part on ideas about beatific vision which rely on later thirteenth-century discussions of the *lumen gloriae*. Yet there are no precedents in scholastic writings for his identification of the tenth heaven, the heaven of fire or Empyrean, with the "pure, intellectual light, full of love and transcending all sweetness," which irradiates from God, and establishes a personal contact between human souls and the deity.

APPENDIX[1]

Principal Optical Works Available to Later Thirteenth-Century Europe

1. Optical Works Translated from Arabic to Latin

Alhazen, *De aspectibus* (*Perspectiva* and *Optica*), trans. c. 1190-1200.

Alhazen, *De speculis comburentibus*, trans. Gerard of Cremona (d. 1187).

Alkindi, *De aspectibus* (*De causis diversitatem aspectus*), trans. Cremona.

Euclid, *Optica* (two translations, one known as *De aspectibus* by Cremona, the other as *De radiis visualibus* or *Liber de fallacia visus* by an unknown translator).

Pseudo-Euclid, *De speculis* (*Compositio speculorum mirabilium*) trans. Cremona.

Ptolemy, *Optica* (*De aspectibus*), trans. Eugene of Sicily, c. 1156-1160.

Tideus, *De speculis* (also circulated as *De speculis comburentibus, De qualitate eius quod videtur in speculo,* and *De aspectibus*), trans. Cremona.

2. Optical Works Translated from Greek to Latin

Euclid, *Catoptrica* (*De speculis*), trans. anonymous, c. 1150.

Euclid, *Optica* (*Liber de visu*), trans. anonymous, early twelfth century.

Hero of Alexander, *Catoptrica* (known as Ptolemy's *De speculis*), trans. William of Moerbeke, 1269.

3. Thirteenth-Century Optical Treatises

Bacon, Roger, Part V (*Perspectiva*) of his *Opus maius*, c. 1263.

___.*De multiplicatione specierum* (*De radiis* and *De generatione specierum*), c. 1262.

___.*De speculis comburentibus* (*De multiplicatione lucis*), c. 1263-1274.

Pecham, John, *Perspectiva communis*, c. 1270-1279.

___.*Tractatus de perspectiva*, c. 1268.

Witelo, *Perspectiva*, c. 1270-1274.

---

[1]The first two sections of this Appendix are adapted from Lindberg, *Theories of Vision*, pp. 209-213.

BIBLIOGRAPHY

Ancient, Arabic, and Medieval Sources

Alain of Lille (Alanus de Insulis). *Anticlaudianus*. Ed. R. Bossuat. Paris: Vrin, 1955.
—.*Contra haereticos*. In *PL* 210, cols 305-430.
Albert the Great (Albertus Magnus). *Analytica posteriora*. In *Opera omnia*. Ed. Peter Jammy. 21 vols. Lyon, 1651. In vol. I, pp. 513-658.
—.*De anima*. Ed. Clemens Stroick. In *Opera Omnia*. Ed. Bernhard Geyer et al. 40 vols. In Progress. Münster: Aschendorff, 1968.
—.*De animalibus*. 2 vols. Ed. H.J. Stadler. *Albertus Magnus de animalibus libri XXVI nach der Cölner Urschrift*. In *Beiträge* 15-16 (1916 and 1921).
—.*De caelo et mundo*. Ed. Paul Hossfeld. Münster: Aschendorff, 1971.
—.*De causis et processu universitatis a prima causa*. Ed. Winfried Fauser. Münster: Aschendorff, 1993.
—.*De intellectu et intelligibili*. Ed. Jammy. In vol. V, pp. 239-262.
—.*De meteoris*. Ed. Jammy. In vol. II, pp. 1-209.
—.*De natura loci* and *De causis proprietatum elementorum*. Ed. Paul Hossfeld. Münster: Aschendorff, 1980.
—.*De quattuor coaequaevis*. Ed. Jammy. In vol. XIXa, pp. 1-235.
—.*De resurrectione*. Ed. Wilhelm Kübel. Münster: Aschendorff, 1958, pp. 236-354.
—.*De sensu et sensato*. Ed. Cemil Akdogan. In *Optics in Albert the Great's "De Sensu et Sensato": An Edition, English Translation, and Analysis*. Unpublished Ph.D. dissertation. University of Wisconsin, 1978.
—.*De sensu et sensato*. Ed. Jammy. In vol. V, pp. 1-51.
—.*In evangelium secundum Ioannem expositio*. Ed. Jammy. In vol. XI.
—.*Liber de natura et origine animae*. Ed. Bernhard Geyer. Münster: Aschendorff, 1955.
—.*Metaphysica*. 2 vols. Ed. Bernhard Geyer. Münster: Aschendorff, 1964.
—.*Physica*. 2 vols. Ed. Paul Hossfeld. Münster: Aschendorff, 1987-93.
—.*Quaestiones de animalibus*. Ed. Ephrem Filthaut. Münster: Aschendorff, 1955.
—.*Summa theologiae, sive De mirabili scientia Dei*. Ed. Siedler Dionysius, Wilhelm Kübel, and Heinrich Georg Vogels. Münster: Aschendorff, 1978.
—.*Super Dionysium De caelesti hierarchia*. Ed. Wilhelm Kübel and Paul Simon. Münster: Aschendorff, 1993.

—.*Super Dionysium De divinis nominibus*. Ed. Paul Simon. Münster: Aschendorff, 1972.

—.*Super Dionysii mysticam theologiam et epistulas*. Ed. Paul Simon. Münster: Aschendorff, 1978.

—.*Super Isaiam*. Ed. Ferdinand Siepmann. Münster: Aschendorff, 1952.

—.*Super Matthaeum*. 2 vols. Ed. Bernhard Schmidt. Münster: Aschendorff, 1987.

—.*Super quattuor Sententiarum*. Ed. Jammy. In vols. XIV-XVI.

—.*Super quattuor Sententiarum*. Ed. F.M. Henquinet, "Une pièce inédite du commentaire d'Albert le Grand sur le IVe livre des sentences," *RThAM* 7 (1935), 263-293.

Alberti, Leon Battista. *"On Painting" and "On Sculpture": The Latin Texts of "De pictura" and "De statua."* Ed. and trans. Cecil Grayson. London: Phaidon, 1972.

Alcher of Clairvaux. *Liber de spiritu et anima*. In *PL* 40, cols 779-832.

Alexander of Hales. *Summa theologica*. 4 vols. Florence: Collegium S. Bonaventura, 1924-48.

Alexander Neckham. *De naturis rerum*. Ed. Thomas Wright. London: Longman, 1863.

Alfarabi. *De ortu scientiarum*. Ed. Clemens Baeumker. *Alfrarabi, Über den Ursprung der Wissenschaften*. In *Beiträge* 19/3 (1916), pp. 17-24.

—.*De scientiis*. Ed. Angel González Palencia. In *Catálogo de las ciencias*. Madrid, 1932.

Alhazen (Ibn-al Haytham). *De aspectibus*. Ed. Friedrich Risner. In *Opticae thesaurus Alhazeni Arabis libri septem. Eundem liber de Crepusculis et Nubium Ascensionibus. Item Vitellonis Thuringopoloni libri X*. Basel, 1572.

—.*The Optics of Ibn-al Haytham: Books I-III, On Direct Vision*. 2 vols. Ed. and trans. A.I. Sabra. London: Warburg Institute, 1989.

Alkindi. *De aspectibus*. Ed. Axel Anthon Björnbo and Sebastian Vogl. In "Alkindi, Tideus und Pseudo-Euklid. Drei optische Werke," *Abhandlungen zur Geschichte der mathematischen Wissenschaften* 26/3 (1912), pp. 3-41.

—.*De radiis*. Ed. M.-Th. d'Alverny and F. Hudry, "Alkindi, *De Radiis*," *AHDLMA* 41 (1974), pp. 215-259.

Ambrose, St. *Hexaemeron*. In *PL* 14, cols 123-174.

—.*Saint Ambrose: Hexaemeron, Paradise, and Cain and Abel*. Trans. John J. Savage. Washington, DC: Catholic University of America Press, 1961; reprint 1977.

Andreas Capellanus, *De amore*. Ed. E. Trojet. München: Eidos Verlag, 1964.

(Anonymous). *Liber de causis*. Ed. Adriaan Pattin. In "Le *Liber de causis*. Édition établie à l'aide de 90 manuscrits avec introduction et notes," *Tijdschrift voor filosofie* 28 (1966), 90-203.

(Anonymous). *Liber de causis primis et secundis*. In *Notes et textes sur l'avicennisme latin aux confins des XIIe-XIIIe siècles*. Ed. Roland de Vaux. Paris: Vrin, 1934.

(Anonymous). *Liber de intelligentiis*. Ed. Clemens Baeumker. *Witelo, ein Philosoph und Naturforscher des XIII. Jahrhunderts*. In *Beiträge* 3/2 (1908), pp. 1-71.

Anselm of Canterbury, St. *De veritate*. In *PL* 158, cols 467-486.

—*Monologion*. In *PL* 158, cols 141-222.

Aristotle. *The Complete Works of Aristotle*. 2 vols. Ed. Jonathan Barnes. Princeton, NJ: Princeton University Press, 1984.

Augustine, St. *Ad Paulinum de videndo Deo*. In *PL* 33, cols 596-622.

—*.Confessionum, libri XIII*. Ed. Lucas Verheijen. In *CCSL* 26, Turnhout: Brepols, 1981.

—*.De Genesi ad litteram*. In *PL* 34, cols 245-486.

—*.De Trinitate*. 2 vols. Ed. W.J. Mountain. In *CCSL* 50-50A, Turnhout: Brepols, 1968.

—*.In Johannis evangelium tractatus*. In *PL* 35, cols 1579-1976.

—*.Sermones*. In *PL* 38-39.

Averroës (Ibn Rushd). *Aristotelis opera cum Averrois commentariis*. 12 vols. Venice, 1562-74; reprint, Frankfurt-am-Main: Minerva, 1962.

—*.Commentarium magnum in Aristotelis De anima libros*. Ed. F. Stuart Crawford. Cambridge, MA: Mediaeval Academy of America, 1953.

—*.De sensu*. In *Compendia librorum Aristotelis qui Parva naturalia vocantur*. Ed. A.L. Shields and H. Blumberg. Cambridge, MA: Mediaeval Academy of America, 1949.

Avicebron (Ibn Gebirol). *Fons vitae*. Ed. Clemens Baeumker. *Avecenbrolis (Ibn Gebirol) Fons vitae ex Arabico in Latinum translatus ab Johanne Hispano et Dominico Gundissalino*. In *Beiträge* I/2-4 (1892-95).

Avicenna (Ibn Sina). *Avicenna latinus: liber de anima seu sextus de naturalibus I-III*. Ed. S. Van Riet. Louvain and Leiden: E.J. Brill, 1972.

—*Avicenna latinus: liber de anima seu sextus de naturalibus IV-V*. Ed. S. Van Riet. Louvain and Leiden: E.J. Brill, 1968.

—*Avicenna latinus: liber de philosophia prima sive scientia divina*. 3 vols. Ed. S. Van Riet. Louvain and Leiden: E.J. Brill, 1977-83.

—*.Compendium de anima*. Venice, 1546; reprint, Farnborough, 1969.

—*.Sufficientia*. In *Avicenne perhypatetici philosophi: ac mediorum facile primi opera in lucem redacta: ac nuper quantum ars niti potuit per canonicos emendata. Logyca, Sufficientia, De celo et mundo, De anima, De animalibus, De intelligentiis, Alpharabius de intelligentiis, Philosophia prima*. Venice, 1508; reprint Frankfurt-am-Main: Minerva, 1961.

Bartholomew of Bologna (Bartholomaeus di Bononia). *Tractatus de luce*. Ed. I. Squadrani. In "Tractatus de luce Fr. Bartholomaei di Bononia inquisitiones et textus," *Antonianum* 7 (1932), 201-238, 337-376, 465-494.

Bartholomew the Englishman (Bartholomaeus Anglicus). *Liber de rerum proprietatibus*. Ed. Wolfgang Richter. Frankfurt, 1601; reprint Frankfurt-am-Main: Minerva, 1964.

Basil, St. *Homiliae in Hexaemeron*. In *PG* 29, cols 3-208.

Bede. *De natura rerum*. In *PL* 90, cols 187-278.

Benvenuto da Imola. *Comentum super Dantis Aldigherij Comoediam*. 5 vols. Ed. Giacomo Filippo Laicaita. Florence: Barbèra, 1887.

Bernard Silvestris. *Cosmographia (De mundi universitate)*. Ed. Peter Dronke. Leiden: E.J. Brill, 1978.

Bible (Vulgate). *Biblia sacra latina iuxta vulgatam versionem*. 2 vols. Ed. Robert Weber. Stuttgart: Württembergische Bibelanstalt, 1969.

Boethius. *Consolatio philosophiae*. In *PL* 63, cols 579-862.

—.*De Trinitate*. In *PL* 64, cols 1247-1256.

Bonaventure, St. *Opera omnia*. 10 vols. Florence: Collegium S. Bonaventura, 1882-1902.

Buti, Francesco da. *Commento di Francesco da Buti sopra la "Divina Commedia" di Dante Allighieri*. 3 vols. Ed. Giannini di Crescentino. Pisa: Fratelli Nistri, 1858-62.

Chalcidius. *Timaeus a Calcidio translatus commentarioque instructus*. Ed. J.H. Waszink and P.J. Jensen. London and Leiden: Warburg Institute, 1962.

Chartularium Universitatis Parisiensis. 4 vols. Ed. Heinrich Denfile and E. Chatelain. Paris, 1889-97.

Dante Alighieri. *Il Convivio*. 2nd edn. Ed. Giovanni Busnelli and G.V. Vandelli. Florence: Le Monnier, 1967.

—.*Il Convivio*. Ed. Cesare Vasoli and Domenico de Robertis. In *Opere minori*, tomo 1, parte 2. Milan and Naples: Ricciardi, 1988.

—.*De vulgari eloquentia*. Ed. Pier Vincenzo Mengaldo. In *Opere minori*, tomo 2, Milan and Naples: Ricciardi, 1979, pp. 26-237.

—.*La "Divina Commedia."* 3 vols. Ed. Umberto Bosco and Giovanni Reggio. Florence: Le Monnier, 1979.

—.*La "Divina Commedia."* 3 vols. Ed. Tommaso Casini, revised by S.A. Barbi, with a new presentation by Francesco Mazzoni. Florence: Sansoni, 1965.

—.*La "Divina Commedia."* Ed. Pietro Fraticelli. Florence: Barbèra, 1887.

—.*La "Divina Commedia."* Ed. Giuseppe Giacalone. Rome: Signorelli, 1988.

—.*Die "Göttliche Komödie": Kommentar*. 3 vols. Ed. H. Gmelin. Stuttgart: Klett, 1954-57.

—.*La "Divina Commedia."* Ed. Attilio Momigliano. Florence: Sansoni, 1950.

—.*"Commedia."* Ed. Emilio Pasquini and Antonio Quaglio. Milan: Garzanti, 1987.

—.*La "Commedia" secondo l'antica vulgata*. 4 vols. Ed. Giorgio Petrocchi. Milan: Mondadori, 1966-67.

—.*La "Divina Commedia."* Ed. Giacomo Poletto. Rome: Desclée Lefebvre, 1894.

—.*La "Divina Commedia."* 3 vols. Ed. Manfredi Porena. Bologna: Zanichelli, 1953.

—.*La "Divina Commedia."* Ed. Tommaso di Salvo. Bologna: Zanichelli, 1987.

—.*La "Divina Commedia."* 3 vols. Ed. Natalino Sapegno. Milan and Naples, Ricciardi, 1957.

—.*La "Divina Commedia."* 3 vols. Ed. Luigi Scorrano and Aldo Vallone. Naples: Ferraro, 1986.

—.*The "Divine Comedy."* 3 vols. Ed. and trans. Charles S. Singleton. Princeton, NJ: Princeton University Press, 1971-75.

—.*Epistole.* Ed. Giorgio Brugnoli and Arsenio Frugoni. In *Opere minori*, tomo 2, pp. 522-643.

—*Monarchia.* Ed. Bruno Nardi. In *Opere minori*, tomo 2, pp. 280-503.

—.*Dante's "De Monarchia."* Ed. and trans. Prue Shaw. Cambridge: Cambridge University Press, 1995.

—.*Oeuvres complètes.* Ed. and trans. André Pézard. Paris: Gallimard, 1968.

—.*Questio.* Ed. Francesco Mazzoni. In *Opere minori*, tomo 2, pp. 744-783.

—.*Rime.* Ed. Gianfranco Contini. In *Opere minori*, tomo 1, parte 1. Milan and Naples: Ricciardi, 1984.

—.*Vita Nuova.* Ed. Domenico de Robertis. In *Opere minori*, tomo 1, parte 1, pp. 27-247.

Dominic Gundissalinus. *De divisione philosophiae.* Ed. Ludwig Baur, *Dominicus Gundissalinus, De divisione philosophiae.* In *Beiträge* 4/2-3 (1903).

—.*De processione mundi.* Ed. Georg Bülow. *Des Dominicus Gundissalinus Schrift De processione mundi.* In *Beiträge* 24/3 (1925).

—.*De unitate.* Ed. Paul Correns. *Die dem Boethius fälschlich zugesschriebene Abhandlung des Dominicus Gundisalvi de Unitate.* In *Beiträge* 1/1 (1891).

Eriugena, John Scottus. *Commentaire sur l'Évangile de Jean.* Ed. and trans. Edouard Jeauneau. In *SC* 180, Paris: Éditions du Cerf, 1972.

—.*Expositiones in Ierarchiam coelestem.* Ed. Jeanne Barbet. In *CCCM* 31, Turnhout: Brepols, 1975.

—.*Homélie sur le prologue de Jean.* Ed. and trans. Edouard Jeauneau. In *SC* 151, Paris: Éditions du Cerf, 1969.

—.*Periphyseon (De divisione naturae).* 3 vols. Ed. I.P. Sheldon-Williams. Dublin: Dublin Institute for Advanced Studies, 1968-81.

Euclid. *Catoptrica.* Ed. J.L. Heiberg and H. Menge. Vol. 7 in *Opera omnia.* 9 vols. Leipzig: Teubner, 1883-1916.

—.*Liber de visu (Optica).* Ed. Wilfred R. Theisen. In *"Liber de visu*: The Greco-Latin Translation of Euclid's *Optics," Mediaeval Studies* 41 (1979), 44-105.

Galen. *Galen on the Usefulness of The Parts of the Body.* 2 vols. Trans. Margaret T. May. Ithaca: Cornell University Press, 1968-72.

Giacomo da Lentini. *Giacomo da Lentini: poesie.* Ed. Roberto Antonelli. Rome: Bulzoni, 1979.

Hero of Alexandria. *Catoptrica.* Ed. L. Nix and W. Schmidt. In *Opera quae supersunt omnia.* 5 vols. Leipzig: Teubner, 1899-1914, vol. II.

Hildegard of Bingen. *Scivias.* In *PL* 197, cols 383-738.

Honoré d'Autun. *Philosophia mundi.* In *PL* 172, cols 41-102.

Hugh of St Victor. *Commentariorum in hierarchiam coelestem S. Dionysii Areopagitae.* In *PL* 175, cols 923-1154.

—.*De bestiis et aliis rebus.* In *PL* 177, cols 15-164.

—.*De unione corporis et spiritus.* In *PL* 177, cols 285-294.

—*Eruditionis didascalicae*. In *PL* 176, cols 741-838.

Index Thomisticus: Sancti Thomae Aquinatis operum omnium indices et concordantiae. Ed. Roberto Busa. Stuttgart: Frommann-Holzboog, 1975.

Isaac Israeli. *Liber de definicionibus*. Ed. T.J. Muckle. In *AHDLMA* 11-12 (1937-38), 299-340.

Isidore of Seville. *Etymologiarum sive originum, libri XX*. In *PL* 82, cols 73-728.

John Peckham. *Perspectiva communis*. Ed. David C. Lindberg. In *John Pecham and the Science of Optics: "Perspectiva communis," edited with an Introduction, English Translation, and Critical Notes*. Madison, Milwaukee, and London: University of Wisconsin Press, 1970.

—.*Tractatus de perspectiva*. Ed. David C. Lindberg. New York: Franciscan Institute Publications, 1972.

Lateinische Hymnen des Mittelalters. 3 vols. Ed. Franz Joseph Mone. Freiburg: Sumptus Herder, 1853-54.

Leonardo da Vinci. *Il Paragone*. Ed. Claire J. Farago. In *Leonardo da Vinci's Paragone: A Critical Interpretation with a New Edition of the Text in the Codex Urbinas*. Leiden and New York: E.J. Brill, 1992.

Lucretius. *De natura rerum*. 2 vols. Ed. W.H.D. Rouse. London: Heinemann, 1978.

Macrobius. *Commentarii in Somnium Scipionis*. 2 vols. Ed. Mario Regali. Pisa: Giardini, 1983-90.

—.*Commentary on the Dream of Scipio*. Trans. William Harris Stahl. 2nd edn. New York and London: Columbia University Press, 1966.

Ovid. *Metamorphoses*. 2 vols. Ed. and trans. Frank Justus Miller. London: Heinemann, 1921.

Petrus Alfonsi. *Dialogus*. In *PL* 157, cols 537-672.

Plato. *The Republic of Plato*. Trans. and comm. Francis M. Cornford. Reprint. New York and London: Oxford University Press, 1964.

—.*Timaeus*. Latin text to 53B. In Chalcidius, *Timaeus a Calcidio translatus commentarioque instructus*. Ed. J.H. Waszink and P.J. Jensen. London and Leiden: Warburg Institute, 1962.

Pliny. *Historia naturalis*. 10 vols. Ed. and trans. H. Rackham, W.H.S. Jones, and D.E. Eichholz. London: Heinemann, 1938-62.

Plotinus. *The Enneads*. 6 vols. Ed. and trans. A.H. Armstrong. London: Heinemann, 1966-88.

Poeti del Duecento. 2 vols. Ed. Gianfranco Contini. Milan and Naples: Ricciardi, 1960.

Proclus. *The Elements of Theology: A Revised Text, with Translation, Introduction, and Commentary*. 2nd edn. Ed. E.R. Dodds. Oxford: Clarendon Press, 1963.

Pseudo-Aristotle. *Problemata*. Latin trans. Bartholomew of Messina. Ed. Gerardo Marenghi. In *Aristotele: Problemi di fonazione e di acustica*. Naples: Libreria scientifica editrice, 1962.

Pseudo-Bede. *De mundi constitutione*. In *PL* 90, cols 881-910.

Pseudo-Bernard. *Tractatus de interiori domo*. In *PL* 184, cols 507-552.

Pseudo-Dionysius the Aeropagite. *Dionysiaca: Recueil donnant l'ensemble des traditions latines des ouvrages attribués au Denys de l'Aréopage.* 2 vols. Ed. Philippe Chevalier. Paris: Desclée de Brouwer, 1937-50.

Pseudo-Euclid. *De speculis.* Ed. Axel Anthon Björnbo and Sebastian Vogl. In "Alkindi, Tideus und Pseudo-Euklid. Drei optische Werke," pp. 97-119.

Ptolemy (Claudius Ptolomaeus). *Optica.* Ed. and trans. Albert Lejeune. In *L'optique de Claude Ptolémée dans la version latine d'après l'arabe de l'émir Eugène de Sicile.* New edition. New York and Leiden: E.J. Brill, 1989.

—.*Optics.* In *Ptolemy's Theory of Visual Perception: An English Translation of the "Optics" with Introduction and Commentary.* Trans. A. Mark Smith. Philadelphia: American Philosophical Society, 1996.

—.*Tetrabiblos (Opus quadripartitum).* Ed. and trans. F.E. Robbins. London and Cambridge, MA: Harvard University Press, 1940.

Restoro d'Arezzo. *La composizione del mondo colle sue cascioni.* Ed. A. Morino, Florence: Accademia della Crusca, 1976.

Rhaban Maur. *De universo.* In *PL* 111, cols 9-614.

Richard of St. Victor. *De Trinitate.* Ed. Gaston Salet. *Richard de St. Victor: La Trinité.* In *SC* 63, Paris: Éditions du Cerf, 1959.

Robert Grosseteste. *Commentarius in VIII libros Physicorum Aristotelis.* Ed. Richard C. Dales. Boulder, CO: Colorado University Press, 1963.

—.*Commentarius in Posteriorum analyticorum libros.* Ed. Pietro Rossi. Florence: Olschki, 1981.

—. *De colore; De generatione sonorum; De iride; De lineis, angulis et figuris seu fractionibus et reflexionibus radiorum; De luce seu de inchoatione formarum; De motu corporali; De natura locorum.* Ed. Ludwig Baur. *Die philosophischen Werke des Robert Grosseteste, Bischofs von Lincoln.* In *Beiträge* 9 (1912).

—.*Hexaëmeron.* Ed. Richard C. Dales and Servus Gieben. London: Oxford University Press, 1982.

Roger Bacon. *De multiplicatione specierum.* Ed. David C. Lindberg. In *Roger Bacon's Philosophy of Nature: A Critical Edition, with English Translation, Introduction, and Notes, of "De multiplicatione specierum" and "De speculis comburentibus."* Oxford: Clarendon Press, 1983.

—.*Opera quaedam hactenus inedita.* Ed. J.S. Brewer. London: Longman, 1859.

—.*The "Opus Majus" of Roger Bacon.* 3 vols. Ed. John Henry Bridges. Oxford: Oxford University Press, 1897-1900.

—.*Perspectiva.* Ed. David C. Lindberg. In *Roger Bacon and the Origins of "Perspectiva" in the Middle Ages: A Critical Edition and English Translation of Bacon's "Perspectiva" with Introduction and Notes.* Oxford: Clarendon Press, 1996.

Sacrobosco (John of Halifax). *De sphaera.* Ed. and trans. Lynn Thorndike. In *The Sphere of Sacrobosco and Its Commentators.* Chicago: Chicago University Press, 1949.

Seneca, Lucius Annaeus. *Naturales quaestiones*. 2 vols. Ed. and trans. Thomas H. Corcoran. London: Heinemann, 1971-72.

A Source Book in Medieval Science. Ed. Edward Grant. Cambridge, MA: Harvard University Press, 1974.

Statius. *Thebaid*. 2 vols. Ed. and trans. J.H. Mozley. Reprint. London: Heinemann, 1982.

Thomas Aquinas. *Commentum in quattuor libros sententiarum*. In *Opera omnia*. 25 vols. Parma, 1852-73; reprint, New York: Musurgia, 1948. Tomus 6-7.

—.*In Aristotelis libros De caelo et mundo, De generatione et corruptione, meteorologicorum*. Ed. Raymund M. Spiazzi. Turin and Rome: Marietti, 1952.

—.*In Aristotelis libros De sensu et sensato commentarium*. 2nd edn. Ed. Raymund M. Spiazzi. Turin: Marietti, 1973.

—.*In Aristotelis libros Peri hermeneias et Posteriorum analyticorum expositio*. 2nd edn. Ed. Raymund M. Spiazzi. Turin: Marietti, 1964.

—.*In Aristotelis librum De anima commentarium*. 4th edn. Ed. Angeli M. Pirotta. Turin: Marietti, 1959.

—.*In duodecim libros Metaphysicorum Aristotelis expositio*. Reprint. Ed. M.-R. Cathala. Turin: Marietti, 1965.

—.*In librum Beati Dionysii De divinis nominibus expositio*. Ed. Ceslai Pera. Rome and Turin: Marietti, 1950.

—.*In librum de causis expositio*. 2nd edn. Ed. Ceslai Pera. Turin: Marietti, 1972.

—.*In octo libros Physicorum Aristotelis expositio*. Reprint. Ed. M.-P. Maggiòlo. Turin: Marietti, 1965.

—.*Quaestiones disputatae de potentia*. 10th edn. Ed. P. Bazzi. Turin and Rome: Marietti, 1965.

—.*Quaestiones quodlibetales*. 9th edn. Ed. Raymund M. Spiazzi. Turin: Marietti, 1956.

—.*Summa contra Gentiles*. In *Liber de veritate catholicae fidei contra errores infidelicum, seu Summa contra Gentiles*. 3 vols. Ed. Ceslai Pera, et al. Turin: Marietti, 1961-67.

—.*Summa theologiae (ST)*. 3 vols. Ed. P. Caramello. Turin: Marietti, 1963.

—.*Super evangelium S. Ioannis lectura*. 6th edn. Ed. R. Raphaelis Cai. Rome and Turin: Marietti, 1972.

Thomas of Cantimpré. *Liber de natura rerum*. Ed. H. Boese. Berlin and New York: De Gruyter, 1973.

Vincent of Beauvais (Vincentius Bellovacensis). *Speculum quadruplex sive speculum maius*. 4 vols. Douai, 1624; reprint, Frankfurt-am-Main: Minerva, 1964.

Virgil. Vol. 1, *Eclogues, Georgics, Aeneid I-VI*. Vol. 2, *Aeneid VII-XII, the Minor Poems*. Trans. H. Rushton Fairclough. Revised edn. London: Heinemann, 1986.

William of Auvergne. *Opera omnia*. 2 vols. Paris, 1674; reprint, Frankfurt-am-Main: Minerva, 1963.

William of Conches. *De philosophia mundi*. In *PL* 172, cols 41-102.

—.*Glosae super Platonem.* Ed. Edouard Jeauneau. Paris: Vrin, 1965.
William of Lorris and Jean de Meun. *Roman de la Rose.* 3 vols. Ed. Félix Lecoy. Paris: H. Champion, 1965-70.
William of St. Thierry. *De natura corporis et animae.* In *PL* 180, cols 695-726.
Witelo. *Perspectiva.* Ed. Friedrich Risner. In *Opticae thesaurus Alhazeni Perspectiva libri septem...* Basel, 1572.

Secondary Literature

Ageno, Brambilla Franca. "Interpretazione e punteggiatura in passi danteschi," *Studi danteschi* 52 (1979-80), 165-192.
Agrelo, Sebastiano. "El tema biblico de la luz," *Antonianum* 50 (1975), 353-417.
Akbari, Suzanne Conklin. "Medieval Optics in Guillaume de Lorris' *Roman de la Rose*," *Medievalia et Humanistica* n.s. 21 (1994), 1-15.
Alonso, J.M. "Teofanía y visón beata en Escoto Erigena," *Revista Española de Teología* 10 (1950), 361-389.
Altmann, A. and Stern, S.M. *Isaac Israeli: A Neoplatonic Philosopher of the Early Tenth Century.* London and Oxford: Oxford University Press, 1958.
Andriani, Beniamino. *Aspetti della scienza in Dante.* Florence: Le Monnier, 1981.
Armstrong, A.H. Ed. *The Cambridge History of Later Greek and Early Medieval Philosophy.* Reprint. Cambridge: Cambridge University Press, 1970.
—."'Emanation' in Plotinus," *Mind* n.s. 46 (1937), 61-66.
Austin, Herbert D. "Dante and Mirrors," *Italica* 21 (1944), 13-17.
—."Dante Notes XI: The Rainbow Colors," *Modern Language Notes* 44 (1929), 315-318.
Badawi, Abd al-Rahman. *Histoire de la philosophie en Islam.* 2 vols. Paris: Vrin, 1972.
Baeumker, Clemens. *Witelo, ein Philosoph und Naturforscher des XIII. Jahrhunderts.* In *Beiträge* 3/2 (1908).
Baig, Bonnie Pavlis. *Vision and Visualization: Optics and Light Metaphysics in the Imagery and Poetic Form of Twelfth- and Thirteenth-Century Secular Allegory, with Special Attention to the "Roman de la Rose."* Unpublished Ph.D. dissertation. University of California, Berkeley, 1982.
Baranski, Zygmunt G. "Dante fra 'sperimentalismo' e 'enciclopedismo'." In *L'enciclopedismo medievale.* Ed. Michelangelo Picone. Ravenna: Longo, 1994, pp. 383-404.
Barasch, Moshe. *Light and Colour in the Italian Renaissance Theory of Art.* New York: New York University Press, 1978.
Barolini, Teodolinda. *Dante's Poets: Textuality and Truth in the "Comedy."* Princeton, NJ: Princeton University Press, 1984.
Baxandall, Michael. *Painting and Experience in Fifteenth-Century Italy.* 2nd edn. Oxford and New York: Oxford University Press, 1988.

Beare, J.I. *Greek Theories of Elementary Cognition from Alcmaeon to Aristotle.* Oxford: Clarendon, 1906.

Beierwaltes, Werner. Ed. *Eriugena Redivivus: Zur Wirkungsgeschichte seines Denkens im Mittelalter und im Übergung zur Neuzeit.* Heidelberg: Carl Winter, 1987.

Bellonzi, Fortunato. "Arti figurative." In *ED* I, pp. 400-403.

Bemrose, Stephen. *Dante's Angelic Intelligences: Their Importance in the Cosmos and in pre-Christian Religion.* Rome: Edizioni di storia e letteratura, 1983.

—."'Una favilla sol della tua gloria': Dante expresses the inexpressible," *Forum for Modern Language Notes* 27 (1991), 126-137.

Bender, John B. *Spenser and Literary Pictorialism.* Princeton, NJ: Princeton University Press, 1972.

Berg, William J. *The Visual Novel: Emile Zola and the Art of His Times.* University Park, PA: Pennsylvania State University Press, 1992.

Berretta, G. "Il Canto XIV del *Paradiso,*" *Filologia e letteratura* 11 (1965), 254-269.

Berti, Enrico. The following entries from the *ED*: "De Anima" (II, pp. 325-326); "De Coelo" (II, pp. 330-332); "De Generatione et Corruptione" (II, pp. 336-337); "De Generatione Animalium" (II, pp. 335-336); "De Meteoris" (II, pp. 364-365); "De Partibus Animalium" (II, p. 378); "De Sensu" (II, pp. 387-388); "Fisica" (II, p. 934); "Metafisica" (III, pp. 924-925).

Bertola, E. "La dottrina dello «spirito» in Alberto Magno," *Sophia* 19 (1951), 306-312.

—."Le fonti medico-filosofiche della dottrina dello «spirito»," *Sophia* 26 (1958), 48-61.

Bevan, Edwyn. *Symbolism and Belief.* London: Allen & Unwin, 1938.

Bigi, V.Ch. "La dottrina della luce in S. Bonaventura," *Divus thomas* 64 (1961), 395-423.

Bigongiari, Dino. Review of Bruno Nardi's *Saggi di filosofia dantesca.* In *Speculum,* 7 (1932), 146-153.

Birkenmajer, Aleksjander. "Witelo est-il l'auteur de l'opuscule 'De intelligentiis'?" In *Études d'histoire des sciences en Pologne.* Wroclaw: Ossolineum, 1972, pp. 259-339.

Block, Irvine. "Truth and Error in Aristotle's Theory of Sense Perception," *Philosophical Quarterly* 11 (1961), 1-9.

Boissard, Edmond. "St. Bernard et le Pseudo-Aréopagite," *RThAM* 26 (1959), 214-263.

Boitani, Piero. "The Sibyl's Leaves: A Study of *Paradiso* XXXIII," *Dante Studies* 96 (1978), 83-126.

Bono, James S. "Medical Spirits and the Medieval Language of Life," *Traditio* 40 (1984), 91-130.

Botterill, Steven. *Dante and the Mystical Tradition: Bernard of Clairvaux in the "Commedia."* Cambridge: Cambridge University Press, 1994.

Bougerol, Jacques G. "Le rôle de l'*influentia* dans la théologie de la grâce chez S. Bonaventure," *Revue de théologie de Louvain* 5 (1974), 273-300.

—."Saint Bonaventure et la hiérarchie dionysienne," *AHDLMA* 36 (1969), 131-167.

—."St. Bonaventure et le Pseudo-Dionysius l'Aréopagite," *Études Franciscaines* 28 suppl. (1968), 33-123.

Boyde, Patrick. *Dante Philomythes and Philosopher: Man in the Cosmos.* Cambridge: Cambridge University Press, 1981.

—.Ed. with Vittorio Russo. *Dante e la scienza: Atti del convegno internazionale di studi su Dante e la scienza* (28-30 May 1993). Ravenna: Longo, 1995.

—.*Perception and Passion in Dante's "Comedy."* Cambridge: Cambridge University Press, 1993.

—."Perception and the Percipient in *Convivio*, III. ix and the *Purgatorio*," *Italian Studies* 35 (1980), 19-24.

Boyer, Carl B. "Aristotelian References to the Law of Reflection," *Isis* 36 (1945-46), 92-105.

—.*The Rainbow: From Myth to Mathematics.* With new colour illustrations and commentary by Robert Greenler. Princeton, NJ: Princeton University Press, 1987.

—."The Theory of the Rainbow: Medieval Triumph and Failure," *Isis* 49 (1958), 378-390.

Bradley, Ritamary. "Backgrounds of the Title *Speculum* in Mediaeval Literature," *Speculum* 29 (1954), 100-115.

Brandeis, Irma. *A Ladder of Vision: A Study of Dante's "Comedy."* New York: Anchor, 1962.

Bremer, Dieter. "Licht als univerales Darstellungsmedium: Materialien und Bibliographie," *Archiv für Begriffsgeschichte* 18 (1974), 185-206.

Brown, Peter. *Chaucer's Visual World: A Study of His Poetry and the Medieval Optical Tradition.* 2 vols. Unpublished D.Phil. dissertation. York University, 1981.

Brownlee, Kevin. "Dante and Narcissus (*Purg.* XXX, 76-99)," *Dante Studies* 96 (1978), 201-206.

Bufano, Antonietta. "Sole." In *ED* V, pp. 296-304.

Busnelli, Giovanni. *Cosmogonia e antropogenesi secondo Dante Alighieri e le sue fonti.* Rome, 1922.

Buti, Giovanni and Bertagni, Renzo. *Commento astronomico della "Divina Commedia."* Florence: Sandron, 1966.

Cantarino, Vincent. "Dante and Islam: Theory of Light in the *Paradiso*," *Kentucky Romance Quarterly* 15 (1968), 3-35.

Caparello, Adriana. *La «perspettiva» in Sigieri di Brabante.* Vatican City: Libreria editrice Vaticana, 1987.

—."Il termine «tunica» e la sua portata scientifico-storica nella dottrina aristotelico-tomista della visione," *Divus thomas* 79 (1976), 369-399.

Capasso, Ideale. *L'astronomia nella Divina Commedia.* Pisa: Domus Galilaeana, 1967.

—.and Tabarroni, Giorgio. The following entries from the *ED*: "Astrologia," (I, pp. 427-431); "Astronomia," (I, pp. 431-435); "Cielo," (I, pp. 1000-1004).

Cappuyns, Maïeul J. *Jean Scot Érigène: Sa vie, son oeuvre, sa pensée*. Reprint. Brussels: Culture et Civilisation, 1964.

Carruthers, Mary J. *The Book of Memory: A Study of Memory in Medieval Culture*. Cambridge: Cambridge University Press, 1990.

Catenazzi, Flavio. *L'influsso dei provenzali sui temi e immagini della poesia siculo-toscanca*. Brescia: Morcelliana, 1977.

Chenu, M.-D. *Nature, Man and Society in the Twelfth Century, Essays on New Theological Perspectives in the Latin West*. Trans. Jerome Taylor and Lester K. Little. Chicago and London: University of Chicago Press, 1968.

Chiamenti, Massimiliano. "The Representation of the Psyche in Cavalcanti, Dante and Petrarch: The 'Spiriti'," *Neophilologus* 82 (1998), 71-81.

Chiavacci Leonardi, Anna Maria. "«Le bianche stole»: Il tema della resurrezione nel *Paradiso*." In *Dante e la Bibbia*. Ed. Giovanni Barblon. Florence: Olschki, 1988, pp. 249-271.

Chimenz, S.A. "Per il testo e la chiosa della *Divina Commedia*," *GSLI* 133 (1956), 161-188.

Chioccioni, Pietro. *L'agostinismo nella "Divina Commedia."* Florence: Olschki, 1952.

Clark, David L. "Optics for Preachers: The 'De oculo morali' by Peter of Limoges," *Michigan Academician* 9 (1977), 329-343.

Cline, Ruth H. "Heart and Eyes," *Romance Philology* 25 (1972), 263-297.

Cocking, J.M. *Imagination: A Study in the History of Ideas*. Ed. Penelope Murray. New York and London: Routledge, 1991.

Collette, Carolyn. "Seeing and Believing in the *Franklin's Tale*," *The Chaucer Review* 26 (1992), 395-410.

Contenson, P.-M., de. "Avicennisme latin et vision de Dieu au début du XIIIe siècle," *AHDLMA* 26 (1959), 29-97.

Copleston, Frederick C. *A History of Medieval Philosophy*. London: Menthuen, 1972.

Cornish, Alison. "Dante's Moral Cosmology." In *Cosmology: Historical, Literary, Philosophical, Religious and Scientific Perspectives*. Ed. Norriss S. Hetherington. New York and London: Garland, 1993, pp. 201-215.

Corti, Maria. *La felicità mentale: Nuove prospettive per Cavalcanti e Dante*. Turin: Einaudi, 1983.

—.*Percorsi dell'invenzione: Il linguaggio poetico e Dante*. Turin: Einaudi, 1993.

Cristiana, Marta. The following entries from the *ED*: "Platonismo" (IV, pp. 550-555); "Scoto Eriugena" (V, pp. 90-92).

Crivelli, E. "Il vetro, gli specchi e gli occhiali ai tempi di Dante," *Giornale dantesco* n.s. 12 (1939), 79-90.

Crombie, A.C. "Expectation, Modelling and Assent in the History of Optics: Part 1. Alhazen and the Medieval Tradition," *Studies in the History and Philosophy of Science* 21 (1990), 605-632.

——.*Robert Grosseteste and the Origins of Experimental Science, 1100-1700*. Oxford: Clarendon Press, 1953.

——.*Science, Optics and Music in Medieval and Early Modern Thought*. London: Hambledon, 1990.

d'Acona, Cristina Costa. "La doctrine de la création «mediante intelligentia» dans le *Liber de Causis* et dans ses sources," *Revue des sciences philosophiques et théologiques* 76 (1992), 209-232.

——."Le fonti e la struttura del *Liber de Causis*," *Medioevo* 15 (1989), 1-38.

——.*Recherches sur le Liber de Causis*. Paris: Vrin, 1995.

Dales, Richard C. "The De-Animation of the Heavens in the Middle Ages," *Journal of the History of Ideas* 41 (1980), 531-550.

d'Alverny, M-T. "Pseudo-Aristotle, *De elementis*." In *Pseudo-Aristotle in the Middle Ages*. Ed. Jill Kraye, W.F. Ryan, Charles B. Schmitt. London: Warburg Institute, 1986, pp. 63-83.

Dauphiné, James. *Le cosmos de Dante*. Paris: Les Belles Lettres, 1984.

Davidson, Herbert A. *Alfarabi, Avicenna and Averroes: Their Cosmologies, Theories of Active Intellect and Theories of Human Intellect*. Oxford and New York: Oxford University Press, 1992.

de Bonfils Tempier, Margherita. "«La prima materia de li elementi»," *Studi danteschi* 58 (1986), 275-291.

——."La prima visione della «Vita Nuova» e la dottrina dei tre spiriti," *Rassegna della letteratura italiana* 76 (1972), 303-316.

de Bruyne, Edgar de. *Études d'esthétique médiévale*. 3 vols. Reprint. Geneva: De Tempel, 1975.

de Mottoni Faes, Barbara. *Il "Corpus Dionysianum" nel Medioevo: Rassegna di studi: 1900-1972*. Rome: Il Mulino, 1977.

——."La dottrina dell' «anima mundi» nella prima metà del secolo XIII: Guglielmo d'Alvernia, «Summa halensis», Alberto Magno," *Studi medievali* ser. 3a/22 (1981), 253-297.

——.*Il platonismo medioevale*. Turin: Einaudi, 1979.

——."Il problema della luce nel commento di Bertoldo Moosburg all'«Elementatio Theologica» di Proclo," *Studi medievali* ser. 3/16 (1975), 325-352.

Devons, Samuel. "Optics through the Eyes of the Medieval Churchmen." In *Science and Technology in Medieval Society*. Ed. Pamela O. Long. New York: New York Academy of Sciences, 1985, pp. 205-224.

Dictionnaire de spiritualité ascétique et mystique doctrine et histoire (DS). Ed. Marcel Viller et al. Paris: Beauchesne, 1937-95.

Di Pino, Guido. *La figurazione della luce nella "Divina Commedia."* Florence: La Nuova Italia, 1952.

Di Scipio, Giuseppe. "Dante and St. Paul: The Blinding Light and Water," *Dante Studies* 98 (1980), 151-157.

—.and Scaglione, Aldo. Ed. *The Divine Comedy and the Encyclopaedia of the Arts and Sciences*. Amsterdam and Philadelphia: John Benjamins, 1988.

—.*The Presence of Pauline Thought in the Works of Dante*. Lewiston, Queenston, and Lampeter: Edwin Mellen Press, 1995.

Doherty, K.F. "St. Thomas and the Pseudo-Dionysian Symbol of Light," *New Scholasticism* 34 (1960), 170-189.

Dondaine, H.-F. *Le corpus dionysien de l'Université de Paris au XIIIe siècle*. Rome: Edizioni di storia e letteratura, 1953.

—."L'objet et le 'medium' de la vision béatifique chez les théologiens du XIIIe siècle," *RThAM* 19 (1952), 60-130.

Dragonetti, Roger. "Dante et Narcisse ou les faux-monnayeurs de l'image," *Revue des études italiennes* 11 (1965), 85-146.

Dronke, Peter. *Dante and Medieval Latin Traditions*. Cambridge: Cambridge University Press, 1986.

—.Ed. *A History of Twelfth-Century Western Philosophy*. Cambridge: Cambridge University Press, 1988.

—."Tradition and Innovation in Medieval Western Colour-Imagery," *Eranos Jahrbuch* 41 (1972), 51-106.

Druart, T.-A. "Alfarabi and Emanationism." In *Studies in Medieval Philosophy*. Ed. John F. Wippel. Washington, DC: Catholic University of America Press, 1987, pp. 23-44.

Dunphy, W.B. "St. Albert and the Five Causes," *AHDLMA* 33 (1966), 7-21.

Durantel, J.S. *St. Thomas et le Pseudo-Denys*. Paris: F. Alcan, 1919.

Durling, Robert M. and Martinez, Roland L. *Time and the Crystal: Studies in Dante's "Rime Petrose."* Berkeley, Los Angeles, and London: University of California Press, 1990.

Eastwood, Bruce S. "Averroes' View of the Retina – A Reappraisal," *Journal of the History of Medicine and Allied Sciences* 24 (1969), 77-82.

—."Mediaeval Empiricism: The Case of Grosseteste's Optics," *Speculum* 43 (1968), 306-321.

Eberle, Patricia J. "The Lovers' Glass: Nature's Discourse on Optics and the Optical Design of the *Romance of the Rose*," *University of Toronto Quarterly* 46 (1977), 241-262.

Edgerton, Samuel Y. Jr. "Alberti's Colour Theory: A Medieval Bottle without Renaissance Wine," *Journal of the Warburg and Courtauld Institutes* 22 (1969), 109-134.

—.*The Renaissance Rediscovery of Linear Perspective*. New York: Harper and Row, 1975.

Enciclopedia dantesca (ED). 6 vols. Ed. Umberto Bosco and Giorgio Petrocchi. Rome: Istituto dell'enciclopedia italiana, 1970-78. Relevant entries are listed by author.

Evans, David. *Mediaeval Optics and Stained Glass*. Unpublished Ph.D. dissertation. University of Birmingham, 1979.

Fallani, Giovanni. *Dante e la cultura figurativa medievale*. Bergamo: Minerva Italica, 1971.

—."Visio beatifica." In *ED* V, pp. 1070-1071.

Fengler, Christie K. and Stephany, William A. "The Visual Arts: A Basis for Dante's Imagery in the *Purgatory* and the *Paradise*," *Michigan Academician* 10 (1977), 127-141.

Ferwerda, Rein. *La signification des images et des métaphores dans la pensée de Plotin*. Groningen: J.B. Wolters, 1965.

Finamore, John J. "Iamblichus on Light and the Transparent." In *The Divine Iamblichus: Philosopher and Man of the Gods*. Ed. H.J Blumenthal and E.G. Clark. London: Bristol Classical Press, 1993, pp. 55-64.

Foster, Kenelm. The following entries from the *ED*: "Summa Contra Gentiles," (V, pp. 479-480); "Tommaso d'Aquino," (V, pp. 626-649).

—.Review of Joseph Anthony Mazzeo's *Medieval Cultural Tradition*. In *Modern Language Notes* 76 (1961), 941-943.

—.*The Two Dantes and Other Studies*. London: Darton, Longman & Todd, 1977.

Frappier, Jean. *Histoire, mythes et symboles: Étude de littérature française*. Geneva: Droz, 1976.

Frugoni, Arsenio. "Gioachino da Fiore." In *ED* III, pp. 165-167.

Führer, M.L. "The Contemplative Intellect in the Psychology of Albert the Great." In *Historia philosophiae Medii Aevi: Studien zur Geschichte der Philosophie des Mittelalters*. Ed. Burkhard Mojsisch and Olaf Pluta. Amsterdam and Philadelphia: Grüner, 1991, pp. 305-319.

—."The Theory of Intellection in Albert the Great and its Influence on Nicholas of Cusa." In *Nicholas of Cusa in Search of God and Wisdom: Essays in Honor of Morimichi Watanabe*. Ed. Gerald Christianson and Thomas M. Izbicki. New York and Leiden: E.J. Brill, 1991, pp. 45-56.

Fujitani, Michio. "Dalla legge ottica alla poesia: la metamorfosi di «Purgatorio» XV, 1-27," *Studi danteschi* 61 (1989), 153-186.

Gage, John. *Colour and Culture: Practice and Meaning from Antiquity to Abstraction*. London: Thames and Hudson, 1993.

—."Gothic Glass: Two Aspects of the Dionysian Aesthetic," *Art History* 5 (1982), 36-58.

Gagné, Joan. "Du 'Quadrivium' aux 'scientiae mediae'." In *Arts libéraux et philosophie au moyen âge*. Paris and Montreal: Vrin, 1969, pp. 975-986.

Gamba, Ulderico. "'Il lume di quel cero ...': Dionigi Areopagita fù l'ispiratore di Dante?," *Studia Patavina* 32 (1985), 101-114.

Gardner, Edmund G. *Dante and the Mystics: A Study of the Mystical Aspects of the Divine Comedy and its Relation with Some of its Mediaeval Sources*. London: Dent, 1913.

Gay, John H. "Four Medieval Views of Creation," *Harvard Theological Review* 66 (1963), 243-273.

Gersh, Stephen. *From Iamblichus to Eriugena: An Investigation of the Prehistory and Evolution of the Pseudo-Dionysian Tradition.* Leiden: E.J. Brill, 1978.

Ghisalberti, Alessandro. "La cosmologia del Duecento e Dante," *Letture classensi* 13 (1984), 33-48.

Giacon, Carlo. "Avicenna." In *ED* I, pp. 481-482.

Giannantonio, Pompeo. "Struttura e allegoria nel *Paradiso*," *Letture classensi* 11 (1982), 63-80.

Gilson, Etienne. "À la recherche de l'Empyrée," *Revue des études italiennes* 10 (1965), 147-161; reprinted in *Dante et Béatrice: Études dantesques.* Paris: Vrin, 1974, pp. 67-77.

—."Notes pour l'histoire de la cause efficiente," *AHDLMA* 29 (1962), 7-31.

—.Review of Bruno Nardi's *Dal "Convivio" alla "Commedia."* In *GSLI* 138 (1961), 562-573.

—."Sur la problématique thomiste de la vision béatifique," *AHDLMA* 31 (1964), 67-88.

Gilson, Simon A. "Light Reflection, Mirror Metaphors, and Optical Framing in Dante's *Comedy*: Precedents and Transformations," *Neophilologus* 83 (1999), 241-252.

Gizzi, Corrado. *L'astronomia nel poema sacro.* 2 vols. Naples: Loffredo, 1974.

Grabes, Herbert. *The Mutable Glass: Mirror-Imagery in Titles and Texts of the Middle Ages and the English Renaissance.* Trans. Gordon Collier. Cambridge: Cambridge University Press, 1982.

Grant, Edward. "Cosmology." In *Science in the Middle Ages.* Ed. David C. Lindberg. Chicago and London: Chicago University Press, 1976, pp. 265-302.

—."Medieval and Renaissance Scholastic Conceptions of the Influence of the Celestial Region on the Terrestrial," *Journal of Medieval and Renaissance Studies* 17 (1987), 1-23.

—.*Planets, Stars and Orbs: The Medieval Cosmos, 1200-1687.* Cambridge: Cambridge University Press, 1994.

Gregory, Tullio. *Anima Mundi: La filosofia di Guglielmo di Conches e la scuola di Chartres.* Florence: Sansoni, 1955.

—."Intenzione." In *ED* III, pp. 480-482.

—."Note sulla dottrina delle «teofanie» in Giovanni Scotto Eriugena," *Studi medievali* ser. 3/4 (1963), 75-91.

—."The Platonic Inheritance." In *A History of Twelfth-Century Western Philosophy.* Ed. Dronke, pp. 54-80.

—.*Platonismo medievale: Studi e ricerche.* Rome: Istituto storico italiano per il medio evo, 1958.

Guardini, Romano. *Studi su Dante.* 2nd edn. Brescia: Morcelliana, 1979.

Guglielminetti, Marziano. "*Paradiso* 13." In *L'arte dell'interpretare: Studi critici offerti a Giovanni Getto.* Cuneo: L'Arciere, 1984, pp. 67-95.

Guidubaldi, Egidio. The following entries from the *ED*:
"Bartolomeo da Bologna," (I, pp. 526-527);"Roberto Grossatesta," (V, pp. 1005-1006).

—."Il canto II del *Paradiso*." In *Nuove Letture Dantesche*. 6 vols. Ed. Enzo Esposito. Florence: Le Monnier, 1972, V, pp. 285-299.

—.Ed. *Dal "De Luce" di R. Grossatesta all'islamico "Libro della Scala": Il problema delle fonti una volta accettata la mediazione oxfordiana*. Florence: Olschki, 1978.

—.*Dante Europeo*. 3 vols. Florence: Olschki, 1965-68.

Gyekye, Kwame. "The Terms 'Prima Intentio' and 'Secunda Intentio' in Arabic Logic," *Speculum* 46 (1971), 32-38.

Hackett, Jeremiah M.G. "The Attitude of Roger Bacon to the *Scientia* of Albertus Magnus." In *Albertus Magnus and the Sciences*. Ed. James A. Weisheipl, pp. 53-72.

Hamesse, Jacqueline. *Les "Auctoritates Aristotelis": Un florilège médiéval*. Louvain and Paris: Peeters and Vrin, 1974.

Harwood, Sharon. "Moral Blindness and Freedom of Will: A Study of Light Images in the *Divine Comedy*," *Romance Notes* 16 (1974), 205-221.

—.*A Study of the Theology and the Imagery of Dante's Divine Comedy*. Lewiston, Queenston, and Lampeter: Edwin Mellen Press, 1991.

Hedwig, Klaus. *Sphaera Lucis: Studien zur Intelligibilität der Seienden im Kontext der mittelaterlichen Lichtspekulation*. In *Beiträge* n.s. 18 (1980).

Hills, Paul. *The Light of Early Italian Painting*. New Haven and London: Yale University Press, 1987.

Hirsch-Reich, Barbara and Reeves, Margaret. *The "Figurae" of Joachim of Flora*. Oxford: Clarendon Press, 1972.

Hollander, Robert. "The Invocations of the *Commedia*." In *Studies in Dante*. Ravenna: Longo, 1980, pp. 31-38.

—.*"Vita Nuova*: Dante's Perceptions of Beatrice," *Dante Studies* 92 (1974), 1-18; reprinted in *Studies in Dante*. Ravenna: Longo, 1980, pp. 11-30.

Holley, Linda Tarte. *Chaucer's Measuring Eye*. Houston, Texas: Rice University Press, 1990.

Hugedé, Norbert. *La métaphore du miroir dans les Épîtres de saint Paul aux Corinthiens*. Neuchâtel: Delachaux and Niestlé, 1957.

Hyman, Arthur. "Maimonides on Creation and Emanation." In *Studies in Medieval Philosophy*. Ed. John F. Wippel. Washington, DC: Catholic University of America Press, 1987, pp. 45-61.

Jeauneau, Edouard. "Macrobe, source du platonisme chartrain," *Studi medievali* ser. 3/1 (1960), 3-24.

—."The Neoplatonic Themes of 'Processio' and 'Reditus' in Eriugena," *Dionysius* 15 (1991), 3-29.

Kay, Richard. "Astrology and Astronomy." In *The Divine Comedy and the Encyclopaedia of the Arts and Sciences*. Ed. di Scipio and Scaglione, pp. 147-162.

—.*Dante's Christian Astrology*. Philadelphia: University of Pennsylvania Press, 1994.

Kemp, Martin. "In the Light of Dante: Meditations on Natural and Divine Light in Piero della Francesca, Raphael and Michelangelo." In *Sonderdruck aus Ars naturam adiuvans: Festschrift für Matthias Winner*. Mainz am Rhein: Philipp von Zabern, 1996, pp. 160-177.

—.*The Science of Art: Optical Themes in Western Art from Brunelleschi to Seurat*. New Haven and London: Yale University Press, 1990.

Klassen, Norman. *Chaucer on Love, Knowledge and Sight*. Cambridge: D.S. Brewer, 1995.

Kleinhenz, Christopher, "Dante and the Tradition of Visual Arts in the Middle Ages," *Thought* 65 (1990), 17-26.

Klibansky, Raymund. *The Continuity of the Platonic Tradition during the Middle Ages*. London: Warburg Institute, 1939.

Knowles, David. *The Evolution of Medieval Thought*. London: Longman, 1962.

Knudsen, Christian. "Intentions and Impositions." In *The Cambridge History of Later Medieval Philosophy*. Ed. Kretzmann, pp. 479-495.

Kogan, Bary S. "Averroes and the Theory of Emanation," *Mediaeval Studies* 43 (1981), 384-404.

Kraye, Jill, with W.F. Ryan, and Charles B. Schmitt. Ed. *Pseudo-Aristotle in the Middle Ages: The Theology and Other Texts*. London: Warburg Institute, 1986.

Kretzmann, Norman, Kenny, Anthony and Pinborg, Jan. Ed. *The Cambridge History of Later Medieval Philosophy: From the Rediscovery of Aristotle to the Disintegration of Scholasticism, 1100-1600*. Cambridge: Cambridge University Press, 1982.

Kucharski, Paul. "Sur la théorie des couleurs et des saveurs dans le "De sensu" aristotélicien," *Revue des études grecques* 67 (1954), 355-390.

Laird, W.R. "Robert Grosseteste on the Subalternate Sciences," *Traditio* 43 (1987), 147-169.

Lansing, Richard H. *From Image to Idea: A Study of Simile in Dante's Comedy*. Ravenna: Longo, 1978.

Leclerq, Jean. "Influence and Non-Influence of Dionysius in the Western Middle Ages." In *Pseudo-Dionysius: The Complete Works*. Trans. Colm Luibheid, notes by Paul Rorem. London and New York: Paulist Press, 1987, pp. 25-32.

Lee, Desmond. "Science, Philosophy and Technology in the Greco-Roman World," *Greece and Rome* ser. 2/19-20 (1972-73), 65-78, 180-193.

Lee, Jonathon Scott. "The Doctrine of Reception According to the Capacity of the Recipient in *Ennead* VI. 4-5," *Dionysius* 3 (1979), 79-99.

Lee, Patrick. "Aquinas and Avicenna on the Active Intellect," *The Thomist* 45 (1981), 41-53.

Leff, Gordon. *Medieval Thought from Saint Augustine to Ockham*. London: Penguin, 1958.

Leisegang, Hans. "La connaissance de Dieu au miroir de l'âme ou de la nature," *Revue d'histoire et de philosophie religieuses* 17 (1937), 145-171.

Lejeune, Albert. *Euclide et Ptolémée: Deux stades de l'optique géométrique grecque*. Louvain: Bibliothèque de l'Université de Louvain, 1948.

Lemay, Richard. *Abu Ma'shar and Latin Aristotelianism in the Twelfth Century: The Recovery of Natural Philosophy through Arabic Astrology*. Beirut: American University of Beirut, 1962.

Lepschy, Antonio. "Osservazioni sul vocabolario cromatico della *Commedia* di Dante," *Atti dell'istituto veneto di scienze ed arti: Classe di scienze fisiche, matematiche e naturali* 152 (1993-94), 1-14.

Lewis, C.S. *The Discarded Image: An Introduction to Medieval and Renaissance Literature*. 2nd edn. Cambridge: Cambridge University Press, 1967.

Leyerle, John. "The Rose-Wheel Design and Dante's *Paradiso*," *University of Toronto Quarterly* 46 (1977), 280-308.

Libera, Alain de. "Albert le Grand et Thomas d'Aquin, interprètes du *Liber de causis*," *Revue des sciences philosophiques et théologiques* 74 (1990), 347-378.

—."Le sens commun au XIIIe siècle: De Jean de la Rochelle à Albert le Grand," *Revue de métaphysique et de morale* 96 (1991), 476-496.

Lindberg, David C. "Alhazen's Theory of Vision and Its Reception in the West," *Isis* 58 (1967), 321-341.

—."Alkindi's Critique of Euclid's Theory of Vision," *Isis* 62 (1971), 469-489.

—.*The Beginnings of Western Science: The European Scientific Tradition in Philosophical, Religious, and Institutional Contexts, 600 B.C. to A.D. 1450*. Chicago and London: Chicago University Press, 1992.

—.*A Catalogue of Medieval and Renaissance Optical Manuscripts*. Toronto: Pontifical Institute of Mediaeval Studies, 1975.

—."The Cause of Refraction in Medieval Optics," *British Journal for the History of Science* 4 (1968-69), 23-38.

—."The Genesis of Kepler's Theory of Light: Light Metaphysics from Plotinus to Kepler," *Osiris* n.s. 2 (1986), 5-42.

—."The Intromission-Extramission Controversy in Islamic Visual Theory: Alkindi versus Avicenna." In *Studies in Perception: Interrelations in the History and Philosophy of Science*. Ed. Machamer and Turnbull, pp. 137-159.

—."Lines of Influence in Thirteenth-Century Optics: Bacon, Witelo, and Pecham," *Speculum* 46 (1971), 66-83.

—."The Science of Optics." In *Science in the Middle Ages*. Ed. David C. Lindberg. Chicago and London: Chicago University Press, 1978, pp. 338-368.

—.*Studies in the History of Medieval Optics*. London: Variorum, 1983.

—.*Theories of Vision from al-Kindi to Kepler*. Chicago and London: Chicago University Press, 1976.

—."The Western Reception of Arabic Optics." In *Encyclopaedia of the History of Arabic Science*. 3 vols. Ed. Roshdi Rashed. London and New York: Routledge, 1996, II, pp. 716-729.

Litt, Thomas. *Les corps célestes dans l'univers de St. Thomas d'Aquin*. Louvain and Paris: Peeters and Vrin, 1963.

Lloyd, A.C. "The Principle that the Cause is Greater than the Effect," *Phronesis* 21 (1976), 146-156.

Lovejoy, Arthur. *The Great Chain of Being*. Cambridge, MA: Harvard University Press, 1933.

Machamer, Peter K. and Turnbull, Robert G. Ed. *Studies in Perception: Interrelations in the History and Philosophy of Science*. Colombus, OH: Ohio State University Press, 1978.

Maggini, Francesco. Review of Contini's *Poeti del Duecento*. In *GSLI* 116 (1940), 40-45.

Malgarini, Patrizia Bertini. "Il linguaggio medico e anatomico nelle opere di Dante," *Studi danteschi* 61 (1989), 29-108.

Mamura, Michael E. "The Metaphysics of Efficient Causality in Avicenna (Ibn Sina)." In *Islamic Theology and Philosophy: Studies in Honor of George F. Hournai*. Ed. Michael E. Mamura. Albany: State University of New York Press, 1984, pp. 172-187.

Marrone, Steven P. *William of Auvergne and Robert Grosseteste: New Ideas of Truth in the Early Thirteenth Century*. Princeton, NJ: Princeton University Press, 1983.

Marshall, Peter. "Nicholas Oresme on the Nature, Reflection and Speed of Light," *Isis* 72 (1981), 357-374.

Martinelli, Bortolo. "La dottrina dell'Empireo nell'Epistola a Cangrande (capp. 24-27)," *Studi danteschi* 57 (1985), 49-143.

—."*Esse* ed *essentia* nell'Epistola a Cangrande (capp. 19-23)," *Critica letteraria* 12 (1984), 627-672.

—."Poesia e scienza in Dante," *Critica letteraria* 9 (1981), 623-667.

Mathieu, Dominique. "Lumière: Étude biblique." In *DS* 9 (1976), pp. 1142-1149.

Mattioli, Mario. *Dante e la medicina*. Naples: Edizione scientifiche italiane, 1965.

Mazzeo, Joseph Anthony. "Light, Love and Beauty in the *Paradiso*," *Romance Philology* 11 (1957-58), 1-18.

—."Light Metaphysics, Dante's *Convivio* and the Letter to Cangrande della Scala," *Traditio* 14 (1958), 191-229.

—*Medieval Cultural Tradition in Dante's Comedy*. Ithaca: Cornell University Press, 1960.

—.*Structure and Thought in the Paradiso*. Ithaca: Cornell University Press, 1958.

Mazzoni, Francesco. "Il Canto XXIX del *Purgatorio*." In *Lectura Dantis Scaligera*. Ed. Giovanni Getto. Florence: Le Monnier, 1965, pp. 5-102.

McEvoy, James. "La connaissance intellectuelle selon Robert Grosseteste," *Revue de philosophie de Louvain* 75 (1977), 5-48.

—."Metaphors of Light and Metaphysics of Light in Eriugena." In *Begriff und Metapher: Sprachform des Denkens bei Eriugena*. Ed. Werner Beierwaltes. Heidelberg: Carl Winter, 1990, pp. 149-167.

—."The Metaphysics of Light in the Middle Ages," *Philosophical Studies* 26 (1979), 124-140.

—."Microcosm and Macrocosm in the Writings of St. Bonaventure." In *S. Bonaventura 1274-1974*. 5 vols. Rome: Collegium S. Bonaventura, 1973, II, pp. 309-343.

—.*The Philosophy of Robert Grosseteste*. Oxford: Clarendon Press, 1982.

—."The Sun as *res* and *signum*: Robert Grosseteste's Commentary on *Ecclesiasticus*, ch. 43, vv. 1-5," *RThAM* 41 (1974), 38-92.

McKeon, C.K. *A Study of the "Summa philosophiae" of the Pseudo-Grosseteste*. New York: Columbia University Press, 1948.

McKirahan, R.D. "Aristotle's Subordinate Sciences," *British Journal for the History of Science* 11 (1978), 197-220.

McNair, Philip. "Dante's Vision of God: An Exposition of *Paradiso* XXXIII." In *Essays in Honour of John Humphreys Whitfield*. Ed. H.G. Davis and others. London: St. George's Press, 1975, pp. 13-29.

Mellone, Attilio. The following entries from the *ED*: "Empireo," (II, pp. 668-671); "Luce," (III, pp. 706-713).

—."Il concorso delle creature nella produzione delle cose secondo Dante," *Divus thomas* 56 (1953), 273-286.

—."Emanatismo neoplatonico di Dante per le citazioni del 'Liber de causis'?" *Divus thomas* 54 (1951), 205-212.

Michaud-Quantin, Pierre. "Albert le Grand et les puissances de l'âme," *Revue du moyen âge latin* 11 (1955), 59-86.

—."La classification des puissances de l'âme au douzième siècle," *Revue du moyen âge latin* 5 (1949), 15-34.

—.*Études sur le vocabulaire philosophique du moyen âge*. Rome: L'Ateneo, 1970.

—."Les petites encyclopédies du XIIIe siècle," *Cahiers d'histoire mondiale* 9 (1966), 580-596.

Mineo, Nicolò. *Profetismo e apocalittica in Dante: Strutture e temi profetico-apocalittici in Dante: dalla Vita Nuova alla Divina Commedia*. Catania: Università di Catania, 1968.

Minio-Paluello, L. "Dante's Reading of Aristotle." In *The World of Dante*. Ed. Cecil Grayson. Oxford: Clarendon Press, 1980, pp. 61-79.

Montgomery, Robert L. *The Reader's Eye: Studies in Didactic Literary Theory from Dante to Tasso*. Berkeley, Los Angeles, and London: University of California Press, 1979.

Moran, Dermot. *The Philosophy of John Scottus Eriugena: A Study of Idealism in the Middle Ages*. Cambridge: Cambridge University Press, 1989.

Morgan, Alison. *Dante and the Medieval Otherworld*. Cambridge: Cambridge University Press, 1990.

Mullahy. B.I. "Liturgical Use of Light." In *New Catholic Encyclopaedia* 8 (1967), 751-754.

Muscatine, Charles. "Locus of Action in Medieval Narrative," *Romance Philology* 17 (1963), 115-122.

Nardi, Bruno. "Alla illustrazione del 'Convivio' dantesco. A proposito dell'edizione di Giorgio Rossi," *GSLI* 95 (1930), 73-114.

—.*Dal "Convivio" alla "Commedia" (Sei saggi danteschi)*. Rome: Istituto storico italiano per il Medio Evo, 1960.

—.*Dante e la cultura medievale*. 2nd edn. Ed. Paolo Mazzantini. Rome and Bari: Laterza, 1983.

—.*«Lecturae» e altri studi danteschi*. Ed. Rudi Abardo. Florence: Le Lettere, 1990.

—."L'origine dell'anima umana secondo Dante." In his *Studi sulla filosofia medievale*. Rome: Edizioni di storia e letteratura, 1960, pp. 9-68.

—.*Il punto sull'Epistola a Cangrande*. Florence: Le Monnier, 1960.

—.*Saggi di filosofia dantesca*. 2nd edn. Ed. Paolo Mazzantini. Florence: La Nuova Italia, 1967.

—.*Saggi e note di critica dantesca*. Milan and Naples: Ricciardi, 1966.

—.*Sigieri di Brabante nella Divina Commedia di Dante Alighieri e le fonti della filosofia di Dante*. Spianate, 1912.

Negri, Luigi. "La luce nella filosofia naturale del '300 e nella «Commedia»," *GSLI* 82 (1923), 328-336.

Newman, F.X. "St. Augustine's Three Visions and the Structure of the *Commedia*," *Modern Language Notes* 82 (1967), 56-78.

Niccoli, Alessandro. "Stimativa." In *ED* V, p. 445.

North, J.D. "Celestial Influence – the Major Premiss of Astrology." In *"Astrologi hallucinati": Stars and the End of the World in Luther's Time*. Ed. Paola Zambelli. Berlin and New York: Walter de Gruyter, 1986, pp. 45-100.

—."Medieval Concepts of Celestial Influence: A Survey." In *Astrology, Science, and Society: Historical Essays*. Ed. Patrick Curry. Woodbridge, Suffolk: Boydell, 1987, pp. 5-17.

Oliva, Gianni. *Per altre dimore: Forme di rappresentazione e sensibilità medievale in Dante*. Rome: Bulzoni, 1991.

Oliviero, Adriana. "La composizione dei cieli in Restoro d'Arezzo e in Dante." In *Dante e la scienza*. Ed. Boyde and Russo, pp. 351-362.

Olschki, Leonardo. "Sacra dottrina e theologia mystica: Il Canto XXX del *Paradiso*," *Giornale dantesco* n.s. 6 (1933), 1-25.

O'Meara, Dominic J. "Eriugena and Aquinas on Beatific Vision." In *Eriugena Redivivus: Zur Wirkungsgeschichte seines Denkens im Mittelalter und im Ubergung zur Neuzeit*. Ed. Werner Beierwaltes. Heidelberg: Carl Winter, 1987, pp. 224-236.

Orestano, F. "Discontinuità dottrinali nella *Divina Commedia*," *Sophia* 1 (1933), 3-17.

Ostendler, Heinrich. "Dante und Hildegard von Bingen," *Deutsches Dante Jahrbuch* 27 (1948), 158-170.

—."Dantes Mystik," *Deutsches Dante Jahrbuch* 28 (1949), 65-98.

Pagliaro, Antonino. "Similitudine." In *ED* V, pp. 253-259.

—.*Ulisse: Ricerche semantiche sopra la "Divina Commedia."* 2 vols. Milan: D'Anna, 1967.

Palgen, Rudolf. "Gli elementi plotiniani nel «Paradiso»." In *Atti del convegno internazionale sul tema: Plotino e il Neoplatonismo in Oriente e in Occidente* (Rome, 5-9 Oct, 1970). Rome: Accademia nazionale dei Lincei, 1974, pp. 509-524.

—."Il Paradiso platonico," In *Letteratura e critica: Studi in onore di Natalino Sapegno*. 5 vols. Florence: Bulzoni, 1974, I, pp. 197-211.

—."Scoto Eriugena, Bonaventura e Dante," *Convivium* 25 (1957), 1-8.

Pannaria, Francesco. "Dante e la scienza," *Nuova antologia* 519 (1973), 247-261.

Panofsky, Erwin. *Abbot Suger. On the Abbey Church of Saint Denis and its Art Treasures*. 2nd edn. Princeton, NJ: Princeton University Press, 1973.

Parent, J.M. *La doctrine de la création dans l'École de Chartres*. Paris and Ottawa: Vrin, 1938.

Parronchi, Alessandro. "Perspettiva." In *ED* IV, pp. 438-439.

—."La perspettiva dantesca," *Studi danteschi* 36 (1959), 5-103; reprinted in *Studi su la dolce prospettiva*. Milan: Martello, 1964, pp. 3-90.

Pasquini, Emilio. "Specchio." In *ED* V, pp. 366-367.

Pelikan, Jaroslav. *The Light of the World: A Basic Image in Early Christian Thought*. New York: Harper, 1962.

Pertile, Lino. *"Paradiso* XXXIII: 'L'estremo oltraggio'," *Filologia e critica* 6 (1981), 1-21.

Pertusi, Agostino. "Cultura greco-bizantina nel tardo Medioevo nelle Venezie e suoi echi in Dante." In *Dante e la cultura veneta*. Florence: Olschki, 1966, pp. 157-197.

Peterson, J. "Aristotle's Incomplete Causal Theory," *The Thomist* 36 (1972), 420-432.

Peterson, Mark. "Dante's Physics." In *The Divine Comedy and the Encyclopedia of the Arts and Sciences*. Ed. di Scipio and Scaglione, pp. 163-180.

Pézard, André. *Dans le sillage de Dante*. Paris: Société d'études italiennes, 1975.

—.*"La Rotta Gonna": Glosses et corrections aux textes mineurs de Dante*. Florence: Sansoni, 1967.

Philippe, M.-D. "Phantasia in the Philosophy of Aristotle," *The Thomist* 35 (1971), 1-42.

Piana, C. "Le Questioni inedite 'De glorificatione Beatae Mariae Virginis' di Bartolomeo di Bologna, O.F.M., e le concezioni del *Paradiso* dantesco," *L'Archiginnasio* 33 (1938), 247-262.

Plumptre, E.H. "Two Studies in Dante I: Dante and Roger Bacon," *The Contemporary Review* 40 (December, 1881), 843-864.

Portelli, J. "The 'Myth' that Avicenna Reproduced Aristotle's Concept of the 'Imagination' in *De anima*," *Scripta mediterranea* 3 (1982), 122-134.

Pouillon, Henri. "La beauté, propriété transcendentale chez les scolastiques (1220-1270)," *AHDLMA* 15 (1946), 263-328.

Quillet, Jeannine. "'Soleil' et 'lune' chez Dante." In *Le soleil, la lune et les étoiles au moyen âge*. Aix-en-Provence: Publications du CUERMA, 1983, pp. 329-337.

Quinn, John Francis. *The Historical Constitution of St. Bonaventure's Philosophy*. Toronto: Pontifical Institute of Mediaeval Studies, 1973.

Rabuse, Georg. "Macrobio." In *ED* III, pp. 757-759.

Radcliff-Umstead, Daniel. "Dante on Light," *Italian Quarterly* 9 (1965), 30-45.

Ricchi, Gino. "Il meccanismo della visione secondo Dante Alighieri," *Giornale dantesco* 10 (1902), 177-179.

Ronchi, Vasco. "De luce et lumine," *Physis* 8 (1966), 5-22.

—.*The Nature of Light: An Historical Survey*. Trans. V. Barocas, London: Heinemann, 1970.

Rorem, Paul. *Biblical and Liturgical Symbols within the Pseudo-Dionysian Synthesis*. Toronto: Pontifical Institute of Mediaeval Studies, 1984.

—.*Pseudo-Dionysius: A Commentary on the Texts and an Introduction to Their Influence*. Oxford and New York: Oxford University Press, 1993.

Rovighi, Vanni, Sofia. "Dionigi." In *ED*, II, pp. 460-462.

Ruello, Francis. "Le commentaire inédit de saint Albert le Grand sur les Noms Divins. Présentation et aperçus de théologie trinitaire," *Traditio* 12 (1956), 231-314.

—."La *Divinorum nominum reseratio* selon Robert Grosseteste et Albert le Grand," *AHDLMA* 26 (1959), 99-197.

Rutledge, Monica. "Dante, the Body and Light," *Dante Studies* 113 (1995), 151-165.

Sabra, A.I. *Optics, Astronomy and Logic: Studies in Arabic Science and Philosophy*. Aldershot: Variorum, 1994.

—."Sensation and Inference in Alhazen's Theory of Visual Perception." In *Studies in Perception*. Ed. Machamer and Turnbull, pp. 160-185.

—.*Theories of Light: from Descartes to Newton*. New edn. Cambridge: Cambridge University Press, 1981.

Saccaro, Giuseppe Battisti. "Il Grossatesta e la luce," *Medioevo* 2 (1976), 21-75.

Saffrey, H.D. "L'état actuel des recherches sur le *Liber de causis* comme source de la métaphysique au moyen âge," *Miscellanea Mediaevalia* 2 (1963), 267-281.

Sage, Athanase. "La dialectique de l'illumination," *Recherches augustiniennes* 2 (1962), 111-123.

Sambursky, S. "Philoponus' Interpretation of Aristotle's Theory of Light," *Osiris* 13 (1958), 114-126.

Santoro, Luigi. "Dante's *Paradiso* (Canto I) and the Aesthetics of Light." In *Dante Readings*. Ed. Eric Haywood. Dublin: Irish Academic Press, 1987, pp. 107-122.

Sayili, Aydin M. "The Aristotelian Explanantion of the Rainbow," *Isis* 30 (1939), 65-83.

Scazzoso, Piero. "Contemplazione naturale e contemplazione soprannaturale confrontate attraverso Plotino e lo pseudo-Dionigi." In *Lectura Dantis mystica: Il poema sacro alla luce delle conquiste odierne*. Florence: Olschki, 1969, pp. 56-84.

Schlanger, Jacques. *La philosophie de Salomon Ibn Gabirol: Étude d'un néoplatonisme*. Leiden: E.J. Brill, 1968.

Schmidt, Margot. "Miroir." In *DS* 10 (1974), 1290-1303

—."Lumière: Au moyen âge." In *DS* 9 (1973), 1158-1173.

Schweig, Bruno. "Mirrors," *Antiquity* 15 (1941), 257-268.

Scrivano, Riccardo. "Poesia e dottrina nel XXX canto del «Paradiso»," *Critica letteraria* 17 (1989), 3-16.

Sebastio, Leonardo. "*Paradiso* II." In *Strutture narrative e dinamiche culturali in Dante e nel Fiore*. Florence: Olschki, 1990, pp. 55-95.

Shoaf, R.A. *Dante, Chaucer, and the Currency of the Word*. Norman, OK: Pilgrim Books, 1983.

Siegel, Rudolph E. *Galen on Sense Perception*. Basel and New York: S. Krager, 1970.

Simonelli, Maria. "Allegoria e simbolo dal 'Convivio' alla 'Commedia' sullo sfondo della cultura bolognese." In *Dante e Bologna nei tempi di Dante*. Bologna: Facoltà di Lettere e Filosofia dell'Università di Bologna, 1967, pp. 207-226.

—."Convivio." In *ED* II, pp. 200-203.

Simson, Otto, von. *The Gothic Cathedral: Origins of Gothic Architecture and the Medieval Concept of Order*. 2nd edn. Princeton, NJ: Princeton University Press, 1962.

Singleton, Charles S. *Dante Studies 2: Journey to Beatrice*. Cambridge, MA: Harvard University Press, 1957.

Siraisi, Nancy G. "Dante and the Art and Science of Medicine Reconsidered." In *The Divine Comedy and the Encyclopaedia*. Ed. di Scipio and Scaglione, pp. 223-245.

—.*Taddeo Alderotti and His Pupils*. Princeton, NJ: Princeton University Press, 1981.

Skinner, Quentin. "Meaning and Understanding in the History of Ideas," *History and Theory* 8 (1969), 3-53.

Smith, A. Mark. "Getting the Big Picture in Perspectivist Optics," *Isis* 72 (1981), 568-589.

—.Introduction. In *Witelonis Perspectivae liber quintus*. Wroclaw: Ossolineum, 1983, pp. 14-72.

—."The Psychology of Visual Perception in Ptolemy's *Optics*," *Isis* 79 (1988), 189-207.

Sorabji, Richard. *Time, Creation and the Continuum: Theories in Antiquity and the Early Middle Ages*. London: Duckworth, 1983.

Southern, Richard W. *Robert Grosseteste: The Growth of an English Mind in Medieval Europe*. Oxford: Clarendon Press, 1986.

Spedicati, Adele Anna. "Sulla teoria della luce in F. Patrizi," *Bollettino di storia di filosofia dell'università degli studi di Lecce* 5 (1977), 243-263.

Spitzer, Leo. "Geistesgeschichte vs. History of Ideas as Applied to Hitlerism," *Journal of the History of Ideas* 5 (1944), 191-203.

Stabile, Giorgio. "Navigazione celeste e simbolismo lunare in «Paradiso» II," *Studi medievali* ser. 3a/21 (1980), 97-140.

Stansbury, Sarah. *Seeing the Gawain-poet: Description and the Act of Perception*, Philadelphia: University of Pennyslvania Press, 1991.

Steneck, Nicholas H. "Albert on the Psychology of Sense Perception." In *Albertus Magnus and the Sciences*. Ed. Weisheipl, pp. 263-290.

—.*Science and Creation in the Middle Ages: Henry of Langenstein (d. 1397) on Genesis*. London and Notre Dame: Notre Dame University Press, 1976.

Stormon, E.J. "The Problems of the Empyrean Heaven in Dante," *Spunti e ricerche: rivista d'italianistica* 3 (1987), 23-33.

Summers, David. *The Judgment of Sense: Renaissance Naturalism and the Rise of Aesthetics*. Cambridge: Cambridge University Press, 1987.

Sweeney, Theodore. "The Doctrine of Creation in *Liber de causis*." In *An Etienne Gilson Tribute*. Ed. C.J. O' Neil. Milwaukee: Marquette University Press, 1959.

—."'Esse Primum Creatum' in Albert the Great's *Liber de Causis et Processu Universitatis*," *The Thomist* 44 (1980), 599-646.

Tachau, Katharine H. *Vision and Certitude in the Age of Ockham: Optics, Epistemology, and the Foundations of Semantics, 1250-1345*. Leiden and New York: E.J. Brill, 1988.

Thonnard, François-Joseph. "La notion de la lumière en philosophie augustinienne," *Recherches augustiniennes* 2 (1962), 125-175.

Tondelli, Luigi with Hirsch-Reich, Barbara and Reeves, Margaret. Ed. *Il "Libro delle Figure" dell'Abate Gioachino da Fiore*. 2nd edn. Turin: Società editrice internazionale, 1961.

Took, J.F. *Dante: Lyric Poet and Philosopher: An Introduction to the Minor Works*. Oxford: Clarendon Press, 1990.

—. *"L'etterno piacer": Aesthetic Ideas in Dante*. Oxford: Clarendon Press, 1984.

Travi, Ernesto. *Dal cerchio al centro: Studi danteschi*. Milan: Vita e pensiero, 1990.

Varese, Claudio. "Parola e immagine figurativa nei canti del Paradiso Terrestre," *Rassegna della letteratura italiana* 94 (1990), 30-42.

Vasoli, Cesare. "Il canto II del *Paradiso*," *Lectura Dantis Metelliana* 2 (1992), 27-51.

—."Il *Convivio* di Dante e l'enciclopedismo medievale." In *L'enciclopedismo medievale*. Ed. Picone, pp. 363-381.

—."Dante e l'immagine enciclopedica del mondo nel *Convivio*," *Studi sulle imago mundi: Centro di studi sulla spiritualità medievale* 22 (1983), 37-73.

—."Fonti albertine nel *Convivio* di Dante." In *Albertus Magnus und der Albertismus: Deutsche philosophische Kultur des Mittelalters*. Ed. J.F.M. Maarten and Alain de Libera. Leiden and New York: E.J. Brill, 1995, pp. 33-49.

—.Introduction. In *Il Convivio*. Milan and Naples: Ricciardi, 1988, pp. xi-lxxxix.

Vescovini, Graziella Federici. "La *perspectiva* nell'enciclopedia del sapere medievale," *Vivarium* 6 (1968), 35-45.

—.*Studi sulla prospettiva medievale*. Turin: Giappichelli, 1965.

Volpini, Enzo. "Galeno, Claudio." In *ED* III, pp. 85-86.

Wallace, William A. *Causality and Scientific Explanation*. 2 vols. Ann Arbor: University of Michigan Press, 1972.

Wallis, R.T. *Neoplatonism*. London: Duckworth, 1972.

Weisheipl, James A. Ed. *Albertus Magnus and the Sciences: Commemorative Essays, 1980*. Toronto: Pontifical Institute of Mediaeval Studies, 1980.

—."Classification of the Sciences in Medieval Thought," *Mediaeval Studies* 27 (1965), 54-90.

Wetherbee, Winthrop. "Philosophy, Cosmology, and the Twelfth-Century Renaissance." In *A History of Twelfth-Century Western Philosophy*. Ed. Dronke, pp. 21-53.

—.*Platonism and Poetry in the Twelfth Century: The Literary Influence of the School of Chartres*. Princeton, NJ: Princeton University Press, 1972.

Wippel, John F. Ed. *Studies in Medieval Philosophy*. Washington, DC: Catholic University Press of America, 1987.

Wlassics, Tibor. *Dante narratore: Saggi sullo stile della Commedia*. Florence: Olschki, 1975.

—."La percezione limitata nella «Commedia»," *Aevum* 47 (1973), 501-508.

Wolfson, Harry Austryn. "The Identification of *Ex Nihilo* with Emanation in Gregory of Nyssa," *Harvard Theological Review* 63 (1970), 53-60; reprinted in *Studies in the History of Philosophy and Religion*. Ed. Isadore Twersky and George H. Williams. Cambridge, MA: Harvard University Press, 1973, I, pp. 199-206.

—."The Internal Senses in Latin, Arabic, and Hebrew Philosophic Texts," *Harvard Theological Review* 28 (1935), 69-133; reprinted in *Studies in the History of Philosophy and Religion*. Ed. Twersky and Williams, vol. I, pp. 250-314.

—."The Meaning of *Ex Nihilo* in the Church Fathers, Arabic and Hebrew Philosophy, and St. Thomas." In *Mediaeval Studies in Honor of J.D.M. Ford* (Cambridge, MA: Harvard University Press, 1948), pp. 355-370; reprinted in *Studies in the History of Philosophy and Religion*. Ed. Twersky and Williams, vol. I, pp. 207-221.

—."The Meaning of *Ex Nihilo* in Isaac Israeli," *The Jewish Quarterly Review* n.s. 50 (1959), 1-12; reprinted in *Studies in the History of Philosophy and Religion*. Ed. Twersky and Williams, vol. I, pp. 222-233.

—."The Platonic, Aristotelian and Stoic Theories of Creation in Hallevi and Maimonides." In *Essays in Honour of the Very Rev. Dr. J.H. Hertz, Chief Rabbi of Great Britain* (London, 1942), pp. 427-442; reprinted in *Studies in the History of Philosophy and Religion*. Ed. Twersky and Williams, I, pp. 234-249.

Zambelli, Paola. "Albert le Grand et l'astrologie," *RThAM* 49 (1982), 141-158.

Index of Longer Quotations from the Works of Dante

COMEDY

INFERNO (*Inf.*)
III (36), p. 77
  (52-63), pp. 77-8
  (65-71), p. 78
VII (73-76), p. 198n
X (69), p. 220n
XXXI (10-15), p. 97
  (19-27), p. 98
  (20, 21, 31, 41, 43, 107, 136), p. 97n
  (34-39), pp. 99-100

PURGATORIO (*Purg.*)
II (37-40), p. 81
VIII (35-36), p. 82
IX (73-84), p. 82
XV (15), p. 80n
  (16-24), p. 110
  (16-17, 22), p. 133n
  (67-75), pp. 119-20
XVII (45, 52-54), p. 83
XVIII (22-23), p. 63n
  (22-27), p. 90
XXI (50-51), p. 138n
XXV (41, 47, 70, 95, 96), p. 140n
  (72), p. 203
  (74-75), pp. 203-4
  (88-96), p. 138
  (91), p. 136n
  (92, 93), p. 136n
XXVII (58-60), p. 83n
  (88-90), p. 94n
XXIX (43-50), p. 100
  (73-78), p. 141
XXX (38), p. 83
XXXI (9), p. 84
  (10-15), p. 84

PARADISO (*Par.*)
I (1-3), p. 244
  (46-54), p. 123
  (58-63), p. 238n
  (75), p. 102
  (79-81), p. 238n
II (32), p. 199
  (34-36), p. 226

  (93), p. 133n
  (94-96), p. 112n
  (97-105), p. 129
  (112-120, 121-123, 127-132, 133-138, 139-144), pp. 201-3
  (145-148), p. 204
III (129), p. 84n
V (7-12), p. 228
VII (1-3), p. 241
  (79-81), p. 229
  (64-72), p. 206
  (141), p. 186
IX (8-9), p. 237n
  (61-62), p. 242n
  (69), p. 237n
X (64-69), pp. 143-4
  (68), p. 136n
  (83-87), p. 234
XII (10-21), p. 144
  (10), p. 136n
  (11-13), p. 136n
XIII (52-60), p. 208
  (58-60), p. 242
XIV (46-51), p. 255
  (52-53), p. 107
  (76-84), p. 85
XV (24), p. 237n
XVII (121-123), p. 238n
XIX (4-5), p. 237n
  (4-6), p. 133n
  (52-54), pp. 229, 231
  (89-90), p. 206n
XXI (11), p. 85
  (83-87), p. 255
XXIII (28-30), p. 237n
XXVI (31-33), p. 229
XXVI (11-12), p. 86
  (20), p. 80n
  (70-79), p. 86
XXVIII (4-12), pp. 125-6
  (22-27), p. 146
  (24), p. 136n
  (31-33), p. 146
XXIX (13-18, 25-30), pp. 206-7
  (25), p. 237
  (136-138, 142-145), p. 242
XXX (38-42), p. 250

(39-40), p. 251
(46-51), pp. 87-8
(58-60), p. 88
(58, 78, 80-81), p. 103
(100-105), p. 254
(118-123), p. 104
(122-123), p. 105
XXXI (22-24), p. 232
(28-30), p. 227
(70-78), p. 104
XXXIII (76-78), p. 89
(115-120), p. 147
(118-119), p. 136n

CONVIVIO (*Con.*)
II, iii, (6), p. 48
II, iii, (8), p. 250n
II, vi, (9), p. 185
II, ix, (4-5), p. 51
II, xiii, (21), p. 135
(27), p. 49
II, xiv, (15), p. 201
III, ii, (4-5), p. 191
III, iii, (13), p. 51n
(9), p. 128
III, ix (6-7), p. 58
(8), pp. 64-5
(9), p. 68
(12), p. 89
(11-13), p. 70
(14), p. 68n
III, xiv (2-4), p. 185
(5), p. 55
IV, viii, (6-7), p. 94n
IV, xii, (17), p. 52
IV, xx, (7-8), p. 194

DE VULGARI ELOQUENTIA (*Dve*)
I, xvi, (2), p. 139n

EPISTOLE (*Ep.*)
XIII (61), p. 192n

MONARCHIA (*Mon.*)
I, viii, (2), p. 196n
I, ix, (1-2), p. 195
II, ii, (2-5), pp. 196-7
III, xv, (15), p. 196n

**QUESTIO DE SITU ET FORMA
AQUAE ET TERRAE** (*Questio*)
(46), p. 196n
(67), p. 197n
(70), p. 197n
(82), p. 51n

RIME (*Rime*)
CX (1-7, 11-15, 16-23, 24-30, 39-45), pp. 43-46

VITA NUOVA (*VN*)
II (5), p. 40
V (1), p. 42
XI (2), p. 41
XIV (5), p. 41

Name and Subject Index

—A—

Abelard, Peter, 179n

Aether: compact structure of in stars, 199; heavenly spheres and bodies composed of, 15, 199; relationship to diaphanous, 15
See also Aristotle

After-image, 80-1n

Ageno, Franca Brambilla, 138n

Agrelo, Sebastiano, 219n

Akdogan, Cemil, 53n

Alain of Lille, 94n, 179, 180n, 205

Albert the Great: commentator on Aristotle, 31-2, 48, 53-5, 60-1n, 64n, 66, 72, 80n, 99n, 114-5, 258; commentator on the Pseudo-Dionysius, 230n, 245; his astrological views, 186-7; influence on Dante, 32, 53-4, 73, 192-3; on celestial influence, 186-7, 192-3; on colour, 167; on internal senses, 93n; on light (analogies), 229-30, (celestial), 121, 186-7, 199n, (intelligible), 236n, (intentional being of), 64, (propagation of), 56n, 64n, (visible), 60-1; on mirrors, 121n, 130-1; on optics (as *scientia media*), 48-9, (geometrical), 53-4, (illusions), 71, 95n, 96n; on reflection, 114-5, 118; on refraction, 133, 134n, 135; on vision (beatific) 103n, 106, 252-3, (infirmities of), 71, (mirroring in), 66, (physical process of), 31-2, 53-4, 66

Alberti, Leon Battista, 12-3n

Alcher of Clairvaux, 92

Alexander of Hales, 251n

Alexander Neckham, optical doctrines in, 66n, 70n, 95n

Alfarabi, 50

Alhazen: influence in thirteenth century, 28-31, 52; on *intentiones*, 61; on light, 60; on optical illusions, 91n; on post-sensory image-transmission, 69n; on reflection, 112n; on refraction, 26-7, 52-3; on transmission of visual form in time, 64n; synthesis of previous optical traditions, 26-8

Ali ibn al- 'Abbas, 21n

Alkindi: on role of light in causation, 21-2, 188n; on reflection, 112n

Alonso, J.M., 253n

Altmann, A., 183n

Ambrose, St., 221

Andreas Capellanus, 45n

Andriani, Beniamino, 4n

Angels: as mirrors of light in medieval writings, 243n; cosmological role of, 186, 205, 242-3; in Dante (as mirrors), 242, (creation of), 206-7, (effects of angelic light on Dante-protagonist), 81-3, (hierarchy of), 240
See also Bonaventure, Intelligences, Pseudo-Dionysius, Thomas Aquinas

Anselm of Canterbury, 92, 236n

Arabic philosophy: emanation, 180-2; theories of celestial causation, 174-5; theories of intellect, 160

Aristotle: account of colour, 15-6, 59-60, 167; account of light, 15, 44, 59, 207; account of vision, 15-6, 59-60; causal theory, 172-3; Dante's knowledge of, 32, 73, 136, 259; his thought enriched by commentators, 31-2, 53-55, 93-6; on aether, 15, 199; on imagination, 16, 93; on intellection, 16, 160; on optics (illusions), 94-5, (of rainbow), 131-2, 136-9, 141-3, 148n, (relationship to geometry), 17-8, 23; on overwhelming of senses, 80-1; misperception at distance, 94, 98, 102; praise of sight, 45n; rejects Empedocles' theory of vision, 65; rejects extramission, 69

Armstrong, A.H., 176n

Art: contemporary influence on Dante, 76; Dante's influence upon later artists, 75, 135n; relationship to perspective, 7; use of light in, 156n

Astrology: Dante's knowledge of, 184-9; foundation in Aristotle, 172-3
See also Albert the Great, Causes, Jean de Meun, Light (celestial)

Atoms, in ray of sunlight, 95n

294

Augustine, St.: biblical commentator, 221n, 230n; colour as cause of visual delight, 45n; concept and image of light in his works, 155, 221-2, 224, 247n; possible echo in Dante, 86n; three visions in, 92n; vision of God in, 252

Austin, Herbert D., 120n, 143n

Averroës: commentator on Aristotle, 32n; on anatomy, 24n, 72; on *intentio*, 63; on light, 60; on optics (as *scientia media*), 48, (illusions), 95n; on reflection, 115; on refraction in atmosphere, 136n; on vision, 32, (mirroring in), 66

Avicebron, 183, 188n

Avicenna: causal theory, 175n, 180-1; influence on medieval writers, 25, 32; on corruptibility, 211; on emanation, 181; on *intentiones*, 62-3; on internal senses, 24-5, 101; on role of light in sensation, 25-6, 60; on technical terms for light, 25-6, 55-6; on vision (theory of), 24-5, (mirroring in), 66

—B—

Badawi, Abd al-Rahman, 180n

Baeumker, Clemens, 151, 160, 173n, 246

Baig, Bonnie Pavlis, 34n

Baldelli, Ignazio, 118n

Baranski, Zygmunt G., 31n

Barasch, Moshe, 135n, 156n

Barolini, Teodolinda, 41n

Bartholomew of Bologna: light analogies in, 234-8; optical doctrines in, 56, 60n, 62n, 236n; relationship to Dante, 57, 235-7, 260

Bartholomew the Englishman, optical doctrines in, 12-3n, 30, 57n, 90-1n, 137n, 224n

Basilisck, its power to blind humans, 35n

Baxandall, Michael, 106n

Beare, J.I., 14-5n

Beatrice: Dante-protagonist's visions of, 40-2, 83, 123-6; effects of her light on Dante-protagonist, 84-6

Bede, 137n

Bellonzi, Fortunato, 76n

Bemrose, Stephen, 186n, 190n, 210n, 211n, 243n, 250n

Bender, John B., 76n

Benvenuto da Imola, 113, 116

Berg, William J., 76n

Bernard Silvestris, 179, 180n, 205

Berretta, G., 246n

Berti, Enrico, 32

Bertini, Patrizia Malgarini, 3n, 9n

Bertola, E., 40n

Bevan, Edwyn, 219n

Bevington, David, 226n

Bible: contradictory passages on vision of God, 252n; Dante's use of, 86-8, 95n, 142, 259-60; light imagery in, 87-8n, 95n, 142, 148n, 192, 207n, 213n, 219n, 220n, 234, 244n; mirror imagery in, 125; rainbow imagery in, 142, 145, 148

Bigi, V.Ch., 246, 247n

Bigongiari, Dino, 211n, 213n

Birkenmajer, Aleksjander, 153n

Bishop of Paris, errors condemned by, 202n, 252

Block, Irvine, 94

Blindness, theme of in Dante, 79-80, 85-9, 258

Boethius, 17n, 202

Boissard, Edmond, 241

Boitani, Piero, 88n, 240n

Bonaventure, St.: on angels, 243n; on light (analogies), 107n, 154n, 224n, 232-3, 248n, (of Trinity), 227, (theory of), 247-9; optical knowledge, 56-7n, 116, 118n, 127, 134n; relationship to Dante, 246, 260; use of Pseudo-Dionysius, 245, 246n

Bono, James S., 40n

Bosco, Umberto, 113n, 117n

Botterill, Stephen, 73n

Bougerol, Jacques G., 174n, 234n, 246n

Boyde, Patrick, 3n, 56n, 76n, 81n, 105n, 138n, 158n, 184n, 186n, 208n, 209n, 211n

Boyer, Carl B., 112n, 132n, 137n

Bradley, Ritamary, 128n, 141n

Brain: as seat of perception, 19; role in vision, 19, 24, 69

Brandeis, Irma, 218n

Bremer, Dieter, 152n

Brown, Peter, 33n, 76n

Brownlee, Kevin, 145n

Bufano, Antonietta, 200n

Busnelli, Giovanni, 53n, 129n

Buti, Francesco da, 117

—C—

Cantarino, Vincent, 158n
Caparello, Adriana, 48n
Capasso, Ideale, 3n, 184n, 190n
Cappuyns, Maieul J., 178n
Carruthers, Mary, 25n
Casini, Tommaso, 124n
Casus lapidis, 113, 123
Catenazzi, Flavio, 34n
Causes: causation as communication of
likeness, 172-4, 176-7; in Dante (efficient
causality), 185, 189-91, 195-8, (God as
first cause), 198, (relationship to effects),
46, (role of secondary causes), 190-2,
196-7, 210
*See also Alkindi, Arabic philosophy,
Aristotle, Avicenna, Light, Robert
Grosseteste*
Cavalcanti, Guido: light and optical
imagery in, 36-7; *spiritus* doctrine in, 36-
7, 40-1; visual rays in, 69n
Chalcidius: account of vision, 14-5;
mediated creation in, 211; optical
doctrines in, 15n, 121n, 130
Chartres, School of, 15n, 179, 205, 211
Chenu, M-D., 205n
Chiamenti, Massimiliano, 40n
Chiavacci-Leonardi, Anna Maria, 107n,
250n
Chimenz, S.A., 126-7n
Chioccioni, Pietro, 92n
Clark, David L., 33n, 55n, 234n
Cline, Ruth, 34n
Cocking, J.M., 102n
Collette, Carolyn, 76n
Colour: as primary object of vision, 59-60,
61n; cause of delight, 45n; definition of,
167; formation in atmosphere, 134-5,
138-9; relationship to light, 59-61
*See also Albert the Great, Aristotle,
Augustine, Plato, Rainbow, Robert
Grosseteste*
Common sense and sensibles, 16, 20n, 25,
61, 69, 94, 98-9, 100, 101
Conklin, Akbari Suzanne, 34n
Constantine the African, 20-1
Contenson, P.-M., 252n
Convivio: account of vision in, 51-2, 58-72,
258; light terminology, 55-8; references
to *perspectiva*, 47-51

Copleston, Frederick C.,151n
Cornish, Alison, 184n
Corti, Maria, 3n, 56n, 57n, 124n, 158n,
212n, 236n
Cosmos, Dante's conceptions of, 184-205
*See also Angels, Causes, Intelligences,
Light (celestial)*
Creation: in Dante (accounts of), 206-7, (act
of divine will), 209, (act of irradiation),
206-7, 259, (instantaneity of), 207,
(related imagery), 206-7; relationship to
emanation, 180-3, 209-11
*See also Angels, Chalcidius, Plato,
Robert Grosseteste, Thomas Aquinas*
Cristiana, Marta, 57n
Crivelli, E., 120n
Crombie, A.C., 9-10n, 20n, 23n, 28n, 132n,
133n, 152n

—D—

d'Acona, Cristina Costa, 182n
Dales, Richard, 202n
d'Alverny, M.-T., 183n
da Todi, Iacopone, 35n
Dauphiné, James, 184n
Davanzati, Chiaro, 35n
Davidson, Herbert A., 161n
de Bonfils Tempier, Margherita, 41n, 212n
de Bruyne, Edgar, 34n, 56n, 163n
de Libera, Alain, 101n, 183n
della Vigna, Pier, 35n
delle Colonne, Guido, 35
de Mottoni Faes, Barbara, 152n, 153n,
202n, 245n
Devons, Samuel, 132n
Differentiation, Dante on role of heavens in,
197-8, 200-1, 204-5
Di Pino, Guido, 75n, 218
Di Salvo, Tommaso, 116n
Di Scipio, Giuseppe, 88n, 125n
Distance, effect on sight in *Comedy*, 81-2,
89, 90n, 94, 98-107
See also Aristotle
Doherty, K.F., 245n
Dominic Gundissalinus, 50, 183n
Dondaine, H.-F., 245n, 252
Dragonetti, Roger, 145n
Dronke, Peter, 9n, 73n, 148n, 205n
Druart, T.-A., 180n

Dunphy, W.B., 181n
Durantel, J.S., 245n
Durling, Robert M., 216n

—E—

Eastwood, Bruce C., 23n, 65n
Eberle, Patricia J., 33-4n
Edgerton, Samuel Y. Jr., 14n, 139n
Emanation, 21-2, 175-83
  See also Arabic philosophy, Avicenna, Creation, Neoplatonism, Plotinus
Empyrean: associated with concept of beatific vision in *Paradiso*, 251-6, 260; Dante's changing views, 250; scholastic writings on, 251n, 256
  See also Light, Vision
Eriugena, John Scotus, 178-9, 212, 226, 230n, 252
Estimation, 63, 101
Euclid, 10-4, 28, 112n
Evans, David, 156n, 239n
Experiment: involving three mirrors in Dante, 128-31; relationship between experience and experiment, 112n
Eye: anatomy of, 18-9; Dante's references to effect of light upon, 80-9; infirmities of, 70-1; of the mind, 225, 235n; mirroring function of, 65-6; rays and light emitted from (extramission), 11, 14, 15n, 19, 34-6, 42, 69, 84n, 85n, 86, 130; retina, 65n
  See also Tunic

—F—

Fallani, Giovanni, 75n, 76n, 251n
Fengler, Christie K., 76n
Ferwerda, Rein, 155n
Finamore, John J., 59n
Fortune, Dante's conception of, 198
Foster, Kenelm, 88n, 102n, 165n, 190n, 210n, 213n, 245n, 250n, 254n
Frappier, Jean, 34n
Fraticelli, Pietro, 118n
Frugoni, Arsenio, 148n
Führer, M.L., 161n
Fujitani, Michio, 119n, 126n

—G—

Gage, John, 148n, 156n, 235n, 239n

Gagné, Joan, 49n
Galen, 18-20, 66n
  See also Tunic
Gamba, Ulderico, 240
Gardner, Edmund G., 239n
Gay, John, 178n
Gersh, Stephen, 177n
Ghisalberti, Alessandro, 184n
Giacalone, Giuseppe, 113n, 116n
Giacomo da Lentini, 34-5
Giacon, Carlo, 55n
Giannantonio, Pompeo, 158n
Gilson, Etienne, 73n, 148n, 181n, 250n, 254n
Gilson, Simon A., 120n
Giullaume de Lorris, 33
Gizzi, Corrado, 3n
Gmelin, Heinrich, 129n
Grabes, Herbert, 122n, 125n
Grace, as light, 232-4, 235n
Grant, Edward, 121, 171n, 172n, 173n
Gregory, Tullio, 15n, 62n, 122n, 153n, 212n, 252n
Guardini, Romano, 158n
Guglielminetti, Marziano, 209n, 212n
Guidubaldi, Egidio, 9n, 113n, 130n, 164n, 166-7, 213n, 235n, 239, 241n, 246n
Guinizzelli, Guido, 36, 194
Gyekye, Kwame, 62n

—H—

Hackett, Jeremiah M.G., 48n
Hamesse, Jacqueline, 81n
Harwood, Sharon, 218n
Hedwig, Klaus, 152n, 156n
Hero of Alexandria, 28, 65n, 112n
Hildegard of Bingen, 148
Hills, Paul, 9n, 14n, 156n
Hirsch-Reich, Barbara, 148n
Hollander, Robert, 41n, 126n
Holley, Linda Tarte, 76n
Honoré d'Autun, 95n, 137n
Hugedé, Norbert, 125n
Hugh of St. Victor, 35n, 45n, 228
Hyman, Arthur, 180n

—I—

Imagination: as source of error, 93; role in vision, 16, 25, 37n, 41, 46, 93, 98-9, 101, 124

See also Aristotle
Intellect, dependence of senses upon, 16, 93
   See also Arabic philosophy, Aristotle, Light (relationship to knowing)
Intelligences, angelic, 180-2, 185-6, 190, 198, 200-4, 243
   See also Angels, Causes
Intentio, different meanings in medieval philosophy, 62-4
   See also Albert the Great, Alhazen, Avicenna, Averroës, Light, Roger Bacon, Thomas Aquinas
Internal senses: Avicenna's classification of, 24-5, 93; Dante's views on, 51, 68-9, 98-103, 124, 258
   See also Common sense, Estimation, Imagination
Isaac Israeli, 183
Isidore of Seville, 66n, 95n, 121n, 137n

—J—

Jean de Meun: his astrological views, 188n; optical knowledge in, 33n, 94n, 129n, 134n
Jeauneau, Edouard, 122n
John of Halifax (Sacrobosco), 13n
John Peckham, optical doctrines in, 28-9, 48n, 52-3n, 60n, 62n, 64n, 68n, 69n, 70n, 71n, 111n, 112n, 132n, 189n

—K—

Kay, Richard, 3n, 184n, 209n, 212n
Kemp, Martin, 14n, 75n
Klassen, Norman, 222n
Kleinhenz, Christopher, 76n
Klibansky, Raymund, 153n
Knowles, David, 151n
Knudsen, Christian, 62n
Kogan, Bary S., 180n
Kucharski, Paul, 139n

—L—

Laird, W.R., 50n
Lansing, Richard H., 118n
Leclerq, Jean, 245n
Lee, Desmond, 131n
Lee, Jonathon Scott, 244n
Lee, Patrick, 161n
Leff, Gordon, 151n

Leisegang, Hans, 128n
Lejeune, Albert, 12n
Lemay, Richard, 174n
Leonardo da Vinci, 135
Lepschy, Antonio, 139n
Lewis, C.S., 163
Leyerle, John, 213n
Liber de causis, 158, 182, 190-2, 193n, 201, 204, 211-2
Liber de causis primis et secundis, 183n
Liber de intelligentiis, 153-4, 189n, 247
Light: analogies drawn from, 1, 32, 44-7, 57-8, 154-5, 219-48, 260; as causal agent, 21-2, 173-5, 186-9; as *intentio*, 62-4; as penetrating other bodies, 221n, 222n, 223n, 226; as principle of (continuity), 153, 159, 177, (form), 177, 216, (life), 154, 220; as visible, 59-61; in Christian Middle Ages, 1, 224-44; in early Christian thought, 178, 220-4; in Neoplatonic thought, 163-4, 174-83; light of faith (*lumen fidei*), 235; light of glory (*lumen gloriae*), 252-6; metaphysics of, 2, 30, 57, 151-69, 212, 216, 218; modes of propagation (*lux, lumen, radius, splendor*), 25-6, 55-7; multiplication of, 193-4, 213-5; of Christ, 213n, 217, 237n; of Empyrean heaven, 250-6; of God, 32, 192n, 217, 229, 231-2, 244; of Virgin Mary, 35n, 217, 225-6, 248; properties of, 1, 224, 235; relationship to knowing, 154, 160-1, 165, 222, 235n, 259; speed of, 207
   See also Albert the Great, Alhazen, Alkindi, Angels, Aristotle, Art, Atoms, Augustine, Averroës, Avicenna, Bartholomew of Bologna, Beatrice, Bible, Blindness, Bonaventure, Cavalcanti, Colour, Convivio, Creation, Eye, Grace, Moon, Planets, Plato, Plotinus, Pseudo-Dionysius, Rainbow, Rays, Reflection, Resurrection, Robert Grosseteste, Souls, Sun, Thomas Aquinas, Thomas of Cantimpré, Witelo
Limoges, Peter, 93n, 234n
Lindberg, David C., 9-10, 12n, 13n, 14n, 15n, 19n, 20n, 21n, 23n, 24n, 27n, 28n, 29n, 31n, 131n, 133n, 152n, 156n, 177n, 234n
Litt, Thomas, 188n
Lloyd, A.C., 172n

Lovejoy, Arthur, 209
Lucretius, 91n, 95n

—M—

Macrobius, 121-2, 180n
Maggini, Francesco, 46n
Mamura, Michael E., 181n
Marrone, Steven P., 161n
Mars, 135
Marshall, Peter, 207n
Martinelli, Bortolo, 9n, 131n, 160n, 189n, 250-1, 256
Mathieu, Dominique, 219n
Mattioli, Mario, 3n, 131n
Mazzeo, Joseph Anthony, 88n, 161-6, 209n, 211n, 212n, 239, 246n, 247
Mazzoni, Francesco, 117-8n, 124n
McCarthy, Edward Randall, 221n, 247n
McEvoy, James, 23n, 47n, 152n, 161n, 175n, 178-9n, 189n, 215n, 247n
McKeon, C.K., 152n
McKirahan, R.D., 18n
McNair, Philip, 148n
Medium: optical effects produced in, 94-100, 134-5, 141; role in vision, 15-6, 62-3
Mellone, Attilio, 158n, 160n, 208n, 209n, 236n, 250n
Mengaldo, Pier Vincenzo, 243n
Michaud-Quantin, Pierre, 31n, 101n
Miller, James L., 57n
Mineo, Nicolò, 41-2n
Minio-Paluello, L., 32n
Mirrors: as means of self-examination, 141n; errors caused by, 94n; in Dante, 110-1, 120, 123-6, 128-9, 141, 199, 242, 258; lead-backing of, 67; relationship to water, 111n
See also Albert the Great, Angels, Bible, Experiment, Pseudo-Dionysius, Thomas Aquinas
Momigliano, Attilio, 117n
Montgomery, Robert L., 102n
Moon: as reflective body, 130-1, 199; Dante enters body of, 226; Dante's explanation of its spots, 200-5; halo formed around, 132, 136, 146
Moran, Dermot, 178n
Morgan, Alison, 76n
Mostacci, Jacopo, 35

Mullahy, B.I., 156n
Muscatine, Charles, 76n

—N—

Nardi, Bruno, 2, 53n, 73n, 101, 112-3, 118, 145n, 148n, 157-60, 164n, 186n, 190n, 191n, 192n, 200n, 209n, 210n, 211-12, 250n, 254n, 255n
Negri, Luigi, 246n
Neoplatonism: ancient writings, 152-3, 155, 157-60, 175-7; Arabic writings, 180-2; Christian forms of, 178-9, 213; Dante's use of its doctrines, 157-8, 191, 201, 204-5; doctrine of participation, 164, 231, 241n; Jewish writings, 183; procession and return, 177
See also Light, Pseudo-Dionysius
Newman, F.X., 92n, 218n
Niccoli, Alessandro, 101n
North, J.D., 173n

—O—

Oliva, Gianni, 9n, 75n
Oliviero, Adriana, 158n
Olschki, Leonardo, 235-6n
O'Meara, Dominic J., 254n
Optics (*Perspectiva*): as auxiliary science, 17-8, 48-50; as universal science, 21-4, 47-8, 257; geometrical branch of, 10-4; in Arabic thought, 20-8; in Aristotelian commentary, 31-2, 53-4, 59-73, 257-8; in medieval encyclopaedias, 30-1, 54-5; in medieval poetry (other than Dante), 33-7; in medieval theology, 32-3, 105-7, 127, 225, 233, 235; medical branch of, 18-21; non-technical thirteenth-century sources, 13, 29-37; optical illusions, 8, 13, 15n, 70-1, 77, 89-102; physical branch of, 14-6; Latin translations, 13, 20-1, 28-9, 72n, 261; thirteenth-century treatises on, 1, 7-8, 28-9; tripartite division of, 12-3
See also Albert the Great, Alexander Neckham, Alhazen, Alkindi, Aristotle, Averroës, Bartholomew of Bologna, Bartholomew the Englishman, Bonaventure, Cavalcanti, Chalcidius, Convivio, Eye, Jean de Meun, John Peckham, Medium, Mirrors, Plato, Pseudo-Aquinas, Ptolemy, Rainbow, Rays, Reflection, Refraction, Robert

Grosseteste, Roger Bacon, Thomas Aquinas, Thomas of Cantimpré, Vision, Vincent of Beauvais, Witelo
Orestano, F., 246n
Ostendler, Heinrich, 148n, 241n
Ovid, 143n, 145
Owens, Jesse, 161n

—P—

Pagliaro, Antonino, 118n
Palgen, Rudolf, 159n, 209n, 212
Pannaria, Francesco, 131n
Panofsky, Erwin, 239n
Parent, J.M., 212n
Parronchi, Alessandro, 7-9, 32, 43n, 51n, 71n, 91n, 95n, 113n
Pasquini, Emilio, 116-7n, 120n
Paul, St.: on seeing face to face, 125; presence in Dante, 86-8, 125-6
Pelikan, Jaroslav, 221n
Pertile, Lino, 88n
Peter of Limoges, 93n, 234n
Peter Lombard, 32, 56
Peterson, J., 172n
Peterson, Mark, 131n
Petrocchi, Giorgio, 111n
Petrus Alfonsi, 182n
Petrusi, Agostino, 239n
Pézard, André, 65n, 116n, 129n
Philippe, M.-D., 93n
Piana, C., 236n
Picone, Michelangelo, 145n
Planets: derive light from sun, 44-5; in Dante (descriptions of), 199, 217, (relationship with Intelligences), 198-200, 202-4
See also Mars, Moon, Stars, Sun
Plato: on cosmic light, 121; on colour, 139n; on creation, 211-2; on sun, 223n; on transmission of light, 11n; optical and visual doctrines in, 14-5, 69n, 94n, 125n
Pliny, 94n, 137n
Plotinus: doctrine of emanation, 175-7; light motifs in, 154n, 155n, 176
Plumptre, E.H., 130
Poletto, Giacomo, 118n
Porena, Manfredi, 110n, 118n
Portelli, J., 25n
Pouillon, Henri, 239n
Primum mobile, 195, 200

Proclus, 177
Protonotaro da Messina, Stefano, 35n
Pseudo-Aquinas: on celestial influence, 188n; optical doctrines in, 13n, 96n, 99n, 115, 129, 134n
Pseudo-Aristotle, 11n, 130
Pseudo-Bede, 137n
Pseudo-Bernard, 141n
Pseudo-Dionysius: angelic hierarchies, 240; angels as mirrors in, 242-3; influence in Middle Ages, 178-9, 224n, 226, 228, 239, 245; light analogies, 154n, 191n, 222-4, 239; Neoplatonic concepts in, 178-9; presence in Dante, 239-45, 260; theophany in, 252-3
Pseudo-Euclid, 28, 112n
Ptolemy: account of celestial influence, 173n; on refraction, 12; optical doctrines in, 11-4, 65n, 112n
Punctiform, analysis of vision, 21, 27, 67-8, 257

—Q—

Quillet, Jeannine, 200n
Quinn, John Francis, 247n

—R—

Rabuse, Georg, 122n
Radcliff-Umstead, Daniel, 158n
Rainbow: Aristotle's account of, 131-2, 137-9, 141-3, 145; Dante's imagery of, 136n, 137-49, 258; double bow, 129, 136, 144-5; medieval theories of, 132-3; number of colours in, 142-3
Rays, 11-4, 45-7, 52, 207, 217, 223, 225-7, 229, 231, 234-5, 244-5, 248-9
See also Eye
Reflection: Dante's imagery of, 110-28; law of, 111-6
See also Albert the Great, Alhazen, Alkindi, Optics, Averroës, Robert Grosseteste, Thomas Aquinas
Refraction: Dante's limited understanding of, 67, 132-3, 257; role of refraction in vision, 26-7, 52-3, 67, 133n
See also Albert the Great, Alhazen, Averroës, Optics, Ptolemy, Robert Grosseteste
Restoro d'Arezzo, 121

Resurrection: light imagery related to, 107n; visual powers of resurrected according to Dante, 107, 255
Rhaban Maur, 95n, 137n
Ricchi, Gino, 131n
Richard of St. Victor, 227
Robert Grosseteste: on colour, 166-7; on light (causal role), 22-3, 188-9, (of Trinity), 227n, (role in creation), 213-5, (role in motion), 167, 189, (role in sensation), 166-7, 247n, (role in sound), 167, 189; on optics (illusions), 70n; (status of), 23, 50; on reflection, 112n; on refraction, 132, 134n; relationship to Dante, 166-7, 188-9, 213, 215-6, 259
Roger Bacon: his doctrine of multiplication of species, 23-4; on *intentio*, 63-4n; optical doctrines in, 28-9, 50, 52n, 56, 60n, 62n, 63n, 64n, 69-70n, 71n, 91n, 111-2, 130n, 132n, 133n, 134n, 189n, 233-4, 259
Ronchi, Vasco, 57n, 68
Rorem, Paul, 179n, 245n
Ruello, Francis, 246n
Rutledge, Monica, 9n

—S—

Sabra, A.I., 27-8n, 62n, 133n
Saccaro, Giuseppe Battisti, 152n, 183n, 215n
Saffrey, H.D., 182n
Sage, Athanase, 222n
Sambursky, S., 59n
Santoro, Luigi, 158n
Sapegno, Natalino, 75n, 117n
Sayili, Aydin M., 137n
Scazzoso, Piero, 239n, 240n
Schlanger, Jacques, 183n
Schmidt, Margot, 128n, 156n
Schweig, Bruno, 128n
Science, medieval, Dante's knowledge of, 3
Scorrano Luigi, 117n
Scrivano, Riccardo, 246n
Sebastio, Leonardo, 200n
Seneca, 91n, 137n
Sheldon-Williams, I.P., 179n
Shoaf, R.A., 145n
Siegel, Rudolph E., 19n
Simonelli, Maria, 31n, 56n, 236n
Simson, Otto von., 152n, 156n, 239n

Singleton, Charles S., 113n, 218n, 254n
Siraisi, Nancy G., 3n, 101n
Skinner, Quentin, 157n
Smith, A. Mark, 9n, 16n, 26n, 93n
Sorabji, Richard, 172n, 178n
Soul-body, formation of according to Dante, 138, 203-4
Souls, blessed, as light in Dante, 217, 237
Southern, Richard W., 101n
Soverchio (overwhelming of senses): Dante's descriptions of, 79-86, 110, 258; medieval sources on, 80n, 81n
  See also Aristotle
Spedicati, Adele Anna, 177-8n
Spirits, 19, 51, 86, 87
  See also Cavalcanti
Spitzer, Leo, 157n
Stabile, Giorgio, 200n
Stansbury, Sarah, 76n
Stars, role in raising land-mass according to Dante, 197
Statius, as poet of the *Thebaid*, 83n
Steneck, Nicholas H., 101n, 221n
Stormon, E.J., 250n
Summers, David, 17n, 99n, 101n
Sun: as symbol, 185, 223; light of, 45, 135, 185, 191, 200, 232-3, 237, 248-9; role in generation and corruption, 172; scintillation of, 95n
  See also Atoms, Planets, Plato, Vapours
Sweeney, Theodore, 182n, 183-4n

—T—

Tachau, Katharine, 9n, 16n
Tasso, Torquato, 227
Thomas Aquinas: character in *Paradiso*, 234; commentator on Aristotle, 31-2, 48, 54-6, 60, 63-4, 66-7, 72, 80n, 93-4n, 95n, 96, 98-9, 115, 105n, 135, 258; influenced by Pseudo-Dionysius, 231, 241n, 245-6; on angels, 240n, 243n; on celestial influence, 188, 193; on creation and emanative language, 210-1; on eclipse at crucifixion, 243n; on *intentiones*, 64n; on internal senses, 101n; on light (analogies), 164-5, 230-2, (celestial), 188, 199n, (multiplication of), 193-4n, (propagation of), 56, (visible), 60-1; on optics (as *scientia media*), 48-9; (illusions), 95n, 96n, 98-9, (mirrors), 66-

7; on reflection, 115, 133n; on vision (beatific), 253-4, (physical), 54, 93-4n

Thomas of Cantimpré, optical and light doctrines in, 72n, 137n, 216n

Thonnard, François-Joseph, 155n, 222n

Tideus, 28n

Tondelli, Luigi, 148n

Took, J.F., 41n, 240n

Travi, Ernesto, 184n

Trinity: Dante's references to, 147, 208-9, 227; in patristic and medieval sources, 148n, 178, 222n, 226-7
See also Bonaventure, Robert Grosseteste

Tunic (of the eye): in Galen, 18; in "perspectivists," 71; in twelfth- and thirteenth-century sources, 72

—V—

Vanni Rovighi, Sofia, 243n

Vapours, effect on stars and sun, 70, 81, 89, 99, 134-6, 238

Varese, Claudio, 9n

Vasoli, Cesare, 9n, 31n, 56, 65n, 69n, 128n, 135n, 156-7, 157n, 160n, 190n, 193n, 200n, 205n, 236n, 240n, 245n

Vescovini, Graziella Federici, 9n, 21n, 23n, 28n, 50n, 62n, 93n, 152n, 189n, 234n

Vincent of Beauvais: optical doctrines in, 30, 57, 66-7n, 80n, 87n, 94n, 95n, 99n, 116, 118n, 137n, 167n; other references, 154n, 188n, 240n, 243n, 251n

Virgil, as poet of the *Aeneid*, 76, 83n, 143n

Vision: as act of perception, 16, 27-8, 68-9, 101-2; beatific, 251-6; geometry of, 10-4, 104-7; in Dante (as structuring device), 77-102, (geometry of), 51-2, (of God), 125-6, 146-7, 217, (in the Empyrean), 87-9, 102-3, 147, 250-6; physical, 58-64, 77-102; praise of, 15
See also After-image, Albert the Great, Aristotle, Augustine, Averroës, Avicenna, Beatrice, Bible, Brain, Chalcidius, Colour, Convivio, Empyrean, Eye, Imagination, Medium, Optics, Punctiform, Refraction, Thomas Aquinas, Tunic

Volpini, Enzo, 20n

—W—

Wallace, William A., 23n

Wallis, R.T., 176n

Weisheipl, James A., 47n, 50n

Wetherbee, Winthrop, 205n, 245n

William of Auvergne, 106n, 134n

William of Conches, 13n, 72n, 211n

William of St. Thierry, 72n

Witelo: on light as form, 215n; optical doctrines in, 28-9, 47, 48n, 53b, 60n, 62n, 69n, 71n, 91n, 111n, 112n, 132-3n, 189n, 259

Wlassics, Tibor, 75, 76n

Wolfson, Harry Austryn, 25n, 69n, 178n, 181-2n

—Z—

Zambelli, Paola, 188n